Recent Developments in Using Seismic Waves as a Probe for Subsurface Investigations

Seismology has come a long way. Being the scientific study of seismic waves and their allied phenomena, it has entered a multidisciplinary realm. As the main tool, it provides a wealth of information when applied systematically to dig inside the Earth structure. Notwithstanding, the utility of seismic waves has increased manifold. Starting from knowing the epicenter of seismic events, it has influenced mapping of civil engineering structures such as dams and bridges, as well as huge constructions. Although there is no dearth of technical papers in the area of seismic waves, there is an absence of synchronized and recent coherent contents in the direction of seismic waves.

The book will be a unique contribution to the field of seismology, with the aim of assimilating theory and practices. It will provide a comprehensive glimpse of recent advancements in this area with a strong unification of theory and practices. The main objective of the book is to present an in-depth analysis of the theory and real implementations of seismic waves as versatile probes that would be integrated with modern and future perspectives. The current and the future strategies to be discussed in the relevant areas of seismic waves will be another boon for readers.

This book will cater to the needs of novices, researchers and practitioners. Additionally, the contents of the book will be useful for undergraduate as well as postgraduate students of earth science disciplines.

Recent Developments in Using Seismic Waves as a Probe for Subsurface Investigations

Theory and Practices

Edited by
Rajib Biswas

CRC Press
Taylor & Francis Group
Boca Raton London New York

CRC Press is an imprint of the
Taylor & Francis Group, an **informa** business
A BALKEMA BOOK

Cover image: Shutterstock

First published 2023
by CRC Press/Balkema
Schipholweg 107C, 2316 XC Leiden, The Netherlands
e-mail: enquiries@taylorandfrancis.com
www.routledge.com – www.taylorandfrancis.com

CRC Press/Balkema is an imprint of the Taylor & Francis Group, an informa business

ISBN: 9781032012087 (hbk)
ISBN: 9781032025131 (pbk)
ISBN: 9781003177692 (ebk)

DOI: 10.1201/9781003177692

Typeset in Times New Roman
by KnowledgeWorks Global Ltd.

Contents

Contributors

Santanu Baruah, North East Institute of Science and Technology (NEIST), Jorhat, Assam, India

Rajib Biswas, Geophysical Lab, Department of Physics, Tezpur University, Tezpur, Assam, India

Nilutpal Borah, Department of Civil Engineering, IIT Guwahati, Guwahati, India

Vladimir V. Chechelnitsky, Baikal Office of Federal Research Center Joint Geophysical Survey of Russian Academy of Sciences, Irkutsk, Russia

Vyacheslav V. Cheptsov, Institute of the Earth's Crust of the Siberian Branch of Russian Academy of Sciences, Irkutsk, Russia

Bijit K. Choudhury, Geo Sciences and Technology Division (GSTD), CSIR-Northeast Institute of Science & Technology (CSIR-NEIST), Jorhat, Assam, India

Anna A. Dobrynina, Institute of the Earth's crust of the Siberian Branch of Russian Academy of Sciences, Irkutsk, Russia and Geological Institute of the Siberian Branch of Russian Academy of Sciences, Ulan-Ude, Russia

Devajit Hazarika, Wadia Institute of Himalayan Geology, Dehradun, Uttarakhand, India

Jyotima Kanaujia, National Geophysical Research Institute, Hyderabad, India

Sreevalsa Kolathayar, Department of Civil Engineering, National Institute of Technology, Karnataka, India

Manoj Kumar, CMPDI, Coal India Limited, Ranchi, Jharkhand, India

Debasis D. Mohanty, Geosciences and Technology Division, CSIR-North East Institute of Science and Technology, Jorhat, Assam, India

Debjeet Mondal, CMPDI, Coal India Limited, Ranchi, Jharkhand, India

Sambit P. Naik, Pukyong National University, Busan, South Korea

C.R. Parthasarathy, Sarathy Geotechnical Engineering Services, Bangalore, India

Arpita Paul, Wadia Institute of Himalayan Geology, Dehradun, Uttarakhand, India

P.N.S Roy, Department of Geology and Geophysics, Indian Institute of Technology, Kharagpur, West Bengal, India

B.K. Sahoo, Radiological Physics and Advisory Division, Bhabha Atomic Research Centre, Mumbai, Maharashtra, India

Sowrav Saikia, National Center for Seismology (MoES), New Delhi, India

Vladimir A. Sankov, Institute of the Earth's crust of the Siberian Branch of Russian Academy of Sciences, Irkutsk, Russia

Vickey Sarma, Geophysical Lab, Department of Physics, Tezpur University, Tezpur, Assam, India

Anup K. Sutar, Ministry of Earth Sciences, Borehole Geophysics Research Laboratory, Karad, Maharashtra, India

Prashanth Talkad, Sarathy Geotechnical Engineering Services, Bangalore, India

T. Thuamthansanga, Department of Physics, Mizoram University, Aizawl, Mizoram, India

R.C. Tiwari, Department of Physics, Mizoram University, Aizawl, Mizoram, India

Monika Wadhawan, National Center for Seismology, New Delhi, India

Mehdi Zaré, IIEES, Tehran, Iran and Geology Div., Academy of Sciences, Iran

Preface

Seismology has come a long way. Being the scientific study of seismic waves and their allied phenomena, it has entered a multidisciplinary realm. The main tool being seismic waves provides wealth of information when applied systematically to dig inside the Earth structure. Notwithstanding, the utility of seismic waves has increased manifolds. Starting from knowing the epicenter of seismic events, it has influenced mapping of civil engineering structures such as dams and bridges, as well as huge constructions. Although there is no dearth of technical papers in the area of seismic waves, however, an absence of synchronized and recent coherent contents in the direction of seismic waves is duly felt.

In this regard, this book will be a unique contribution, which will fill up this gap. With the aim of assimilating theory and practices, this book is expected to provide a comprehensive glimpse of recent advancements in this area with a strong unification of theory and practices. The main objective of the book is to present an in-depth analysis of the theory and real implementations of seismic waves as versatile probes that would be integrated with modern and future perspectives. The current and the future strategies to be discussed in the relevant areas of seismic waves will be another boon for the readers.

This proposed book will be a unique contribution with the aim of assimilating theory and practices, thereby providing a comprehensive glimpse of recent advancements in this area with a strong unification of theory and practices. The main objective of the book is to present an in-depth analysis of the theory and real implementations of seismic waves as versatile probes that would be integrated with modern and future perspectives

Chapter 1 starts with application of travel time inversion in order to execute deep Earth imaging. It appraises reader about the basics of tomography with illustrative examples. Chapter 2 discuses another strong tool for subsurface investigation, that is, waveform modelling. Chapter 3 covers theory and practices involved in depth and source localization of anisotropy with touch in recent developments in subsurface investigation.

Chapter 4 is about the sedimentary basin configuration of different basin structures of Northeast India via teleseismic body wave imaging. Chapter 5 illustrates a geological tool for future land use code and land use planning in the low seismic area with the objective of studying seismic wave amplification and localized ground effects.

Chapter 6 contemplates HVSR as a versatile tool for subsurface investigation. It also covers two case studies as examples of implementation of this tool. Application of teleseismic receiver function in investigation of crustal thickness and Poisson's ratio are covered in Chapter 7.

Chapter 8 deliberates on Lg wave attenuation in special context to Indo-Burman mobile belt of North-East India. Chapter 9 is a case study on analysis of fluctuation in radon concentration and its correlation with geophysical phenomena of a seismically active region. Chapter 10 details unique seismicity pattern during violent roof movements in longwall mines with rigorous theory and practice approaches. Similarly, Chapter 11 demonstrates estimation of source parameters and scaling relation of the Baikal rift's earthquakes.

Chapter 12 explores seismic anisotropy for understanding the dynamics of crust and upper mantle: A case study in the northwest Himalaya and eastern Ladakh-Karakoram zone. Chapter 13 is about lessons learned from earthquakes. Lastly, Chapter 14 illustrates integrated geotechnical and geophysical approaches to assess integrity of tank foundations for a crude oil terminal project in the Middle East.

The editor firmly believes that this book will be a valuable addition in expanding the knowledge base in the field of subsurface investigation using seismic wave as probes. The book is an amalgamation of geological, geophysical, geodetic and engineering problems pertaining to the subsurface and converging to one-stop solution, that is, seismic waves, which is a novel approach. The book is expected to capture the attention of readers from diverse background of science and engineering. The most unique feature being the chapters are practically oriented with case studies based on realistic nature of problems. The book will be enriched with the rich repertoire of knowledge and experience gained by the authors through their extensive research in their chosen areas. The book will enable the readers to appreciate the interdisciplinary approach to some universal problems pertaining to subsurface investigation.

Chapter 1

Deep earth imaging

An application of seismic wave (travel time inversion)

Jyotima Kanaujia

1.1 INTRODUCTION

Seismic waves represent one of the most valuable resources to study the structure and composition of the Earth's interior. A. Mohorovicic, a Serbian seismologist, was the first to deduce Earth structure from seismic records in 1909. He noticed two branches on the travel time graph from an earthquake occurring at a regional distance and interpreted that one of the branches represents arrival times of direct wave and the other branch represents arrival times of refracted wave from a discontinuity. This discontinuity marked the sharp change in the elastic properties between the crust and upper mantle and is called the Mohorovicic discontinuity. The technique of Herglotz and Wiechart (see, e.g., Gubbins, 1992) was the first to be used in 1910 to construct a one-dimensional model of the whole Earth. The method uses the relationship between angular distance and ray parameter to determine velocity as a function of the Earth's radius (Rawlinson and Sambridge, 2003).

Nowadays, a large number of methods are available to determine Earth structure from seismic waves. These methods use different properties of seismic records: arrival times of seismic phases, amplitudes of seismic waves, spectra of P-, S-, and surface waves, and the entire waveform of the recorded signal. Based on different data, methods, source-receiver geometry, and type of sources, the scale of study can be defined, that is, local, regional, and global scale for crustal, mantle, and core structure of the Earth, respectively. For example, the layout of network stations may be 1-D, 2-D, and 3-D, sources may be located near or far from the stations, sources of seismic waves may be natural or artificial, and the scale

DOI: 10.1201/9781003177692-1

may vary few meters to thousands of kilometers. Finally, there are many ways of translating the seismograms' data to represent seismic structures.

In this chapter, a brief review of seismic travel time tomography to image Earth structure has been presented. This technique is based on inverting travel times of seismic waves to determine the 2-D and 3-D structure of the Earth. The word *tomography* means *slice picture* (from the Greek word *tomos* meaning "slice"). This term was first used in medical imaging to describe mapping the human body's internal density distribution using x-rays (Lee and Pereyra, 1993) (CT scan). The seismological community later adopted the term to describe a similar process to map the Earth's structure using seismic waves. Seismologists now routinely use *tomography* to refer to the Earth's 3-D model; however, the word was initially designed to describe the imaging of 2-D horizontal slices of the Earth only. Although the basic theory behind seismic tomography and CT scan is similar, these techniques differ in their technical details. The CT scan allows measuring data at a large number of points because these points can be controlled, and thus high-resolution images of the human body can be obtained (e.g., Carlson, 2006; Zhao, 2015). However, in seismic tomography, we cannot control the sources and receivers because about 90% of earthquakes occur along the plate boundaries. Further, there is a considerable cost involved in the installation and maintenance of seismic stations. Given this, it is difficult to image the subsurface Earth's structure from seismic tomography as precisely and accurately as the human body's CT scan.

1.2 CLASSIFICATION OF SEISMIC VELOCITY MODELING

For the last about four decades, the area of velocity modeling has witnessed rapid progress. The estimation of velocity structure involves finding the information about the Earth structure in terms of velocities and thickness of the various layers. Accurate velocity structure is essential in locating earthquakes, determining the composition and origin of the Earth's outer layers, interpreting large-scale tectonics, and estimating strong ground shaking. Currently, it is a vast field that covers various aspects of solid earth sciences. Extensive research has been carried out to seek solutions for different tomographic problems. Different terms are used to classify tomography in terms of data used, the study region's lateral extent, the depth range of modeling space, and the relative distance between the seismic station network and the seismic events used.

Based on seismic data used, tomography can be classified into body wave tomography (BWT) and surface wave tomography (SWT). BWT (Figure 1.1) is the most widely used tomography, and when it is based on the inversion of P- and/or S-wave travel time data, it is called travel time tomography. Figure 1.1 shows the paths of seismic rays in four cases of BWT. SWT uses dispersion data of Rayleigh waves or Love waves to estimate the 3-D crustal velocity structure.

According to the scale of study areas, there are three tomography types: local, regional, and global (Figure 1.1). In local tomography, the study area is small and varies typically from a few kilometers to hundreds of kilometers. In regional tomography, the study area varies from about a few hundred kilometers to a few thousands of kilometers. In global

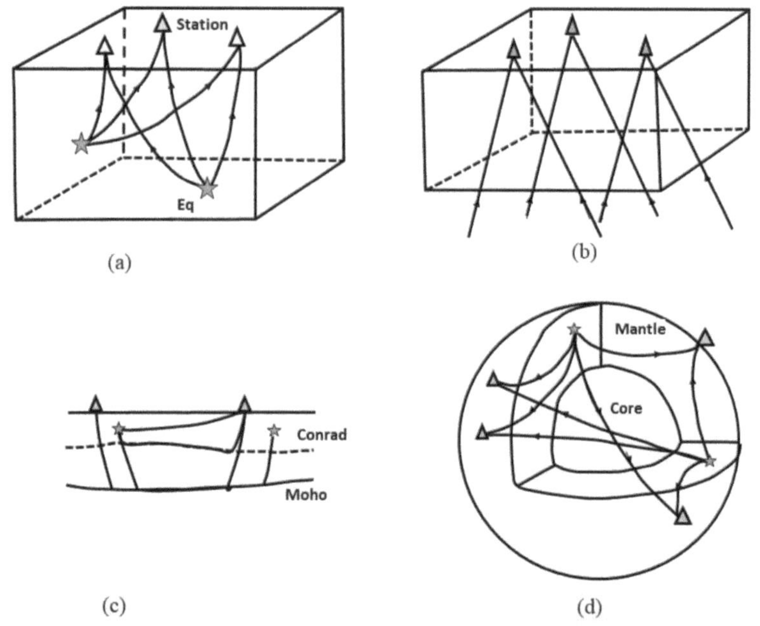

Figure 1.1 Various approaches of body-wave tomography. (a) Local earthquake tomography; (b) teleseismic tomography; (c) determining the Conrad and Moho geometry; (d) global tomography. (Zhao 2009.)

tomography (Figure 1.1d), the area encompasses the entire mantle or whole Earth (Inoue et al., 1990; Vasco et al., 1995; Zhao, 2001).

Based on the depth range of modeling space, there are crustal, mantle, and core tomography. The depth range of crustal tomography varied from surface to Moho depth or up to the upper mantle and was carried out from the travel times of crustal phases. The depth range of mantle tomography extends up to the upper and lower mantle. Crustal tomography usually involves local earthquake tomography, and its range depends upon the diving depth of P- and S waves and epicentral distances. The tomography for depths up to 670 km discontinuity is called upper mantle tomography, and for depths up to core-mantle boundary is called mantle tomography. Core tomography is global and uses core phases to estimate the velocity structure of the Earth's core. (e.g., Lei and Zhao 2006).

Based on the distance between seismic networks and seismic events, tomography can be divided into local earthquake tomography (LET) and teleseismic earthquake tomography (TET). In the LET (Figure 1.1a), seismic stations and events considered fall within the study area (e.g., Aki and Lee, 1976; Thurber, 1983; Zhao et al., 1992). LET is carried out from the inversion of travel times direct crustal phases (e.g., P- and S waves). TET (Figure 1.1b) 3-D velocity structure beneath a seismic network is estimated from P- or/and S-wave data from teleseismic earthquakes and/or artificial explosions. In the TET, the seismic events are located far away from the seismic network. These are called teleseismic events (with epicenter distance 30°–100°), and their travel times can be used to estimate velocity structure up to the lower mantle.

1.3 OVERVIEW OF VARIOUS METHODOLOGIES USED TO STUDY VELOCITY STRUCTURE

Several methods are available to estimate velocity structures, and each method has advantages and shortcomings in specific applications. These methods can be classified based on various factors, such as the type of data available/used and the scale of study (local/regional/global). Some of the methods include explosion seismology, surface wave dispersion, use of crustal phases and teleseismic phases, receiver function/crustal transfer function, waveform modeling, and use of reflection phases. A brief overview of these methods is given below.

1.3.1 Explosion seismology

Explosion seismology has been used to map both continental and oceanic regions' velocity structures. Most of the work done using explosion seismology is based on measurements of arrival times of direct and refracted seismic phases. Some of the difficulties and limitations in interpreting the explosion-generated data are well known (Carpenter, 1965; Kosminskaya et al., 1969). For example, initial arrivals may fail to indicate transitional layering and unable to detect low-velocity layers. The incorporation of arrival times of refracted and reflected phases and their amplitudes, as well as wide-angle reflections, have been attempted to overcome these shortcomings (Mueller and Landisman, 1966; Landisman et al., 1971). However, substantial difficulties were encountered while assessing the reliability of the different types of the data set (Healy, 1971).

1.3.2 Reflection measurements

Measurements of timings of reflections from explosions have been used to determine shallow layering, but it has been recognized that additional information of velocities is customarily required to improve the subsurface structure's interpretation. Deep crustal reflection measurements became more familiar with the increased sophistication of recording equipment (Perkins and Phinney, 1971). Drill hole data provides detailed knowledge about the shallow velocity structure (e.g., sonic velocity log), but such investigations are restricted in scope because they provide only localized information from very shallow depths.

1.3.3 Dispersion of surface waves

In the period range from 1 to 30 sec, dispersion of surface waves has been used to study the crustal structure (Ewing and Press, 1959; Bucher and Smith, 1971). Theoretically, surface waves are sensitive to low-velocity zones, but these waves cannot map detailed crustal structures in practice. However, studies based on P wave delay times from teleseismic earthquakes in conjunction with gravity or surface wave data may provide additional constraints on average crustal velocity and layer thickness (Press and Biehler, 1964).

1.3.4 Waveform inversion tomography

These techniques use the original waveforms and wave equation modeling to obtain high-resolution images. The full waveform inversion technique involves constructing earthquake and Earth structure models from analyzing the complex waveforms recorded by modern broadband seismographs. Techniques dealing with waveform inversion can be broadly divided into two categories: (a) direct waveform-fitting techniques that map waveform mismatches directly to model parameter perturbations (Tarantola, 1986; Nolet, 1990; Akçelik et al., 2003; Tromp et al., 2005), and (b) data functional techniques that isolate the waveform-fitting step, which is nonlinear, from the inversion step, which is linear or can be linearized. The data functionals are estimated from seismograms that, on the one hand, can quantify waveform misfit and, on the other hand, can be used as linear constraints on the model parameters (Gee and Jordan 1992; Gaherty et al., 1996; Katzman et al., 1998).

1.3.5 Travel time inversion

This is the most popular and robust technique that depends on the Ray theory and Fermat's principle. Travel time inversion has been widely used to image subsurface structure at all scales (global, local, and regional) from the inversion of source-receiver travel times of seismic waves. For imaging crustal and lithospheric structure, this method was introduced in the mid-1970s (Aki, 1976), widely known as travel time tomography. Travel time tomography seeks to estimate the velocity of propagation of seismic waves from the first arrival times. The inverse problem is formulated in terms of the difference between observed and estimated travel times from known sources (earthquake events) to seismic receivers' location (recording stations).

1.3.6 Finite-frequency travel time tomography

These methods aim at obtaining a higher resolution, and their mathematical formulations include the effects of wave diffraction. Ray paths are replaced by volumetric sensitivity kernels. Methods stem from the banana-doughnut sensitivity kernels and consider finite-frequency effects. The formulation considers frequency-dependent travel time and amplitude anomalies to achieve increased resolution (e.g., Dahlen et al., 2000; Pollitz, 2007; Zhang et al., 2011). The technique provides an accurate estimate of amplitudes of velocity anomalies and their spatial resolution because the method considers wave propagation physics. This method helps map near-surface structures where infinite-frequency (ray) theory has limitations because of the length scales of heterogeneities comparable to or shorter than the wavelengths of seismic waves.

1.3.7 Receiver function analysis

Burdick and Langston (1977) found that some energy traces in the SV component near the P wave arrival times inferred that it was attributed to converted seismic waves because of crustal discontinuity near the station. They carried out forward waveform modeling by taking different crustal velocity structures and obtained similarities between computed and observed waveforms. The computed time series based on three component seismograms provided information on Earth structure beneath a seismic station. Thus, it was recognized that P to S converted phases contain information about the Earth structure between discontinuity and station. An algorithm to compute RF was developed by Park and Levin (2000) using multiple-taper spectral correlation. This method was further improved by Helffrich (2006) to be applied to a more extended time series. Ammon (1991) introduced an inversion method based on spectral deconvolution (Clayton and Wiggins, 1976) to estimate the velocity structure without a prior knowledge of the local velocity model. The method involved deconvolution of the radial component from the vertical component and resulted in the retrieval of responses whose peaks indicate positive or negative velocity anomalies. Because this provides structure beneath the recording station (receiver), it is called a "Receiver Function" (Figure 1.2).

Several other techniques have also been used to obtain crustal structure, such as studying the particle motion of Rayleigh wave (Boore and Toksoz, 1969); the use of crustal transfer function, and the spectral ratio between observation at different localities (e.g., Ellis and Basham, 1968; Phinney,

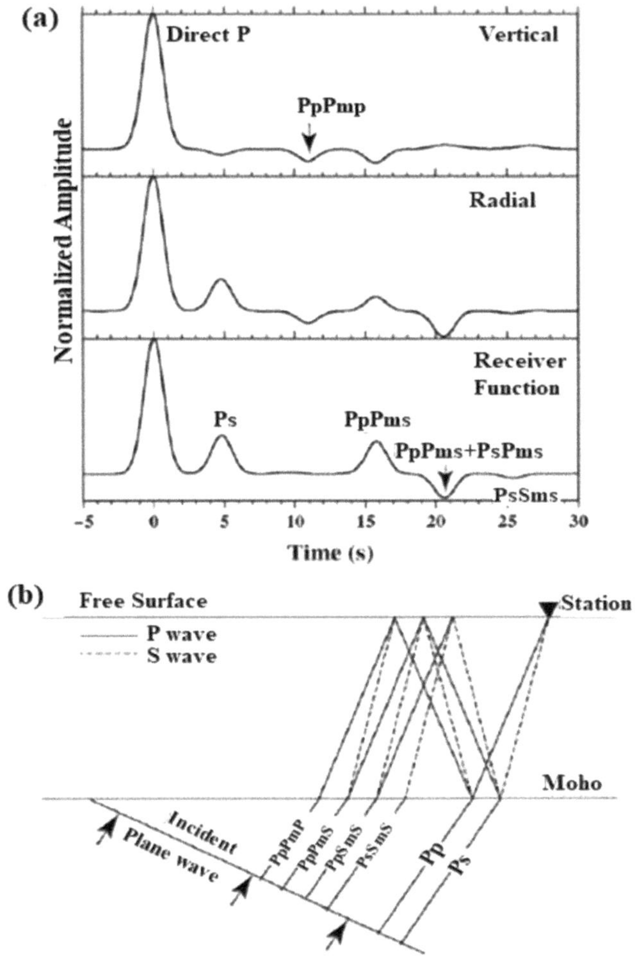

Figure 1.2 (a) Radial receiver function for a single layer over a half-space computed from deconvolution of vertical seismogram with the radial seismogram. (b) Ray diagram of different crustal phases. (Ammon, 1991.)

1969; Bakun, 1971); and determination of lateral velocity variations using least-squares inversion of refraction measurements (Wesson, 1971). However, these and many other techniques have specific advantages and disadvantages, and no single techniques are universally applicable to estimate the crustal structure.

Further limiting the scope of this chapter to the travel time tomography, the following sections are developed to the mathematical formulation of the problem and procedure to perform the travel time inversion.

1.4 DEVELOPMENT OF SEISMIC TOMOGRAPHY

The propagation of seismic waves from the source to the receiver is affected by the medium's physical properties, such as elastic, anelastic, anisotropy, and density. Seismic tomography is a promising technique to map the 3-D distribution of these properties. Tomographic models have helped analyze the subsurface–lithological structure, temperature, fracturing and faulting process, fluid content, and other factors. Seismic tomography began in the mid-1970s and has advanced to become a critical mapping technique. The roots of this technique can be traced back to the research work of Keiiti Aki and coworkers. This work on regional- and local-scale BWT began in 1974. Further, the work of BWT on a global scale was started by Dziewonski and coworkers in 1975 (Dziewonski et al., 1977). These early works were referred to as 3-D inversion and 3-D perturbations of velocity structure instead of seismic tomography.

Radon (1917) developed the theoretical background necessary to solve tomography problems. Reagan (1978) and Clayton and Hearn (1982) used seismic tomography for the first time in the study of seismic reflections and seismology, respectively.

Many different tomography types find applications in science and technology, such as medicine, biology, materials science, geology, and geophysics. The primary objective of tomography is to obtain a cross-section of an object's internal structure to infer its various properties (Nowack and Li, 2009).

The way medical tomography uses radiation and magnetic fields to image the structure of the human body's parts; on similar principles, seismic tomography can produce Earth images. The main aim of seismic tomography is to study the subsurface structure of the Earth on a different scale and to advance the knowledge of the Earth's internal structure. Based on the scale, seismic tomography can be divided into global and near-surface tomography. As stated earlier, global tomography is based on the data obtained from teleseismic earthquakes. Near-surface tomography deals with mapping the shallow subsurface structure a few kilometers deep, covering parts of the crust and upper mantle. The focus of the further sections is on near-surface tomography.

1.5 LOCAL EARTHQUAKE TOMOGRAPHY

Aki and Lee (1976) were the first to study seismic tomography from the inversion of first arrival P-wave travel times from local earthquakes to estimate the velocity structure and hypocenter locations in Bear Valley, California. The source-receiver geometry for this type of study is shown schematically in Figure 1.1a; the earthquake sources lie beneath the receiver array within the model volume. The study suggested that hypocenter coordinates, which are not accurately known, must be included in the inversion. Since this work, this branch of seismic tomography is popularly known as local earthquake tomography or LET (Thurber, 1993). Aki and Lee (1976) used a constant velocity block parameterization using a homogeneous initial model and a linear inversion. Thurber (1983) used an approximate ray tracing (ART) method to facilitate an iterative nonlinear inversion of P-wave travel times for continuously varying velocity structure and hypocenter locations. In many studies, S-wave data is also included in their analysis to allow inverting for V_P/V_S ratio (Graeber and Asch, 1999). These studies suggested that V_P/V_S ratio is a better indicator of hydration levels than the V_P.

The drawback of LET can only be conducted in seismically active regions, usually located at or near the plate boundaries. LET has been carried out in many different regions, including California (Aki and Lee, 1976; Thurber, 1983; Eberhart-Phillips and Michael, 1993), Taiwan (Kao and Rau, 1999), Japan (Zhao et al., 1992), South America (Bosch, 1997; Graeber and Asch, 1999), South Australia (Greenhalgh et al., 1989), Taiwan (Ustaszewski et al., 2012; Huang et al., 2014), and the Mediterranean (Luca et al., 1997; Papazachos and Nolet, 1997).

LET has been able to image the lithosphere and upper asthenosphere up to depths of 200 km in subduction zone settings (e.g., Abers, 1994; Graeber and Asch, 1999). High-resolution images of the crust have also been obtained using shallow earthquakes (e.g., Thurber, 1983; Chiarabba et al., 1997). Advantages of LET include more significant depth of penetration and the added structural information provided by the relocated hypocenters, for example, the existence of double seismic zones (Hasegawa et al., 1978; Kawakatsu, 1985; Kao and Rau, 1999). However, the relocated hypocenters add to the solution's non-uniqueness, and other phases (except first P and S arrivals) are challenging to incorporate. Thus, LET models rarely include interfaces, although Zhao et al. (1992) included interfaces in their LET model of NE Japan by using observed SP waves converted at the Moho and PS/SP waves converted at the upper boundary of the subducted Pacific Plate.

1.6 TRAVEL TIME TOMOGRAPHY

In simple terms, travel time tomography is a more accurate way of referring to seismic tomography. In this technique, the critical parameter to study the subsurface structure is the travel time of seismic waves. The term *travel time* refers to the time taken by a seismic wave from a source to the recording station after propagating through different subsurface structures, including the reflecting boundaries. Travel time tomography seeks to estimate seismic waves' propagation velocities from the measurements of first arrival times. Generally speaking, this data rarely fits a straight line precisely because of scattered data, but we attempt to "fit" a straight line—called a line of best fit. This allows the estimation of an initial velocity model, which can then be utilized for preliminary interpretation of the subsurface structure and can serve as an initial velocity model to initiate inversion. When the 1-D model is estimated, the next step is to parameterize the 3-D velocity structure into uniform blocks of velocity perturbations. The block parameterization includes dividing a three-dimensional medium represented by a geometrical model into similar-size blocks, as illustrated in Figure 1.3. The model contains, in general, all paths connecting the sources and receivers. The next challenging step is to trace the exact ray path, and this is not a straightforward exercise. Several methods have been developed for solving the problem of two-point ray tracing. The well-known techniques include the shooting method—in which the source is sampled to converge on the correct receiver location and the bending method—where the source and receiver are kept fixed and try to bend the ray. Other than these finite differences or graph theory, techniques have also been used. These techniques require a grid of points for ray tracing rather than block parameterization.

The travel time data required for travel time tomography become available from the analysis of suit of seismograms collected from various recording stations. The accurate seismic phase picking requires considerable

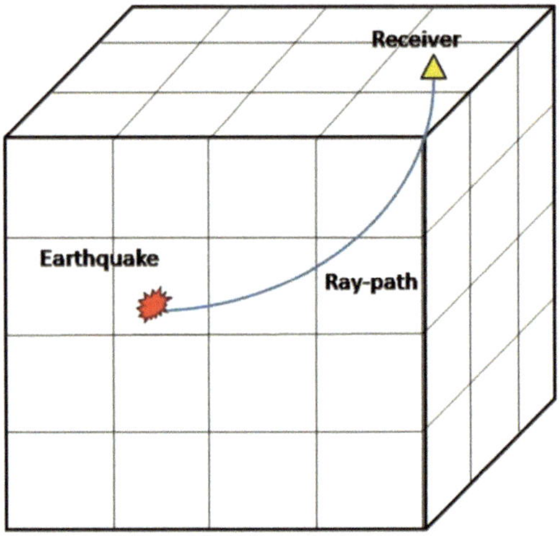

Figure 1.3 An example of ray path in 3-D media with different velocity blocks for tomographic inversion.

insight and experience. Picking of the first P- and S phases are considerably simple and straightforward compared to other seismic phases. An initial velocity model is estimated from this observed travel time data adopting linear regression and an appropriate inversion technique. The data predicted from the estimated model cannot precisely match the predicted data because of uncertainties in phase picking and various other factors such as signal-to-noise ratio, instruments measurements, and site effects. Due to this, we define the travel time residuals (δt) as the difference between the observed and predicted travel times, and these residuals can be represented as the sum of three terms.

$$\delta t = \delta t_{3D} + \delta t_{mislocation} + \delta t_{noise}.$$

Considering that seismic waves are analogous to acoustic sound waves—the propagation velocity of which is known for a particular medium—we can translate the velocity information to characterize the properties of the substance, structure present below the Earth's surface and thus gain a better understanding of its structure and composition. The general procedure adopted in travel time tomography can be summarized as follows:

- The first step involves identifying and reading the arrival times of seismic phases of the earthquake records collected from various seismic stations.

- The next step involves understanding the distances traveled by seismic waves from sources to the receivers. For this, we need to make assumptions on velocities. This step is accomplished by ray tracing—based on the theory of seismic wave propagation within various media. The ray-tracing results allow approximating the distances traveled from sources to receivers adopting different paths from various reflecting layers.
- In the third step, using the data from the previous two steps, the travel time equations are solved for velocity parameters. This step is also known as travel time inversion because we start with observational data and infer the model parameters that could explain this data.

1.6.1 Forward and inverse modeling

The travel time inversion used in the seismic tomography is called inverse modeling. It is appropriate to mention the differences between forward and inversion modeling to understand these modeling techniques better.

Forward modeling involves predicting the data that an experiment would likely produce. In the context of Earth Sciences, we should have prior knowledge of the subsurface geology of the study area. After deciding the type of experiment we are likely to conduct, the model parameters for data acquisition can be inferred for such an experiment. Based on the theory and physical processes, we can then predict the data from the geological model's parameters. However, inversion deals with how to estimate model parameters after physically running an experiment. The known data here consists of the data acquisition parameters and the data produced from the conducted experiment. We then apply a series of mathematical methods to obtain a subsurface geological model from the experiment.

While the two methods work at opposite ends, they can be combined with one working as a test for the other. Any geophysical inversion can allow us to obtain a subsurface structure model, which fits observed data. Forward modeling is used to establish the relation between the subsurface

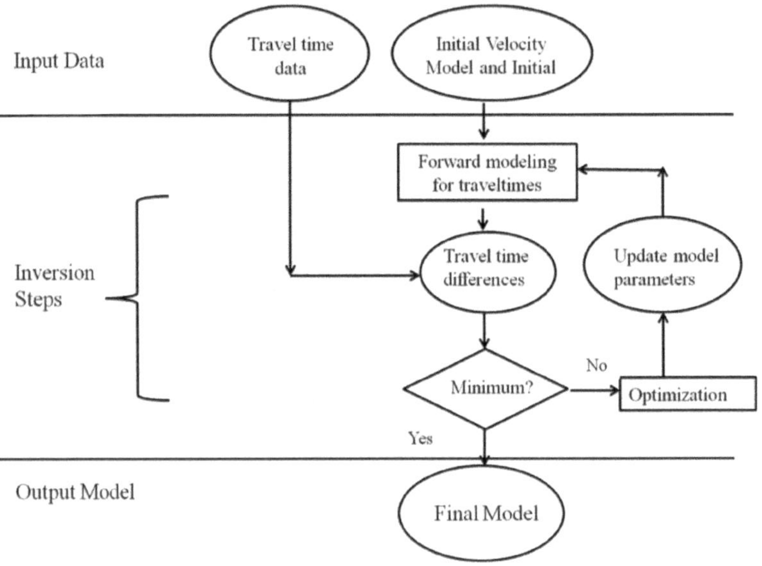

Figure 1.4 Flowchart of tomographic inversion.

model and the observed data. If M is a model and D is observed data, then forward modeling can be expressed as:

$$F(M) = D \tag{1.1}$$

where F is forward modeling, operator, which calculates geophysical data D for a given M. Then inversion can be described as solving equation (1.1) for unknown model M. As the inverse problem is usually ill-conditioned, additional constraints are imposed and play an essential role in travel time inversion. The objective function of equation (1.1) can be minimized using L2 (the least-square method) or L1 norm. L1 norms are usually stable in case of spikes and significant errors present in travel times.

For seismic travel time inversion, model (1) is the most commonly used. In this model, D is observed travel times, M is a depth, velocity model, and F is a forward-modeling operator, which calculates travel times of seismic phases (see Figure 1.4).

1.6.2 The nonlinearity of tomographic inversion

If the relation between the model parameters and the data would be linear, a result could be easily obtained. However, the tomographic problem in principle is nonlinear because it depends on the model parameters and the ray path geometry. The tomographic problem needs to be linearized around an approximate ray path to invert for the model parameters. Approximate ray paths can be defined by the introduction of an initial model of the subsurface. In this initial model, the approximate ray paths of the focusing operators can be calculated. The difference between these modeled travel times and the observed travel times of a particular ray path becomes the data to be inverted.

Consequently, the output of the inversion represents the difference between the initial and actual subsurface models. This relation is assumed to be linear and forms the basis of many tomography studies (Tarantola, 1986; Nolet, 1987; Scales, 1987; Phillips and Fehler, 1991). The relation is represented as:

$$\Delta d = A\Delta m \tag{1.2}$$

where Δd is the vector that contains the travel time differences for all measurement locations, Δm is the vector that contains the model parameter updates, and A is the matrix that describes a linear relationship between the differences in travel times and the model parameter updates. In this way, the tomographic problem is linearized, and the inversion of the travel time differences leads to an update of the initial model parameters. It is to be noted that this linear relationship is only an approximation of the proper relation between Δd and Δm. In the updated model, new travel time differences can be calculated, and these differences can be inverted again, and so on. This iterative process is commonly denoted as nonlinear tomography.

1.6.3 Method of travel time inversion

In order to set up the nonlinear tomographic system for the focusing operators, the following steps should be followed:

1. Model parameterization: A set of model parameters is defined to represent the study area's seismic structure. In tomographic methods, typically, the preliminary estimates of model parameters have to be specified.

2. Forward modeling: Procedure has been defined to calculate travel times in the initial velocity model for the source-receiver pairs. The travel time differences between the observed and estimated travel times are calculated.
3. Optimization or inverse modeling: the differences between the observed and estimated travel times are minimized to optimize the model parameters. Furthermore, the adjustments of parameters are then automated to the initial parameters.

Once the velocity model is revised in the third step, we return to the second step to further optimize the velocity model. This procedure is repeated till an accurate velocity model is obtained. In the subsequent sections, further details are provided regarding parameterization, forward modeling, and optimization.

1.6.3.1 Parameterization of structure

1.6.3.1.1 1-D VELOCITY STRUCTURE

One-dimensional crustal velocity models find applications in many research areas in seismology. These models are adopted to estimate hypocenter parameters and focal mechanism solutions. These models can be adopted as initial models to develop 2-D and 3-D velocity models (Kissling et al., 1994). The travel times of seismic phases have been widely used to develop a 1-D velocity model (Crosson, 1976; VanDecar and Crosson, 1990). The 1-D Earth model represents the Earth composed of constants velocity layers (Figure 1.5), varying with increasing depths. Several forward-modeling and inverse-modeling techniques were developed to estimate the velocity structure as a function of depths from travel times and wavefield. These techniques include the reflection method, Herglotz-Wiechert inversion, and tau-p inversion (Fuchs and Muller 1971; Clayton and McMechan, 1981; Aki and Richards 2002).

1.6.3.1.2 3-D VELOCITY STRUCTURE

The velocity model of the region is defined by several model parameters. In the tomographic inversion, these parameters include the subsurface

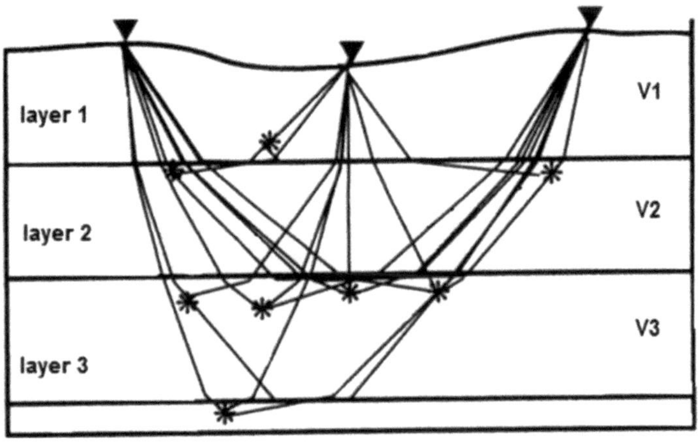

Figure 1.5 A systematic representation of 1-D velocity model having three layers with varying velocity along with traveled ray paths from source-receiver pairs.

parameters that influence travel times, velocities, and sources' locations. Tomographic methods usually need preliminary estimates of the model parameter. The velocity parameterization is more complex and describes a complete area or volume, and the source parameters describe each source's position. The key to success in tomographic inversion is defining an optimum number of parameters that accurately describe the subsurface velocity structure.

The travel time of a seismic wave between source and receiver solely depends on the medium's velocity structure through which the wave propagates. Therefore, subsurface structure in a seismic travel time inversion is represented by variations in P- or S-wave velocity (or slowness). Velocity variations may be defined by a set of interfaces whose geometry is varied to satisfy the data. These variations are represented by a set of constant velocity blocks or nodes with a specified interpolation function or a combination of velocity and interface parameters (Figure 1.6). The most appropriate choice will depend on the *a priori* information (e.g., known faults or other interfaces), whether or not the data indicates the presence of interfaces (e.g., reflections, mode conversions), whether data coverage is adequate to resolve the trade-off between interface position and velocity, and finally, the capabilities of the inversion routine. There are several methods for parameterization of the subsurface model. Some of them are discussed next.

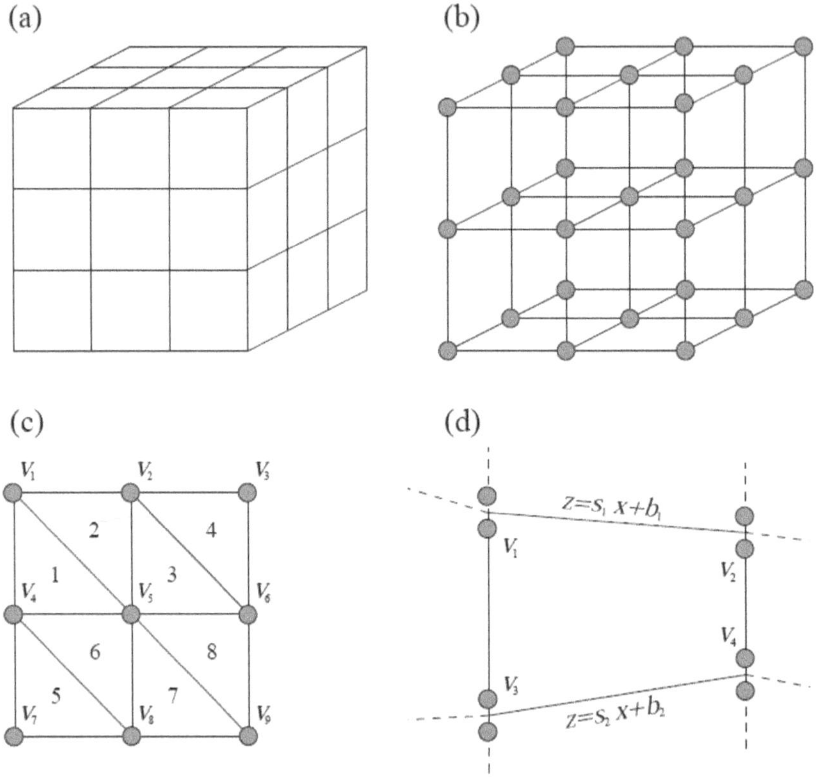

Figure 1.6 Different types of velocity parameterization: (a) constant velocity blocks, (b) a grid of velocity nodes, (c) triangulated velocity grid designed for constant velocity gradient cells (after White, 1989), (d) velocity defined by a trapezoidal block (after Zelt and Smith, 1992). The four corner vertices labeled V_1 to V_4 define the velocity within the block. If adjacent nodes in vertically adjacent blocks are the same, velocity will be vertically continuous; else subhorizontal interface will be defined.

1.6.3.1.3 CONSTANT VELOCITY BLOCKS

(Figure 1.6a) constant velocity blocks are simple to define and allow rays to be traced straightforwardly. Their use in inversion is to recover smooth variations in velocity. Although this parameterization is not optimal for such a purpose, if many blocks are used, and smoothing is applied, an acceptable approximation to a continuously varying velocity field is possible.

1.6.3.1.4 GRID OF VELOCITY NODES

An alternative approach to represent a velocity field is to define a set of discrete velocity nodes on a grid (Figure 1.6b) together with a specified interpolation function. This approach was adopted by Thurber (1983) for local earthquake tomography. Following function describes the velocity at any point (x, y, z) within a rectangular grid of nodes:

$$v(x,y,z) = \sum_{i=1}^{2}\sum_{j=1}^{2}\sum_{k=1}^{2} V(x_i,y_j,z_k)\left(1-\left|\frac{x-x_i}{x_2-x_1}\right|\right)\left(1-\left|\frac{y-y_i}{y_2-y_1}\right|\right)\left(1-\left|\frac{z-z_i}{z_2-z_1}\right|\right) \quad (1.3)$$

where $V(x_i, y_j, z_k)$ is the velocities values of the eight grid points around (x, y, z). Equation 1.3 ensures that the velocity field will be continuous throughout the model volume, although the velocity gradient will be discontinuous from cell to cell. Needless to say, that during the inversion process, too many parameters result in an under-determined system, while too few parameters will be unable to describe the subsurface structure. Therefore, for the velocity parameterization, it is vital to find a method that optimally describes the subsurface.

1.6.3.1.5 CONSTANT VELOCITY GRADIENT CELL

This method aims to bridge the gap between a block approach and a grid approach and uses cells with a constant velocity gradient. White (1989) described a method of 2-D refraction tomography in which a rectangular grid of nodes was used to define triangular regions of constant velocity gradient (see Figure 1.6c). The velocity within each cell is defined as:

$$v(x,y) = v_0 + (x-x_0)\nabla_x v + (z-z_0)\nabla_z v \quad (1.4)$$

where, v_0, $\nabla_x v$ and $\nabla_z v$ are computed from the velocities at the vertices of the triangle (e.g., $v_0 = 1$, $\nabla_x v = (v_2 - v_1)/\nabla x$, $\nabla_z v = (v_5 - v_2)/\nabla z$ in cell 1 of Figure 1.6c). In this method, velocity varies continuously throughout the medium, and the rays are traced analytically within each cell. However, the velocity gradient is discontinuous at each cell boundary which, in conjunction with their triangular shape, can result in difficulties in finding the source-receiver ray path. Similar methods were used by Chapman and Drummond (1982) and Bishop et al. (1985).

1.6.3.1.6 IRREGULAR GRID NODES

The use of irregular blocks or grid nodes to describe the velocity distribution is not very common in travel time tomography. However, irregular-shaped elements can be adapted to suit variations in subsurface data coverage. The 2-D wide-angle inversion method of Zelt and Smith (1992) used such a method. The velocity field is discretized into a series of trapezoidal blocks, with four corner vertices used to specify the velocity within the block. Figure 1.6d shows the design of the trapezoid. The velocity within the trapezoid is given as (Zelt and Smith, 1992):

$$v(x,z) = \frac{c_1 x + c_2 x^2 + c_3 z + c_4 xz + c_5}{c_6 x + c_7} \qquad (1.5)$$

c_1's are linear combinations of the corner velocities (see Zelt and Smith, 1992). Discontinuities are permitted in the velocity field by including pairs of nodes on subhorizontal boundaries (see Figure 1.6d). If these nodal values are equal, the velocity field is continuous. The inherent flexibility of this technique means that a velocity structure with or without layering can be represented. If layers are present, then it is possible to vary velocity within the layers arbitrarily. Pinch out, and isolated bodies can be represented by using three-sided blocks. Finally, the trapezoid's size and shape can be varied to suit the resolving power of the data set.

1.6.3.2 The forward problem of travel time determination

The forward problem involves calculating theoretical travel times of seismic waves—between the source-receiver pairs, traveling through the Earth's 3-D structure for a given model M. The travel times are computed by the ray tracing, form hypocenter to the recording station placed on the Earth's surface. In its simplest form, the ray is represented by a straight line connecting the source and receiver. Nevertheless, this representation is not sufficiently exact in practical applications. A better method is to assume that the seismic velocity increases with depth, and the ray path forms a part of a circle connecting the source and receiver. However, the best approximation of ray tracing includes refraction and reflection at the layer boundaries or the vertical boundaries of the blocks. Some of these parameters and their methods of computation are described below.

We express the travel time of a seismic wave between source S and receiver R the following integral:

$$t = \int_{S}^{R} \frac{1}{v(x)} dl \tag{1.6}$$

where *dl* is differential path length, *x* is the position vector, and *v* is velocity. In order to evaluate this integral, the path taken by the seismic ray should be known. However, we face difficulty performing this integration because the wave's travel path is not known as it depends on the velocity structure. The equation that governs the geometry of ray paths can be derived from the eikonal equation by considering how a slight change in time *dt* affects a point *x* on a wavefront (see Aki and Richards, 2002). The resultant *ray equation:*

$$\frac{d}{dl}\left(\frac{1}{v(x)}\frac{dx}{dl}\right) = \nabla\left(\frac{1}{v(x)}\right) \tag{1.7}$$

In travel time inversion, the traditional method of determining source-receiver travel times is ray tracing. These methods of travel time determination are described below.

1.6.3.2.1 RAY TRACING

Finding a ray path between a source and receiver is a boundary value problem. There are two basic methods to obtain the solution methods, viz, shooting and bending.

1.6.3.2.2 SHOOTING METHOD

In this method, the ray equation (Eq. 1.7) is formulated as an initial value problem, and a complete ray path can be traced. However, the initial data in the form of source coordinates and the takeoff angle of the ray are specified. The boundary value problem is then solved by shooting rays in the medium from the source and using information from the computed paths to update the initial ray trajectories to more accurately target the receivers (see Figure 1.7). Rays may also be shot from receivers towards the sources as the principle of reciprocity applies in ray tracing.

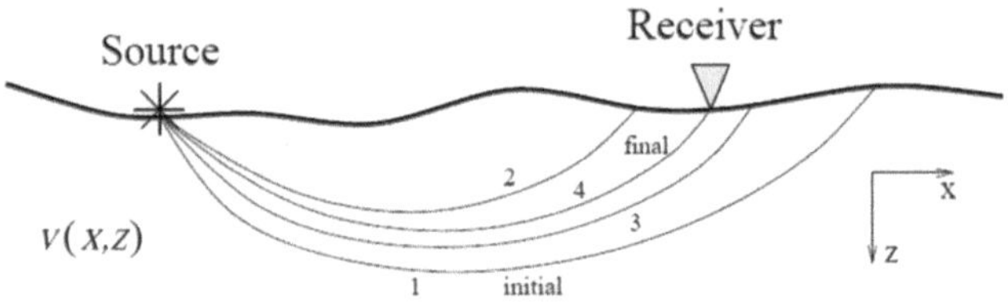

Figure 1.7 Principle of the shooting method. The initial projection angle of ray (1) is iteratively adjusted until the final ray (4) passes sufficiently close to the receiver. (After Rawlinson 2000)

As shown in Figure 1.7, the shooting method fixes one end of the ray path (source point), takes initial incidence angle i_0 and initial azimuth j_0, and then uses ray path equations to find the coordinates of another end point (station).

Equation (1.6) could be rewritten as first-order equations:

$$\frac{\partial x}{\partial l} = v(x)p$$
$$\frac{\partial p}{\partial l} = \frac{\partial}{\partial l}\left(\frac{1}{v(x)}\right) \tag{1.8}$$

where p is the slowness. Equations (1.8) gives a system of six first-order ODE (ordinary differential equation) equations (can be simplified into five independent equations), which must be integrated numerically to find the desired ray path. For *Cartesian Coordinates*, the system becomes

$$x' = c\sin i\cos j;$$

$$y' = c\sin i\sin j;$$

$$z' = c\cos i;$$

$$i' = -\cos i\left[c_x \cos j + c_y \sin j\right] + c_z \sin i;$$

$$j' = \frac{1}{\sin i}\left[c_x \sin j - c_y \cos j\right].$$

The solution should satisfy

$$h(i_0, j_0) = H, \tag{1.9}$$

$$g(i_0, j_0) = G \tag{1.10}$$

Here h and g are the calculated coordinates (latitude and longitude) of the end of the ray. H and G are the coordinates of the end of the desired ray.

Julian and Gubbins (1977) suggested Newton's method and false position method to solve equations (1.9) and (1.10), and both methods have to be applied iteratively.

In Newton's method, we have:

$$\begin{bmatrix} \dfrac{\partial h}{\partial i_0} & \dfrac{\partial h}{\partial j_0} \\ \dfrac{\partial g}{\partial i_0} & \dfrac{\partial g}{\partial j_0} \end{bmatrix} \begin{bmatrix} i_0^{(n+1)} - i_0^n \\ j_0^{(n+1)} - j_0^n \end{bmatrix} = \begin{bmatrix} H - h_n \\ G - g_n \end{bmatrix}$$

However, to apply Newton's method first, we need to calculate partial derivatives $\frac{\partial h}{\partial i_0}, \frac{\partial h}{\partial j_0}, \frac{\partial g}{\partial i_0}, \frac{\partial g}{\partial j_0}$.

In the false position method, we have the following formulation:

$$\begin{vmatrix} k_0 - k_0^{(1)} & k_0 - k_0^{(2)} & k_0 - k_0^{(3)} \\ h^{(1)} - H & h^{(2)} - H & h^{(3)} - H \\ g^{(1)} - G & g^{(2)} - G & g^{(3)} - G \end{vmatrix} = 0, (k = i, j).$$

In these expressions, the superscripts represent estimated values in three previous iterations. Further, this does not require the calculation of the partial derivatives, but convergence is slow.

In 2-D problems, a shooting approach is often used because the source-receiver array lies on a single vertical plane. This makes the shooting of a single fan of rays an effective way of obtaining nearby rays to all targets. Zelt and Smith (1992) use a bisection method to find rays that bound each required phase (e.g., a set of rays that reflect at the surface from a particular interface). Similar methods were used by Cassell (1982) and Langan et al. (1985). Other applications of shooting methods in 2-D reflection and/or refraction travel time inversion include those by Farra and Madariaga (1988), White (1989), Lutter et al. (1990), Williamson (1990), Zelt and Smith (1992), and McCaughey and Singh (1997). Examples of its use in 3-D reflection and/or refraction travel time inversions are harder to find, although several 3-D tomographic studies that combined refraction and local earthquake data (Benz and Smith, 1984; Ankeny et al., 1986; Sambridge, 1990) and some teleseismic tomography studies (Neele et al., 1993; VanDecar et al., 1995) have used shooting methods of ray tracing.

1.6.3.2.3 BENDING METHOD

As shown in Figure 1.8, the bending method fixes the two ends, one at the source end and another at the receiver end, and takes some initial estimate of the ray path and adjusts it until it satisfies minimum travel time criteria (Fermat's criteria). The bending method is derived for a continuous 3-D velocity media, which traces a two-point ray geometry modeled from first-order differential equations (Julian and Gubbins, 1977). Consider a ray traveling between source S and receiver R through an inhomogeneous medium with velocity v and slowness p. If $x = x(q)$ represents this ray path and q can be chosen at any point on this ray path, then

$$\frac{\partial l}{\partial q} = \sqrt{\dot{x}^2 + \dot{y}^2 + \dot{z}^2} \equiv F, \ where\ \dot{x} = \frac{\partial x}{\partial q}$$

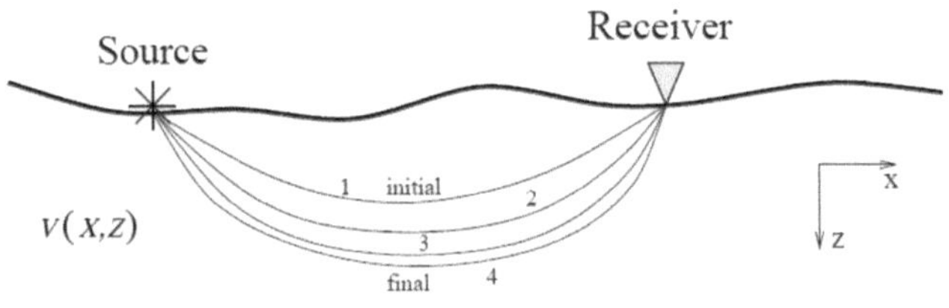

Figure 1.8 Principle of the bending method. The initial path geometry (ray 1) is adjusted until it satisfies Fermat's principle (ray 4). (After Rawlinson 2000)

using these expressions, the equation 1.6 of travel time can be written as

$$t = \int_{qS}^{qR} pFdq \tag{1.11}$$

According to Julian and Gubbins (1977), equation of travel time can be solved by the standard method of calculus of variation and used an Euler's equation to represent the boundary value problem between two points A and B as:

$$\begin{aligned} &\frac{d}{dq}(pF)_{\dot{x}} = (pF)_x \\ &\frac{d}{dq}(pF)_{\dot{y}} = (pF)_y \\ &\frac{d}{dq}(pF)_{\dot{z}} = (pF)_z \\ &\frac{dF}{dq} = 0, \end{aligned} \tag{1.12}$$

these equations are solved iteratively with conditions for $(n+1)$ iterations:

$$\mathrm{x}^{(n+1)}(q) = \mathrm{x}^{(n)}(q) + \xi^{(n)}, \tag{1.13}$$

where $\xi^{(n)}$ is perturbation added in nth iteration, $q = l/L$, L is the total length of the ray path, and $0 \leq l \leq L$. If $l = L$ then q represents single representation of total ray path and boundary condition will be $x(0) = x_S$, $x(1) = x_R$; $y(0) = y_S$, $y(1) = y_R$; and $z(0) = z_S$, $z(1) = z_R$. Equation 1.13 is substituted in equation 1.12, and the resulting equation can be solved by linearizing $\xi^{(n)}$. Iterations are carried out until the solution converges to RMS's specified limit (Julian and Gubbins 1977).

Um and Thurber (1987) used the pseudo-bending technique to solve the two-point ray tracing in the 3-D velocity medium. In this method, after each iteration, the integration step-size reduces to half of the previous step-size. This method is comparatively much faster than the earlier bending technique. Bending methods of ray tracing have been used in many studies based on the inverting that invert teleseismic data (Thomson and Gubbins, 1982; Zhao et al., 1994, 1996; Steck et al., 1998). Recently, Koulakov (2009a) used the bending tracing technique for local earthquake tomography and showed bending works faster than the shooting method.

In comparing between bending and shooting methods, Julian and Gubbins (1977) found that bending is computationally faster than shooting by a factor of 10 or more in media with continuous velocity variations. When discontinuities are present, however, the formulation of the bending problem becomes much more complex.

Two alternative approaches to the ray-tracing problem include the methods based on finite difference (Vidale, 1990) and network theory (Moser, 1991). An essential aspect of these methods is the apparent ability to estimate both the global minimum time path and, in Vidale's method, a crude estimation of the arrival amplitude. Vidale obtained a finite difference solution to the eikonal equations on a 3-D grid to evaluate the entire travel time field, allowing for the subsequent assignment of ray paths and arrival amplitudes (Vidale and Houston, 1993). Moser's method uses network theory to find a global minimum travel time path efficiently. Accurate ray-tracing techniques such as these combined with ever more powerful computers promise to make travel time tomography increasingly practical for large-scale problems.

1.6.3.3 Inversion of travel time and its solution

The inversion involves estimating an adjustment vector that best satisfies the observed data, d_{obs}, through the known relationship D = F(M). In travel time inversion, data parameters are travel times, and model parameters are velocities. The functional F is nonlinear because the ray path's trajectory is a function of the velocity structure of the medium through which it travels. Hence, inversion schemes should account for this nonlinearity unless shown that its effects are not severe. The three techniques are adopted to solve the inversion. These techniques include back projection, conjugate gradient methods, and global optimization techniques and are described below.

1.6.3.3.1 BACK PROJECTION

This is one of the commonly applied methods for the inversion of travel time data. As stated below, the travel time of a seismic ray along a ray path can be expressed as:

$$t = \int_{L(s)} p(\mathrm{x})dl \tag{1.14}$$

where $L(s)$ is the path along which the integration is performed, p in the slowness vector, and x is the position vector.

In forward modeling, with consideration that the region to be modeled is divided into a set of blocks, the forward problem can be written as

$$t = \sum_b l_b p_b$$

where t represents the time delay of the ray, p_b signifies slowness perturbation of the b_{th} block, and l_b indicates the length of the r_{th} ray segment in the b_{th} block.

If $p(\mathrm{x}) = p_o(\mathrm{x}) + \delta p(\mathrm{x})$, then in (eq. 1.14) the modified path is $L_o + \delta L = L(p_o + \delta p,)$ and the adjusted travel time is given as:

$$t = \int_{L_0 \delta L} p_o(\mathrm{x})dl + \int_{L_0 \delta L} \delta p(\mathrm{x})dl \tag{1.15}$$

Fermat's principle can be invoked to replace $Lo + \delta L$ with Lo, so that the travel time perturbation is accurate to first order:

$$\delta t = \int_{L_0} \delta p(\mathrm{x})dl + O\left(\delta p(x)^2\right) \tag{1.16}$$

Note that $\delta t = t - t_o$ where $t_0 = \int_{L_0} p_0(\mathrm{x})dl$ is the path integral for the unperturbed medium. If a medium is represented by M constant slowness blocks, then equation 1.16 (for N rays) can be expressed as:

$$d = Gm \tag{1.17}$$

where d represents the travel time's residual vector components and m is the slowness perturbation vector. G is a matrix with N × M dimensions—N stands for travel time data, and M signifies the number of model blocks. Ray lengths l_{ij} corresponding to the distance traversed by each ray in each m_j. $G = \partial g / \partial m$, where $g(m)$ is the model prediction. As most of the blocks are empty because rays do not pass through these blocks, most of G's elements are zero. The least-squares solution of this equation is given as (e.g., Aki et al., 1977):

$$G^T G m = G^T d \tag{1.18}$$

$G^T G$ has a large dimension, and its direct inversion is difficult. According to the back-projection method's simple scheme, direct inversion can be avoided, and initially, only the diagonals of $G^T G$ (Humphreys and Clayton, 1988) are used to obtain the inversion. Adopting this approximation, the estimates of the model parameters are:

$$m = D^{-1} G^T d, \text{where}, D^{-1} = dig\left(G^T G\right) \tag{1.19}$$

Although this method is fast and needs less storage space, due to which the resulting model is strongly blurred. Humphreys and Clayton (1988) suggested two algorithms—iteration and filtering, to take care of blurring (e.g., Natterer, 1986). Filtering is a deconvolutional scheme implemented either in the space or wavenumber domain. Iteration focuses on an image through repeated back projection of the current residuals and is a stable procedure. However, the iteration has a limitation that focusing can be very slow along paths with few ray crossings. Filtering provided an unstable model when ray distribution is inhomogeneous. The generalized inverse provided an alternative means to get a state solution (Backus and Gilbert 1968). Many studies have been carried out using back-projection methods (Humphreys and Clayton, 1990; Hole, 1992; and Zelt and Barton, 1998).

Two back-projection techniques have been applied to solve equations (1.17). These techniques initially used in medical imaging include algebraic reconstruction technique (ART) and the simultaneous iterative reconstruction technique (SIRT). Kissling (1988) used parameter separation and applied ART approximation to the damped least-square solution of equation 1.17. However, ART suffers from poor convergence (Blundell, 1993). SIRT takes care of some of the convergence problems associated with ART by averaging the perturbations applied to each parameter. SIRT has been used for LET but is generally applied when the hypocenter-velocity structure coupling was not treated explicitly (Walck and Clayton, 1987). Further, SIRT provides a least-squares solution with different weighting than that specified by the 'user'.

1.6.3.3.2 CONJUGATE GRADIENT METHOD

Conjugate gradient method, in particular, the algorithm LSQR (least squares), is an effective alternative to the above reconstruction techniques (Paige and Saunders, 1982). It is based on Lanczos procedures used to diagonalize the system of equations followed by a QR decomposition to obtain the solution. This method is fast and works very well to solve sparse linear systems. This method's basic idea is to formulate the inversion as an optimization problem and then minimize the objective function in terms of the data residuals. The details of the method are outlined in Golub and Van Loan (1983), Spakman and Nolet (1988), Scales (1987), and van der Sluis and van der Vorst (1987).

In this method, the travel time equation (1.15) is linearized about some initial model, and the linearized equation in discrete form is represented as:

$$\Delta t = G\Delta m \tag{1.20}$$

Δt represents the components of the travel time residual vector, and Δm represents components of the differences between the model slowness and the actual slowness (referred to as a cell-based discretization), and G is a matrix whose element G_{ij}, is the distance the ith ray travels in the jth cell. A generalized solution (e.g., a least-squares solution of equation [1.20]) is a vector Δm that minimizes the Euclidean norm of ($G\,\Delta m$- Δd). Further, since each ray travels through only a tiny portion of the medium, the distance matrix will be sparse, with 99% or more of the elements equal to zero for large problems. Singular value decomposition (SVD) (Golub and Reinsch, 1970) and the row-action methods (Censor, 1981) can directly deal with equation (1.1). But SVD is unsuitable for large least-squares problems because of large, dense, intermediate matrices. Scales (1987) defined the application of the classic conjugate gradient algorithm of Hestenes and Stiefel (1952) to solve these large, dense, sparse matrices of travel time inversion.

Solution of equation (1.20) carried out by an iterative process where, at each step k, we have the following linear relation between the vector $t^{(k)}$ of travel times of all rays and the vector $m^{(k)}$ with the current estimated slowness field (Scales, 1987):

$$\Delta t^{(k)} = G^{(k)} \Delta m^{(k)} \tag{1.21}$$

where $\Delta t^{(k)} = t^{obs} - t^{(k)}$ and $\Delta m^{(k)} = m^{(k+1)} - m^{(k)}$.

Under the assumption that error distribution in the relationship $t_{obs} \approx g(m_{true})$ is Gaussian, then L2 norms are applicable on this difference. We want to find the vector $\Delta m^{(k)}$ that minimizes $\left\| G^{(k)} \Delta m^{(k)} - \Delta t^{(k)} \right\|^2$. This can be formally stated as finding the minimum of the following objective function:

$$\Phi\left(\Delta m^{(k)}\right) = \left(\Delta t^{(k)} - G^{(k)} \Delta m^{(k)}\right)^T \left(\Delta t^{(k)} - G^{(k)} \Delta m^{(k)}\right) \tag{1.22}$$

The estimated solution, also called least-squares solution, is

$$\left[G^{(k)}\right]^T G^{(k)} \Delta m^{(k),est} = \left[G^{(k)}\right]^T \Delta t^{(k)} \tag{1.23}$$

Least-squares solutions very often do not provide good results, and sometimes they do not even exist. Thus, eq. (1.23) now becomes

$$G^T G m = G^T t, \tag{1.24}$$

$$m = \left(G^T G\right)^{-1} G^T t, \; or \; m = G^{-g} t,$$

where G^{-g} is known as generalized inversion operator which act on the data to yield the model. The solution $m = m^{est}$ will be obtained through conjugate gradient schemes as presented in the next section.

The classical conjugate gradient method (Hestenes and Stiefel, 1952) to solve the least-squares linear system (24) is often referred to as CGLS (Paige and Saunders 1982; Santos and Pierro 2006; Bazan et al., 2014) and is as follows, for a given initial guess m_0 and a maximum number of iterations k^{max},

$$r_0 = t - Gm_0, z_0 = G^T r_0, \text{and } p_0 = z_0;$$
$$\text{repeat}$$
$$w_k = Gp_k;$$
$$\alpha_k = \left(z_k^T z_k\right) / \left(w_k^T w_k\right);$$
$$s_{k+1} = sk + \alpha_k p_k;$$
$$z_{k+1} = G^T r_{k+1}$$
$$\beta_k = \left(z_{k+1}{}^T z_{k+1}\right) / \left(z_k^T z_k\right);$$
$$p_{k+1} = z_{k+1} + \beta_k p_k;$$
$$\text{until } \Phi(k) \text{ reaches a local minimum or k} > \text{k}_{\text{max}}$$

The matrix $G^T G$ is not explicitly computed since it usually has a higher number of non-zero entries and a higher condition number than G (Scales, 1987).

Other gradient optimization methods to solve the inverse problem include the Gauss-Newton method (Aki et al., 1977, Zhao et al., 1994; Steck et al. 1998), steepest descent method (Blundell 1993), and subspace method (Kennett et al., 1988, Sambridge 1990; Williamson 1990). These methods are explained in detail by Rawlinson (2000).

The principal advantage of the steepest descent and conjugate gradient methods over the Newton and DLS methods is that a large system of potentially ill-conditioned linear equations need not be solved. Consequently, these methods are much more rapid for the inversion step. However, since they only minimize one dimension at each iteration, they tend to converge more slowly. If the procedure for solving the forward step is fast compared to solving the M × M matrix equation, then steepest descent or conjugate gradients may be preferable. However, if the forward step is slow compared to solving the matrix inverse, then a Newton or DLS approach may be quicker.

1.6.3.3.3 GLOBAL OPTIMIZATION

The above inversion methods provide local minimum because they span the model space near the initial model estimates. Therefore, they are unable to provide a global minimum solution. There is a renewed interest in using global optimization techniques to solve geophysical inverse problems because of unprecedented computer power advancements. Several types of global optimization techniques have been applied to the inversion of seismic data using random processes to search model space and find better models. These methods include Monte Carlo, genetic algorithms, and simulated annealing.

The Monte Carlo approach, defined as a purely random search of model space, is the simplest of these global methods. The misfit function is evaluated at a set of points in the model space that are randomly chosen. The preferred solution is the model with a minimum misfit. The selection of points in model space has no dependence on previous points. A basic Monte Carlo approach is not suited to most seismic travel time inversion problems because it requires significant computation time, even when model space is not very large. Therefore, it becomes difficult to apply this method if many model parameters are involved (Sambridge and Drijkoningen, 1992; Rawlinson, 2000).

Based on the concept of biological evolution, genetic algorithms are used to develop new models from the parent generation of the randomly picked initial models. Producing a new set of models from a preexisting set involves four basic steps: forward step, reproduction, mutation, and crossover. Such techniques are adopted in the 1-D inversion of marine seismic data (Sambridge and Drijkoningen, 1992; Drijkoningen and White, 1995) and the 2-D inversion of refraction travel times (Boschetti et al., 1996). The first step involves determining travel times for the set of input models. The next step of reproduction involves selecting the parent population and assigning a measure of fitness to each model in the pool based on the misfit function's magnitude at the corresponding point in model space. The crossover step creates a pool of offspring, each of which is a "mix" of its two parents. Sambridge and Drijkoningen (1992) represented models by binary strings and performed the crossover by simply cutting and transposing two segments at a randomly chosen point along the two strings. Mutation involves randomly changing values of some parameter (or bits if a binary string representation is used) in selected models, which ensures that some "freshness" remains in the model pool, although the mutation rate should be kept low (Boschetti et al., 1996, use a mutation rate of 1%) so that "good" models are not corrupted. The basic principle behind genetic algorithms is that models with a high fitness index will pass their characteristics to subsequent generations, while those aspects of the cycle that introduce random changes of the population permit new parts of the model space to be tested.

An artificial neural network (ANN) method was used to compute the least-squares fit to the measured travel time data (Ning et al., 1996). ANN are nonlinear data-driven, self-adaptive approaches compared to the traditional methods. This approach is quite different from the other methods since the ANN computes the Earth Model that provides the best explanation (travel times). Indeed, in the other works (e.g., Roth, 1993; Roth and Tarantola 1994), the ANN learns to invert the function that allows obtaining velocities from travel times, from data resulting from several randomly generated Earth Models. In these works, different Earth Model structures and ray-tracing algorithms were used. Neural networks can be viewed as applications that map one space, the input space, into some output space (Roth and Tarantola, 1994). Baraez et al. (2002) have used modular neural networks for seismic tomography and showed that the modular approach is best suited for the inversion problem. This method was advocated and used by various researchers (e.g., Roth and Tarantola, 1994; Langer et al., 1996; Calderon-Macias et al., 1998; Baraez et al., 2002).

The technique of simulated annealing is analogous to physical annealing in thermodynamic systems. This technique can be applied to optimize the model parameters. Weber (2001) demonstrated a method for finding an optimal, irregular triangular cell parameterization for 2-D travel time tomography that best suits the ray-path geometry (Figure 1.9). He defined a cost function, whose minimization is equivalent to the minimization of the tomographic matrix's null space, and used a nonlinear optimization algorithm of simulated annealing to find the optimal parameterization. They proved that the optimal triangulated model provides more accurate and reliable inversion results than the regular grid model in 2-D travel time tomography.

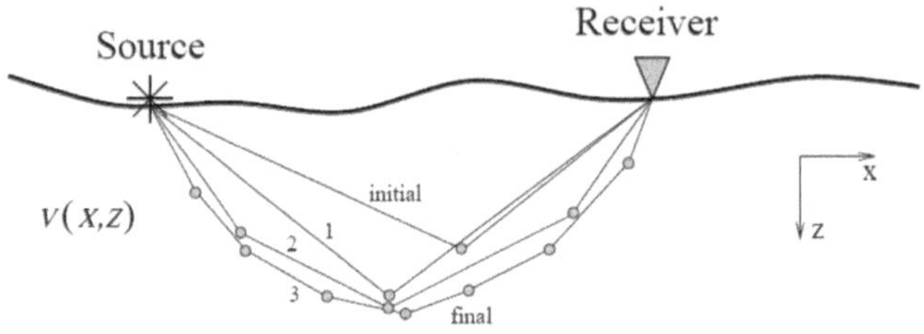

Figure 1.9 Principle of the pseudo-bending method of Um and Thurber (1987). An initial guess ray defined by three points is provided. The center point is perturbed to satisfy the ray equation best. Then the number of segments is doubled, and the process is repeated. This figure schematically represents three such iterations. (After Rawlinson 2000)

Global optimization techniques based on stochastic methods are a fast-developing field of optimization. However, current applications of these techniques in the travel time inversion are restricted because of high computation costs. Global optimization techniques can find global minimum solutions to highly nonlinear inverse problems, but the computational expense when many parameters are involved currently limits their use in seismic data inversion.

1.6.4 Analysis of solution quality

A solution obtained from inversion using the above methods has to be checked for its robustness or quality. A solution without knowledge of resolution or non-uniqueness may lead to erroneous interpretation. Two methods are commonly used to assess the solution robustness in travel time tomography. The first method under the assumption of local linearity estimates model covariance and resolution matrices and the second method involves testing resolution by reconstructing a synthetic model from the same source-receiver geometry as in the actual experiment.

1.6.4.1 Data resolution matrix

After computing a generalized inverse G^{-g}, we solve for the model parameters $m^{est} = G^{-g} d$. We can calculate the data resolution matrix by using m^{est} into the equation $Gm = d$. The relation between the observed and predicted data (d^{obs} and d^{pre}) may be written as:

$$d^{pre} = Gm^{est} = G\left[G^{-g}d^{obs}\right] = \left[GG^{-g}\right]d^{obs} = Hd^{obs} \tag{1.25}$$

where H is the data resolution matrix that describes how well the prediction matches the data. If $H = I$ (identity matrix), then $d^{pre} = d^{obs}$ and the prediction error is zero. In case H is not equal to I, the prediction error is non-zero.

The matrix H's diagonal elements provide essential information regarding the data resolution, and their values always lie between 0 and 1. The row of the matrix H describes how well the neighboring data can be independently predicted or resolved.

1.6.4.2 Model resolution matrix

The model resolution matrix determines whether the model parameters can be independently predicted or resolved. Let us assume a true but unknown set of model parameters m^{true} and $Gm^{true} = d^{obs}$. We seek the transformation that links the estimated model parameters m^{est} with the actual model m^{true}.

$$m^{est} = G^{-g}d^{obs} = G^{-g}\left(Gm^{true}\right) = \left(G^{-g}G\right)m^{true} = Rm^{true} \tag{1.26}$$

where R is known as the model resolution matrix. In the case of $R = I$, then this implies that each model parameter is uniquely resolved. A small number in the ith diagonal element of R indicates the ith model parameter's poor resolution. A non-zero off-diagonal element indicates a trade-off between two model parameters.

1.6.4.3 Model and data covariance matrices

The covariance of model parameters depends on the data's covariance and how the error is mapped from data space to model space. The unit model covariance matrix describes the degree of error amplification that occurs in the mapping. If the data are uncorrelated and have equal variance σ^2, the unit model covariance matrix $\mathrm{cov}_u\, m$ is defined as

$$\left[\mathrm{cov}_u\, m\right] = \sigma^2 G^{-g} G^{-gT} \tag{1.27}$$

σ denotes the standard error in the data.

In case of unequal variance, $\text{cov}_u\, m$ it can be written as

$$[\text{cov}_u\, m] = \sigma^2 G^{-g} [\text{cov}_u\, d] G^{-gT} \quad (1.28)$$

$\text{cov}_u\, m$ is a unit data covariance matrix. Square roots of the model covariance matrix's diagonal elements are a measure of the standard errors in the model parameters.

1.6.4.4 Synthetic parameterization

The parameterizations that describe continuous velocity fields often opt for resolution tests that attempt to reconstruct a synthetic model using the same source-receiver geometry as the actual experiment. The rationale behind this approach is that if a known structure with similar length scales to the solution model can be recovered using the same (for linearized solutions) or similar (for nonlinear solutions) ray paths, and then the solution model should be reliable. The quality criterion is the similarity between the recovered model and the synthetic model. The is the standard test model "checkerboard test", in which the synthetic model is divided into alternating regions of high and low velocity with a length scale equal to the smallest wavelength structure recovered in the solution model. Regions in which the checkerboard pattern is recovered are those regions where the solution model structure can be considered well resolved. Lévéque et al. (1993) demonstrated that such an approach is not necessarily as reliable as it might seem. In the checkerboard test, the small-scale structure is well retrieved while the large-scale structure is poorly retrieved. Furthermore, if the nonlinearity of the problem is accounted for, the ray paths through the synthetic model may have significantly different geometries to those through the solution model. Under these situations, testing several different synthetic models is advisable, especially if the solution model is complex.

Teleseismic travel time tomography studies that used checkerboard resolution tests include those by Glahn and Granet (1993), Achauer (1994), and Seber et al. (1991). Ritsema et al. (1998) used several different synthetic models rather than checkerboards and analyzed their recovered model's accuracy. In local earthquake tomography, checkerboard tests have been implemented by Walck and Clayton (1987), Walck (1988),

Chiarabba et al. (1997), Graeber and Asch (1999), Dinc et al. (2010), Koulakov (2009a), Kanaujia et al. (2016). Koulakov et al. (2009) defined various synthetic representations, including checkerboard, to test the local earthquake tomography algorithm.

1.7 COUPLED HYPOCENTER VELOCITY MODEL PROBLEM

Many investigators adopt a general methodology involving first determining the 3-D structure by fixing the earthquake locations and then obtaining the earthquake locations by fixing the velocity structure. However, theoretical and numerical studies have demonstrated that this type of methodology introduces bias in both locations and structure (e.g., Thurber, 1992; Roecker et al., 2006). To substantiate this problem's nature, we provide the following analysis given by Roecker et al. (2006). This problem can be stated as:

$$Hdh + Sds = \mathrm{r} \tag{1.29}$$

where H and S represent the derivative matrices, and dh and ds signify perturbation vectors for hypocenters and slowness, respectively. The SVD of H is

$$H = U\Lambda V^T = [U_P \,|\, U_0]\Lambda V^T \tag{1.30}$$

where, U_p is the range space of H and U_0 is the null space of H (Aster et al., 2004). Multiplying the original equation by U^T and separation of the p and 0 components provides:

$$\begin{bmatrix} U_P^T & H \\ U_0^T & H \end{bmatrix} dh + \begin{bmatrix} U_P^T & S \\ U_0^T & S \end{bmatrix} ds = \begin{bmatrix} r_p \\ r_0 \end{bmatrix} \tag{1.31}$$

Note that $U_0^T H = 0$ by definition and, if we initially relocate the earthquakes, then $\boldsymbol{r}_p = 0$ also, we have

$$\begin{bmatrix} U_P^T \ H \\ 0 \end{bmatrix} dh + \begin{bmatrix} U_P^T & S \\ U_0^T & S \end{bmatrix} ds = \begin{bmatrix} 0 \\ r_0 \end{bmatrix} \tag{1.32}$$

Thus, we have two sets of equations.

$$U_p^T Hdh + U_p^T Sds = 0 \tag{1.33}$$

and

$$U_0^T Sds = r_0 \tag{1.34}$$

The second set of equations allow calculating *ds*. This is the decoupled problem (the method of separation of parameters of Pavlis and Booker [1980]). In case we seek a solution for the entire system involving *ds* by ignoring *dh* in equation (1.32), we solve the reduced system represented by the equation:

$$\begin{bmatrix} U_P^T & S \\ U_0^T & S \end{bmatrix} ds = \begin{bmatrix} 0 \\ r_0 \end{bmatrix} \tag{1.35}$$

This amounts to solving equation (1.34) but with the constraint that the weighted sum of *ds* will be zero. This bias occurs because full simultaneous inversion is not conducted.

The above description provides insight into some features of a joint hypocenter–velocity structure inversion (JHVI) versus a velocity-only inversion (VOI). In a VOI, the added constraints (35) force *ds* to be small. This means that subsequent estimates of *dh* will be small as well. The *ds* term is kept small at the cost of misfits to the data. Because when *ds* is small, the fit to r_0 will be degraded. This property, documented in Thurber (1992), can be observed in VOI tests with actual data—both *dh* and *ds* are too small in a VOI, even when alternated with hypocenter relocation, the data fit is poor.

Thurber (1992) showed by the practical example of a 1-D velocity model that the VOI substantially underestimates the velocity contrast, while the simultaneous inversion recovers the velocity structure to within 1%, and the events are located within 100 m. He concluded that for LET, solving the entire system of equations is essential to allow obtaining unbiased estimates of velocity structure.

Many algorithms utilize the fast sparse-matrix solvers such as LSQR, but the historical use of subspace methods (e.g., Kennett et al., 1988) for efficient inversion is worth mentioning. In particular, these methods deal with large matrix sizes when many earthquakes are included in an inversion. For LET, three research groups published comparable methods for separating hypocenter parameters from velocity model parameters. These methods allow the efficient solution of a smaller matrix in one giant matrix (Pavlis and Booker, 1980; Spencer and Gubbins, 1980; Rodi et al., 1981). Consider a LET problem with 10,000 earthquakes recorded on an average at 50 stations and 20,000 model parameters. The size of the matrix to represent the complete system would be (50 × 10,000) by (4 × 10000 + 20000), or 500000 by 60000. If we adopt the annulling method of Pavlis and Booker (1980), we can decompose the JHVI system of equations for each earthquake i as:

$$H_i dh_i + S_i ds_i = r_i \quad (1.36)$$

where H_i and S_i are the derivatives matrices of arrival times with respect to hypocenter and model parameters, respectively. dh_i and ds_i are the corresponding parameter perturbations. Using the orthogonal matrix

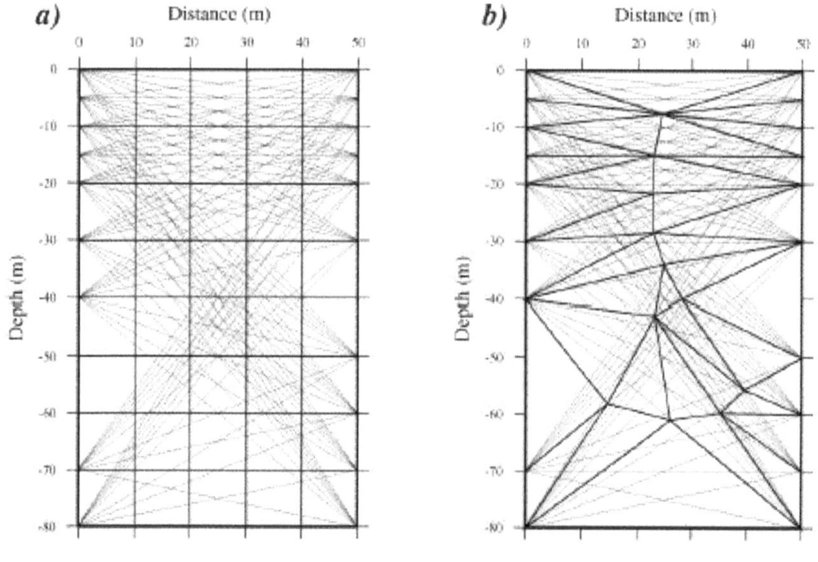

Figure 1.10 An example of optimized model parameterization (a) Ray-path geometry with rectangular grid parameterization. (b) Ray-path geometry with optimized triangular cell parameterization. (After Weber 2001)

U_{0i} that satisfies $U_{0i}H_i = 0$ allows us to recast the decoupled equations in the following form:

$$U_{0i}H_i dh_i + U_{0i}S_i ds_i = U_{0i}r_i, \text{or } S_i' ds_i = r_i' \tag{1.37}$$

The matrix size of the original partial system in eq. (1.36) would be 40 by 20004, whereas the matrix size of the decoupled system in eq. (1.37) would be 36 by 20000. If we consider all the earthquakes in this manner, the size of the resulting system of equations is 460000 by 20000, which shows a substantial size reduction. We can further reduce the problem by incrementally constructing the standard equations $S'^T S' = S'^T r'$ (Spencer and Gubbins, 1980). The resulting system of equations now has a size of 20000 by 20000 only.

Unfortunately, the last step results in squaring the condition number of the system matrix, leading to a loss of matrix sparseness. As stated earlier, the single-step "tomographic" inversion underestimates velocity contrast, while the "simultaneous inversion" recovers the actual structure and event location nearly precisely. Furthermore, ignoring direct hypocenter-velocity structure coupling leads to bias in the derived "tomographic" models (Walck and Clayton, 1987; Lees and Crosson, 1989).

1.8 CONCLUDING REMARKS

This chapter provides a brief description of the various methods employed to model 1-D and 3-D velocity structure (seismic tomography) and mathematical background to solve the travel time tomography problem. In a nutshell, many new ways have been evolved to exploit seismic travel time information and solve tomographic problems. The main application of seismic travel times includes determining the crust and lithosphere's structure and composition from tomographic inversion.

REFERENCES

Abers, G. A. (1994), Three-dimensional inversion of regional P and S arrival times in the East Aleutians and sources of subduction zone gravity highs. *J. Geophys. Res.*, 99, 4395–4412.

Achauer, U. (1994), New ideas on the Kenya rift based on the inversion of the combined dataset of the 1985 and 1989/90 seismic tomography experiments. *Tectonophysics*, 236, 305–329.

Akcelik, V., J. Bielak, G. Biros, I. Epanomeritakis, A. Fernandez, O. Ghattas, E. J. Kim, J. Lopez, D. O., Hallaron, T. Tu and J. Urbanic (2003), High-resolution forward and inverse earthquake modeling on terascale computers. *Proceedings of the 2003 ACM/IEEE Conference on Supercomputing*, page 52. New York.

Aki, K. and P. G. Richards (2002), *Quantitative Seismology*, 2nd edition, University Science Books, Mill Valley, California, pp. 704.

Aki, K., A. Christoffersson and E. S. Husebye (1977), Determination of the three-dimensional seismic structure of the lithosphere. *J. Geophys. Res.*, 82(2), 277–296.

Aki, K. and W. H. K. Lee (1976), Determination of the three-dimensional velocity anomalies under a seismic array using first P arrival times from local earthquakes: A homogeneous initial model. *J. Geophys. Res.*, 81, 4381–4399.

Ammon, C. J. (1991), The isolation of receiver effects from teleseismic P-waveform. *Bull. Seismol. Soc. Am.*, 81(6), 2504–2510.

Ankeny, L. A., L. W. Braile and K. H. Olsen (1986), Upper crustal structure beneath the Jemez Mountains volcanic field, New Mexico, determined by three-dimensional simultaneous inversion of seismic refraction and earthquake data. *J. Geophys. Res.*, 91, 6188–6198.

Aster, R., B. Borchers and C. Thurber (2013), *Parameter Estimation and Inverse Problems,* 2nd edition, Academic Press, Boston, pp. 360.

Backus, G. and F. Gilbert (1968), The resolving power of gross earth data. *Geophys. J. Int.*, 16, 169–205.

Bakun, W. H. (1971), Crustal model parameters from P-wave spectra. *Bull. Seismol. Soc. Am.*, 61, 913–935.

Barraez, D., S. Garcia-Salicetti, B. Dorizzi, M. Padron and E. Ramos (2002), Modular neural networks for seismic tomography. *Proceedings of the 16th International Conference on Pattern Recognition* (*ICPR'02*). Washington, DC: IEEE Computer Society, Volume 3, pp. 30407.

Baźan, F. S. V., M. C. C. Cunha and L. S. Borges (2014), Extension of GKB-FP algorithm to large-scale general form Tikhonov regularization. *Linear Algebra Appl.*, 21(3), 316–339.

Benz, H. M. and R. B. Smith (1984), Simultaneous inversion for lateral velocity variations and hypocenters in the Yellowstone region using earthquake and refraction data. *J. Geophys. Res.*, 89, 1208–1220.

Bishop, T. P., K. P. Bube, R. T. Cutler, R. T. Langan, P. L. Love, J. R. Resnick, R. T. Shuey, D. A. Spindler and H. W. Wyld (1985), Tomographic determination of velocity and depth in laterally varying media. *Geophysics*, 50, 903–923.

Blundell, C. A. (1993), Resolution Analysis of Seismic P-wave Velocity Estimates Using Reflection Tomographic Inversion, Ph.D. thesis. Monash University, Australia.

Boore, D. M. and M. N. Toksoz (1969), Rayleigh wave particle motion and crustal structure. *Bull. Seismol. Soc. Am.*, 59, 331–346.

Bosch, M. (1997), P wave velocity tomography of the Venezuelan region from local arrival times. *J. Geophys. Res.*, 102, 5455–5472.

Boschetti, F., M. K. Dentith and R. D. List (1996), Inversion of seismic refraction data using genetic algorithms. *Geophysics*, 61, 1715–1727.

Bucher, R. L. and R. B. Smith (1971), Crustal structure of the eastern Basin and Range Province and the northern Colorado Plateau from phase velocities of Rayleigh waves. In *The Structure and Physical Properties of the Earth's Crust*. Geophysical Monograph Series, vol. 14, edited by J. G. Heacock, pp. 59–70. Washington, DC: AGU.

Burdick, L. J. and C. A. Langston (1977), Modeling crustal structure through the use of converted phases in teleseismic body waveforms. *Bull. Seismol. Soc. Am.*, 67, 677–692.

Calderon–Macias, C., M. K. Sen and P. L. Stoffa (1998), Hopfield neural networks, and mean field annealing for seismic deconvolution and multiple attenuation. *Geophysics*, 62, 992–1002.

Carlson, W. D. (2006), Three-dimensional imaging of earth and planetary materials. *Earth Planet. Sci. Lett.*, 249(3–4), 133–147.

Carpenter, E. W. (1965), Explosion Seismology: Capabilities and limitations of long-range methods for detecting and recognizing explosions are discussed. *Science*, 147 (3656), 363–373.

Cassell, B. R. (1982), A method for calculating synthetic seismograms in laterally varying media. *Geophys. J. Int.*, 69, 339–354.

Censor, Y. (1981), Row action methods for huge and sparse systems and their applications. *SIAM Review*, 23, 444–466.

Chapman, C. H. and R. Drummond (1982), Body-wave seismograms in inhomogeneous media using Maslov asymptotic theory. *Bull. Seismol. Soc. Am.*, 72, S277–S317.

Chiarabba, C., A. Amoto and M. Meghraoui (1997), Tomographic images of the El Asnam fault zone. *J. Geophys. Res.*, 102(B11), 24485–24498.

Clayton, R. W. and G. A. McMechan (1981), Inversion of refraction data by wavefield continuation. *Geophysics*, 46, 860–868.

Clayton, R. W. and R. A. Wiggins (1976), Source shape estimation and deconvolution of teleseismic body waves. *Geophys. J. Int.*, 103, 301–320.

Clayton, R. W. and T. M. Hearn (1982), A tomographic analysis of lateral velocity variations in Southern California. *EOS Trans. AGU*, 63, 1036.

Crosson, R. S. (1976), Crustal structure modeling of earthquake data 2. Velocity structure of the Puget Sound region, Washington. *J. Geophys. Res.*, 81, 3047–3054.

Dahlen, F. A., S.-H. Hung and N. Guust (2000), Frechet kernels for finite-frequency traveltimes—I. Theory. *Geophys. J. Int.*, 141, 157–174.

Dinc, A. N., I. Koulakov, M. Thorwart, W., Rabbel, E. R. Flueh, I. Arroyo, W. Taylor and G. Alvarado (2010), Local earthquake tomography of central

Costa Rica: Transition from seamount to ridge subduction. *Geophys. J. Int.*, 183 (1), 286–302.

Drijkoningen, G. G. and R. S. White (1995), Seismic velocity structure of oceanic crust by inversion using genetic algorithms. *Geophys. J. Int.*, 123, 653–664.

Dziewonski, A. M., B. H. Hager and R. J. O'Connell (1977), Large-scale heterogeneities in the lower mantle. *J. Geophys. Res.*, 82, 239–255.

Eberhart-Phillips, D. and A. J. Michael (1993), Three-dimensional velocity structure, seismicity and fault structure in the Parkfield Region, central California. *J. Geophys. Res.,* 98(B9), 15737–15758.

Ellis, R. M. and P. W. Basham (1968), Crustal characteristics from short-period P waves. *Bull. Seismol. Soc. Am.*, 58, 1681–1700.

Ewing, M. and F. Press (1959), Determination of crustal structure from phase velocity of Rayleigh waves, III, the United States. *Geol. Soc. Am. Bull.*, 52, 1031–1035.

Farra, V. and R. Madariaga (1988), Non-linear reflection tomography. *Geophys. J.*, 95, 135–147.

Fuchs, K. and G. Muller (1971), Computation of synthetic seismograms with reflectivity method and comparison with observations. *Geophys. J. Int.*, 23, 417–433.

Gaherty, J. B., T. H. Jordan and L. S. Gee (1996), Seismic structure of the upper mantle in a Central Pacific corridors. *J. Geophys. Res.*, 101, 22291–22309.

Gee, L. S. and T. J. Jordan (1992), Generalized seismological data functionals. *Geophys. J. Int.*, 111, 363–390.

Glahn, A. and M. Granet (1993), Southern Rhine Graben: Small-wavelength tomographic study and implications for the dynamic evolution of the graben. *Geophys. J. Int.*, 113, 399–418.

Golub, G. H. and C. F. Van Loan (1983), *Matrix computations.* Oxford: North Oxford Academic.

Golub, G. and C. Reinsch (1970), Singular value decomposition and least squares solution. *Numer. Math.*, 14, 403–420.

Graeber, F. M. and G. Asch (1999), Three-dimensional models of P wave velocity and P-to-S velocity ratio in the southern central Andes by simultaneous inversion of local earthquake data. *J. Geophys. Res.*, 104, 20237–20256.

Greenhalgh, S. A., D. Tapley and R. Singh (1989), Crustal heterogeneity in South Australia, earthquake evidence. *Geophys. J. Int.*, 96, 85–99.

Gubbins, D. (1992), *Seismology and Plate Tectonics.* Cambridge: Cambridge University Press.

Hasegawa, A., N. Umino and A. Takagi (1978), Double-planed deep seismic zone and upper-mantle structure in the northeastern Japan arc. *Geophys. J. Int.*, 54, 281–296.

Healy, J. H. (1971), A comment on the evidence for a worldwide zone of low seismic velocity at shallow depths in the earth's crust. In *The Structure and*

Physical Properties of the Earth's Crust. Geophysical Monograph Series, vol. 14, edited by J. G. Heacock, pp. 35–40. Washington, DC: AGU.

Helffrich, G. (2006), Extended-time multitaper frequency domain cross-correlation receiver function estimation. *Bull. Seismol. Soc. Am.*, 96 (1), 344–347.

Hestenes, M. and E. Stiefel (1952), Methods of conjugate gradients for solving linear systems. *Nat. Bur. Standards J. Res.*, 49, 409–436.

Hole, J. A. (1992), Nonlinear high-resolution three-dimensional travel-time tomography. *J. Geophys. Res.*, 97, 6553–6562.

Huang, H. H., Y. M. Wu, X. Song, C. H. Chang, S. J. Lee, T. M. Chang and H. H. Hsieh (2014), Joint Vp and Vs tomography of Taiwan: Implications for subduction-collision orogeny. *Earth Planet. Sci. Lett.*, 392, 177–191.

Humphreys, E. and R. W. Clayton (1988), Adaption of back projection tomography to seismic travel time problems. *J. Geophys. Res.*, 93, 1073–1085.

Humphreys, E. D. and R. W. Clayton (1990), Tomographic image of the Southern California Mantle. *J. Geophys. Res.*, 95, 19725–19746.

Inoue, H., Y. Fukao, K. Tanabe and Y. Ogata (1990), Whole mantle P-wave travel time tomography. *Phys. Earth Planet. Inter.*, 59, 294–328.

Julian, B. R. and D. Gubbins (1977), Three-dimensional seismic ray tracing. *J. Geophys.*, 43, 95–113.

Kanaujia, J., A. Kumar and S. C. Gupta (2016), 3-D velocity structure around Tehri region of the Garhwal Lesser Himalaya: Constraints on geometry of the underthrusting Indian plate. *Geophys. J. Int.*, 205(2), 900–914.

Kao, H. and R. J. Rau (1999), Detailed structures of the subducted Philippine Sea plate beneath northeast Taiwan: A new type of double seismic zone. *J. Geophys. Res.*, 104, 1015–1033.

Katzman, R., L. Zhao, and T. H. Jordan (1998), High-resolution, two-dimensional vertical tomography of the central Pacific mantle using *ScS* reverberations and frequency-dependent travel times. *J. Geophys. Res.*, 103 (B8), 17933–17971.

Kawakatsu, H. (1985), Double seismic zone in Tonga. *Nature*, 316, 53–55.

Kennett, B. L. N., M. S. Sambridge and P. R. Williamson (1988), Subspace methods for large scale inverse problems involving multiple parameter classes. *Geophys. J. Int.*, 94, 237–247.

Kissling, E. (1988), Geotomography with local earthquake data. *Rev. Geophys.*, 26, 659–698.

Kissling, E., W. L. Ellsworth, D. Eberhart and U. Kradolfer (1994), Initial reference models in local earthquake tomography. *J. Geophys. Res.* 99(B10), 19635–19646.

Kosminskaya, I. P., N. A. Belyaevsky and I. S Volvovsky (1969), Explosion seismology in the USSR. In *The Earth's Crust and Upper Mantle.* Geophysical Monograph Series, vol. 13, edited by P. J. Hart, pp. 195–208. Washington, DC: AGU.

Koulakov, I. (2009), Out-of-network events can be of great importance of improving results of local earthquake tomography. *Bull. Seismol. Soc. Am.*, 99(4), 2556–2563.

Koulakov, I., T. Yudistira, B. G. Luehr and Wandono (2009), P, S velocity and V_P/V_S ratio beneath the Toba caldera complex (Northern Sumatra) from local earthquake tomography. *Geophys. J. Int.*, 177(3), 1121–1139.

Landisman, M., S. Muller and B. J. Mitchell (1971), Review of evidence for velocity inversions in the continental crust. In *The Structure and Physical Properties of the Earth's Crust*. Geophysical Monograph Series, vol. 14, edited by J. G. Heacock, pp. 11–34. Washington, DC: AGU.

Langan, R. T., I. Lerche and R. T. Cutler (1985), Tracing of rays through heterogeneous media: An accurate and efficient procedure. *Geophysics*, 50, 1456–1465.

Langer, H., G. Nunnari and L. Occhipinti (1996), Estimation of seismic waveform governing parameters with neural networks. *J. Geophys. Res.*, 101, 20109–20118.

Lee, W. H. K. and V. Pereyra (1993), Mathematical introduction to seismic tomography. In *Seismic Tomography: Theory and Practice*, edited by H. M. Iyer and K. Hirahara, pp. 9–22. London: Chapman & Hall.

Lees, J. M. and R. S. Crosson (1989), Tomographic imaging for three dimensional velocity structure of Mount St. Helens using earthquake data. *J. Geophys. Res.*, 94(B5), 5716–5728.

Lei, J. and D. Zhao (2006), Global seismic tomography: On the effect of various mantle and core phases. *Phys. Earth Planet. Inter.*, 154(1), 44–69.

Lèvèque, J. J., L. Rivern and G. Wittlinger (1993), On the use of the checker-board test to assess the resolution of tomographic inversions. *Geophys. J. Int.*, 115, 313–318.

Luca, G. D., L. Filippi, D. Caccamo, G. Neri and R. Scarpa (1997), Crustal structure and seismicity of southern Tyrrhenian basin. *Phys. Earth Planet. Inter.*, 103, 117–133.

Lutter, W. J. and R. L. Nowack (1990), Inversion for crustal structure using reflections from the PASSCAL Ouachita experiment. *J. Geophys. Res.*, 95, 4633–4646.

McCaughey, M. and S. C. Singh, (1997), Simultaneous velocity and interface tomography of normal incidence and wide-aperture seismic traveltime data. *Geophys. J. Int.*, 131, 87–99.

Moser, T. J. (1991), Shortest path calculation of seismic rays. *Geophysics*, 56, 59–67.

Mueller, St. and M. Landisman (1966), Seismic studies of the earth's crust in continents, 1, Evidence for a low-velocity zone in the upper part of the lithosphere. *Geophys. J. Int.*, 10, 525–538.

Natterer, F. (1986), *The Mathematics of Computerized Tomography*. Hoboken, NJ, USA, Wiley, pp. 222.

Neele, F., J. Van Decar and R. Snieder (1993), The use of P wave amplitude data in a joint inversion with travel times for upper mantle velocity structure. *J. Geophys. Res.*, 98, 12033–12054.

Ning, M., W. Yanping, H. Zhengyi and B. Zongdi (1996), An iterative algonthrn using a neural network for nonlinear traveltime tomography. *Proceedings of ICSP'96*, pp. 130–136.
Nolet, G. (1987), Waveform tomography. In *Seismic Tomography: With Applications in Global Seismology and Exploration Geophysics*, edited by G. Nolet, pp. 301–322. Dordrecht: D. Reidel.
Nolet, G. (1990), Partitioned waveform inversion and two-dimensional structure under the network of autonomously recording seismographs. *J. Geophys. Res.*, 95, 8499–8512.
Nowack, R. L. and C. Li (2009), Methods and applications of seismic tomography. In *Handbook of Signal Processing in Acoustics*, edited by D. Havelock, S. Kuwano and M. Vorlander, pp. 1633–1653. New York: Springer-Verlag.
Paige, C. and M. Saunders (1982), LSQR: An algorithm for sparse linear equations and sparse least squares. *ACM Transactions on Mathematical Software*, 8(1), 43–71.
Papazachos, C. and G. Nolet (1997), P and S deep velocity structure of the Hellenic area obtained by robust nonlinear inversion of travel times. *J. Geophys. Res.*, 102, 8349–8367.
Park, J. and V. Levin (2000), Functions from multiple-taper spectral correlation estimates. *Bull. Seismol. Soc. Am.*, 90(6), 1507–1520.
Pavlis, G. L. and J. R. Booker (1980), The mixed discrete continuous inverse problem: Application to the simultaneous determination of earthquake hypocenters and velocity structure. *J. Geophys. Res.*, 85(B9), 4801–4810.
Perkins, W. E. and R. A. Phinney (1971), *A Reflection Study in the Wind River Uplift, Wyoming*. Geophysical Monograph Series, 14, 14–50.
Phillips, W. S. and M. C. Fehler (1991), Traveltime tomography: A comparison of popular methods. *Geophysics*, 56(10), 1639–1649.
Phinney, R. A. (1969), Structure of the earth's crust from spectral behavior of long-period body wave. *J. Geophy. Res.*, 69, 2997–3017.
Pollitz, F. F. (2007), Finite-frequency traveltime tomography of San Francisco Bay region crustal velocity structure. *Geophys. J. Int.*, 171(2), 630–656.
Press, F. and S. Biehler (1964), Inference on crustal velocities and densities from P-wave delays and gravity anomalies. *J. Geophys. Res.*, 69, 2979–2995.
Radon, J. (1917), On the determination of functions from their integrals along certain manifolds (Translation), Ber, Verbhandl. *Sachs. Akad. Wiss, Leipzig, Math-Phys*, 69, 262–277.
Rawlinson, N. (2000), Inversion of Seismic Data for Layered Crustal Structure, Ph.D. thesis, Monash University, Australia.
Rawlinson, N. and M. Sambridge (2003), Seismic traveltime tomography of the crust and lithosphere. *Adv. Geophys.*, 46(C), 81–198.
Reagan, R. L. (1978), A Finite-Difference Study of Subterranean Cavity Detection and Seismic Tomography. Ph.D. thesis, University of Missouri-Rolla.

Ritsema, J., A. A. Nyblade, T. J. Owens, C. A. Langston and J. C. VanDecar (1998), Upper mantle seismic velocity structure beneath Tanzania, East Africa: Implications for the stability of cratonic lithosphere. *J. Geophys. Res.*, 103, 21201–21213.

Rodi W. L., Jordan T. H., Masso J. F. and J. M. Savino (1981), Determination of the three-dimensional structure of Eastern Washington from the joint inversion of gravity and earthquake data. *Report: Systems Sciences and Software SSS-R-80-4516*, La Jolla, California.

Roecker S, C. Thurber, K. Roberts and L. Powell (2006), Refining the image of the San Andreas Fault near Parkfield, California using a finite difference travel time computation technique. *Tectonophysics*, 426(1–2), 189–205.

Roth, G. (1993), Application des reseaux de neurones aux problemes inverres sismiques. Ph.D. thesis, lnstitut de Physique du Globe de Paris, Univrrsite de Paris 7, Paris.

Roth, G. and A. Tarantola (1994), Neural networks and inversion of seismic data. *J. Geophys. Res.*, 99(B4), 6753–6768.

Sambridge, M. S. (1990), Non-linear arrival time inversion: Constraining velocity anomalies by seeking smooth models in 3-D. *Geophys. J. Int.*, 102, 653–677.

Sambridge, M. S. and G. Drijkoningen (1992), Genetic algorithms in seismic waveform inversion. *Geophys. J. Int.*, 109, 323–342.

Santos, R. J. and A. de Pierro (2006), The effect of the nonlinearity on GCV applied to conjugate gradients in computerized tomography. *Comput. Appl. Math.*, 25, 111–128.

Scales, J. (1987), Tomographic inversion via the conjugate gradient method. *Geophysics*, 52(2), 179–185.

Seber, D., M. Barazangi, B. A. Tadili, M. Ramdani, A. Ibenberahim and D. B. Sari (1991), Three-dimensional upper mantle structure beneath the interplate Atlas and interplate Rifmountains of Morocco. *J. Geophys. Res.*, 101, 3125–3138.

Spakman W. and G. Nolet (1988), Imaging algorithms: Accuracy and resolution in delay time tomography. In *Mathematical Geophysics: A Survey of Recent Developments in Seismology and Geodynamics*, edited by N. J. Vlaar, G. Nolet, M. Wortel and S. Cloetingh, vol 3. pp. 155–188. Dordrecht: Springer.

Spencer, C. and D. Gubbins (1980), Travel-time inversion for simultaneous earthquake location and velocity structure determination in laterally varying media. *Geophys. J. Int.*, 63, 95–116.

Steck, L. K., C. H. Thurber, M. Fehler, W. J. Lutter, P. M. Roberts, W. S. Baldridge, D. G. Stafford and R. Sessions (1998), Crust and upper mantle P wave velocity structure beneath Valles Caldera, New Mexico: Results from the Jemez teleseismic tomography experiment. *J. Geophys. Res.*, 103, 24301–24320.

Tarantola, A. (1986), A strategy for nonlinear elastic inversion of seismic reflection data. *Geophysics*, 51, 1893–1903.

Thomson, C. J. and D. Gubbins (1982), Three-dimensional lithospheric modelling at NORSAR: Linearity of the method and amplitude variations from the anomalies. *Geophys. J. Int.*, 71, 1–36.

Thurber, C. H. (1992), Hypocenter–velocity structure coupling in local earthquake tomography. Special Issue: Lateral Heterogeneity and Earthquake Location. *Phys. Earth Planet. Inter.*, 75, 55–62.

Thurber, C. H. (1983), Earthquake locations and three-dimensional crustal structure in the Coyote Lake Area, Central California. *J. Geophys. Res.*, 88(B10), 8226–8236.

Thurber, C. H. (1993), Local earthquake tomography: Velocities and V_P/V_S—theory. In *Seismic Tomography: Theory and Practice*, edited by H. M. Iyer and K. Hirahara, pp. 563–583. London: Chapman & Hall.

Tromp, J., C. H. Tape and Q. Liu (2005), Seismic tomography, adjoint methods, time reversal and banana-doughnut kernels. *Geophys. J. Int.*, 160, 195–216.

Um, J. and C. Thurber (1987), A fast algorithm for two-point seismic ray tracing. *Bull. Seismol. Soc. Am.*, 77, 972–986.

Ustaszewski, K., Y. M. Wu, J. Suppe, H. H. Huang, C. H. Chang and S. Carena (2012), Crust–mantle boundaries in the Taiwan–Luzon arc-continent collision system determined from local earthquake tomography and 1D models: Implications for the mode of subduction polarity reversal. *Tectonophysics*, 578, 31–49.

van der Sluis, A. and H. A. van der Vorst (1987), Numerical solution of large, sparse linear algebraic systems arising from tomographic problems. In *Seismic Tomography. Seismology and Exploration Geophysics*, edited by G. Nolet, vol 5. pp. 49–84. Dordrecht: Springer.

van der Sluis, A. and H. van der Vorst (1990), SIRT- and CG-type methods for the iterative solution of sparse linear least-squares problems. *Linear Algebra Appl.,*, 130, 257–303.

VanDecar, J. C., D. E. James and M. Assumpção (1995), Seismic evidence for a fossil mantle plume beneath South America and implications for plate driving forces. *Nature*, 378, 25–31.

VanDecar, J. C. and R. S. Crosson (1990), Determination of teleseismic relative phase arrival times using multi-channel cross-correlation and least-squares. *Bull. Seismol. Soc. Am.*, 80, 150–169.

Vasco, D. W., L. R. Johnson and J. Pulliam (1995), Lateral variations in mantle velocity structure and discontinuities determined from P, PP, S, SS and SS-SdS travel time residuals. *J. Geophys. Res.*, 100, 24037–24059.

Vidale, J. E. and H. Houston (1993), The depth dependence of earthquake duration and implications for rupture mechanisms. *Nature*, 365, 45–47.

Vidale, J. E. (1990), Finite-difference calculations of traveltimes in three dimensions. *Geophysics*, 55, 521–526.

Walck, M. C., (1988), Three-dimensional Vp/Vs variations for the Coso region, California. *J. Geophys. Res.*, 93, 2047–2052.

Walck, M. C. and R. W. Clayton (1987), P wave velocity variations in the Coso region, California, derived from local earthquake travel times. *J. Geophys. Res.*, 92, 393–405.

Wéber, Zoltán. (2001). Optimizing model parameterization in 2D linearized seismic traveltime tomography. *Phys. Earth Planet. Inter.*, 124, 33–43. 10.1016/S0031-9201(01)00185-6.

Wesson, R. L. (1971) Travel-time inversion for laterally inhomogeneous crustal velocity models. *Bull. Seismol. Soc. Am.*, 61, 729–746.

White, D. J. (1989), Two-dimensional seismic refraction tomography. *Geophys. J.*, 97, 223–245.

Williamson, P. R. (1990), Tomographic inversion in reflection seismology. *Geophys. J. Int.*, 100, 255–274.

Zelt, C. A. and P. J. Barton (1998), Three-dimensional seismic refraction tomography: A comparison of two methods applied to data from the Faeroe Basin. *J. Geophys. Res.*, 103, 7187–7210.

Zelt, C. A. and R. B. Smith (1992), Seismic traveltime inversion for 2-D crustal velocity structure. *Geophys. J. Int.*, 108, 16–34.

Zhang, F.-X., Q.-J. Wu, J.-T. Pan, G.-C. Zhang and Q.-Q. Feng (2011), The computation of a finite-frequency travel time sensitive kernel for P-waves in the AK135 earth model. *Appl. Geophys.*, 8(2), 158–163.

Zhao, D. (2001), New advances of seismic tomography and its applications to subduction zones and earthquake fault zones: A review. *Island Arc*, 10(1), 68–84.

Zhao, D. (2015), *Multiscale seismic tomography,* Springer, Tokyo, pp. 304.

Zhao, D. (2015), *Methodology of Seismic Tomography.* In: Multiscale Seismic Tomography. Springer Geophysics. Springer, Tokyo. https://doi.org/10.1007/978-4-431-55360-1_2

Zhao, D. (2009), Multiscale seismic tomography and mantle dynamics. *Gondwana Res.*, 15(3–4), 297–323.

Zhao, D., A. Hasegawa and S. Horiuchi (1992), Tomographic imaging of P and S wave velocity structure beneath Northeastern Japan. *J. Geophys. Res.*, 97(B13), 19909–19928.

Zhao, D., A. Hasegawa and H. Kanamori (1994), Deep structure of Japan subduction zone as derived from local, regional, and teleseismic events. *J. Geophys. Res.*, 99(B11), 22313–22329.

Zhao, D., H. Kanamori and E. Humphreys (1996), Simultaneous inversion of local and teleseismic data for the crust and mantle structure of southern California. *Phys. Earth Planet. Inter.*, 93, 191–214.

Chapter 2

Waveform modelling

Anup K. Sutar

2.1 INTRODUCTION

Every wiggle of seismogram contains information about (i) the earthquake source, where the signal generated, (ii) path travelled between source to station and (iii) the instrument that recorded the signal. It is possible to model each of the effects numerically and construct a synthetic seismogram (Aki and Richards, 2002), which can be represented as:

$$u(t) = s(t) * g(t) * I(t) \tag{2.1}$$

where $u(t)$ is the seismogram, $s(t)$ is the source term, $g(t)$ is the propagation term, $I(t)$ is the instrument response and '*' represents the convolution operator. The comparison of synthetic seismogram with the observed earthquake record is known as waveform modelling. This is a tool used for understanding the source fault rupture process and earth structure through which the waves travel. It is an iterative process used to minimize the difference between the observed and synthetic seismograms either by adjusting the earth structure or the source parameters.

Out of the three parameters, the path parameter, also known as earth's transfer function, is the most complex and accounts for all propagation effects, such as reflections, refraction, diffraction, scattering, mode conversions, attenuation, amplification and so on. Therefore, it is challenging to model $g(t)$ numerically. Similarly, the source parameter can further be decomposed and modelled by source radiation and fault rupture characteristics (Stein and Wysession, 2003). Thus, having preliminary information about the medium properties,

DOI: 10.1201/9781003177692-2

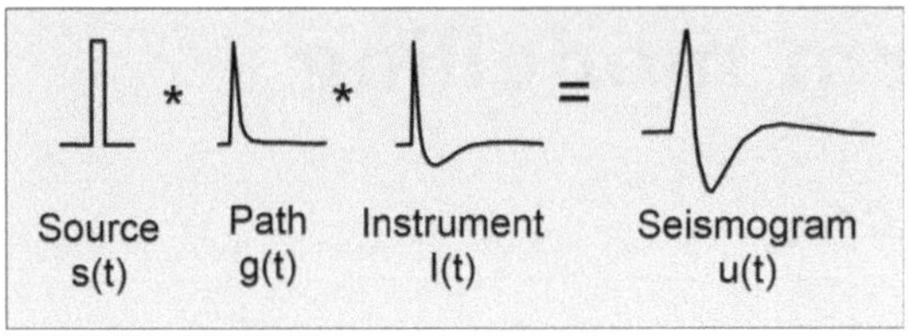

Figure 2.1 Convolution of source, path and instrument response to get synthetic seismogram.

the source/fault rupture characteristics can be extracted from the observed seismogram, and the technique is called moment tensor inversion. Similarly, earth's subsurface physical properties such as, seismic velocity structure, attenuation, amplification can be extracted from the observed seismic waves by a technique called seismic waveform tomography (see Figure 2.1).

2.2 SOURCE PARAMETER USING WAVEFORM MODELLING

In this section, we will understand the procedure adopted to estimate earthquake source parameters using waveform modelling. The displacement field $u(x,t)$ due to a body force $f(x,t)$ applied impulsively (i.e. a spike), in space (homogeneous, unbounded, isotropic elastic medium) and time, to a given particle at position $x = \xi$ and time $t = \tau$ can be expressed,

$$\rho \ddot{u}_i = f_i + (\lambda + \mu) u_{j,ji} + u_{i,jj} \tag{2.2}$$

In vector notation, it can be expressed as:

$$\rho \ddot{u} = f + (\lambda + 2\mu) \nabla (\nabla \cdot u) - \mu \nabla \times (\nabla \times u) \tag{2.3}$$

where λ and μ are lame parameters, $\nabla \cdot u$ and $\nabla \times u$ represents volumetric and shearing deformation, respectively. The body force f is given by $f_i = X_0(t)\delta(x)\delta_{i1}$ where X_0 is the amplitude of force applied, and the initial conditions given are $u(x,0) = 0$ and $\dot{u}(x,0) = 0$ *for* $x \neq 0$. The problem is complicated due to the directionality at the source and receiver.

The solution of equation (2.3) using Green's function technique can be written as: (Aki and Richards, 2002)

$$u_i = \frac{1}{4\pi\rho}(3\gamma_i\gamma_j - \delta_{ij})\frac{1}{r^3}\int_{r/\propto}^{r/\beta} \tau X_0(t-\tau)d\tau$$
$$+\frac{1}{4\pi\rho\alpha^2}\gamma_i\gamma_j\frac{1}{r}X_0\left(t-\frac{r}{\alpha}\right) \tag{2.4}$$
$$-\frac{1}{4\pi\rho\beta^2}(\gamma_i\gamma_j - \delta_{ij})\frac{1}{r}X_0\left(t-\frac{r}{\beta}\right)$$

where r is the source-receiver distance, γ is a unit vector directed from source to receiver, τ is the retard time and expressed as $t-|x-\xi|/\alpha$, ρ is the density, α is the P-wave velocity, β is the S-wave velocity, δ is the delta function.

The equation consists of three terms, namely, (i) near-field term, (ii) far-field P-wave term and (iii) far-field S-wave term.

i. ***The near-field term:***

Let the near-field term is defined by u_N and can be expressed as,

$$u_N = \frac{1}{4\pi\rho}(3\gamma_i\gamma_j - \delta_{ij})\frac{1}{r^3}\int_{r/\propto}^{r/\beta} \tau X_0(t-\tau)d\tau \tag{2.5}$$

The aforementioned equation describes the solution due to point source $X_0(t)$ in the x_j-direction at the origin. u_N is composed of both P- and S-wave motions and it is neither irrotational ($\nabla \times u = 0$) nor solenoidal ($\nabla \cdot u = 0$), thus it is not fruitful to decompose an elastic displacement into its P-wave and S-wave components. u_N can further be decomposed by both longitudinal and transverse motions.

The longitudinal component is expressed as:

$$u_N.\gamma = \gamma_j \frac{1}{2\pi\rho r^3}\int_{r/\alpha}^{r/\beta} \tau X_0(t-\tau)d\tau \tag{2.6}$$

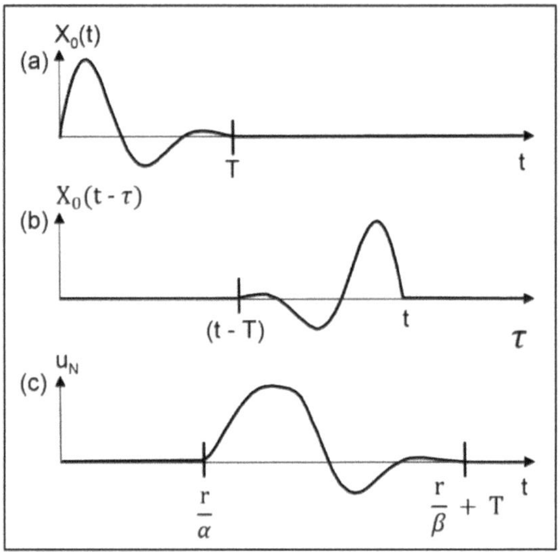

Figure 2.2 Arrival time and duration of the near-field term, (a) Body-force time function, which is non-zero only for time between 0 and T. (b) Variation of $X_0(t - \tau)$ and τ, (c) Variation of $\int_{r/\alpha}^{r/\beta} \tau X_0(t - \tau)d\tau$ with time.

The transverse component is expressed as:

$$u_N.\gamma' = \gamma_j \frac{1}{2\pi\rho r^3} \int_{r/\alpha}^{r/\beta} \tau X_0(t - \tau)d\tau \tag{2.7}$$

ii. ***Far-field P-wave term***

Let the far-field displacement is noted by u_α and can be expressed as,

$$u_\alpha(x,t) = \frac{1}{4\pi\rho\alpha^2} \gamma_i \gamma_j \frac{1}{r} X_0\left(t - \frac{r}{\alpha}\right) \tag{2.8}$$

The far-field P-wave term attenuates at $\frac{1}{r}$ and depends on the space-time function, $\left(t - \frac{r}{\alpha}\right)$ and, therefore, propagates with speed α, where $\alpha^2 = (\lambda + 2\mu)/\rho$ and λ, μ are lame parameters and ρ is density of the medium. The direction of propagation is parallel to the direction of γ from the source; thus, the direction of particle motion is same as the direction of propagation. Therefore, the wave is called as longitudinal or sometimes called as radial waves (Figure 2.2).

iii. ***Far-field S-wave term***

Let the far-field S-wave term is represented by u_β, which can be expressed as,

$$u_\beta = \frac{1}{4\pi\rho\beta^2}(\delta_{ij} - \gamma_i\gamma_j)\frac{1}{r}X_0\left(t - \frac{r}{\beta}\right) \tag{2.9}$$

The S-wave term also attenuates as $\frac{1}{r}$ and propagates at a speed of β, which is defined by $\beta^2 = \mu / \rho$, where μ is lame parameter and ρ is density. The direction of u_β is perpendicular to γ as $u_\beta \cdot \gamma = 0$, thus the far-field S-wave is transverse in nature. The direction of particle motion is perpendicular to the direction of propagation.

2.3 MOMENT TENSOR

Moment tensor is an alternative description of a seismic source. The displacement across a seismic source is described in terms of equivalent body forces, which produce the same radiation as the fault slips (Lay and Wallace, 1995). A moment tensor is defined as:

$$M_{pq} = \begin{bmatrix} M_{11}M_{12}M_{13} \\ M_{21}M_{22}M_{23} \\ M_{31}M_{32}M_{33} \end{bmatrix} \tag{2.10}$$

where M_{12}, for example, represents a couple whose force is in the x-direction and arm in the y-direction defined in an appropriate coordinate system with respect to the seismic source. The summed displacement fields produced by M_{12} and M_{21} is known as a double-couple solution.

The displacement field due to a double couple is represented by $u_n = M_{pq} * G_{np,q}$, where $G_{np,q}$ represents the Green's functions corresponding to each of the respective moment tensor elements.

Using time-dependent seismic moment $M_0(t) = \mu u(t) A$, where μ is the shear modulus, u is the displacement and A is the area of rupture.

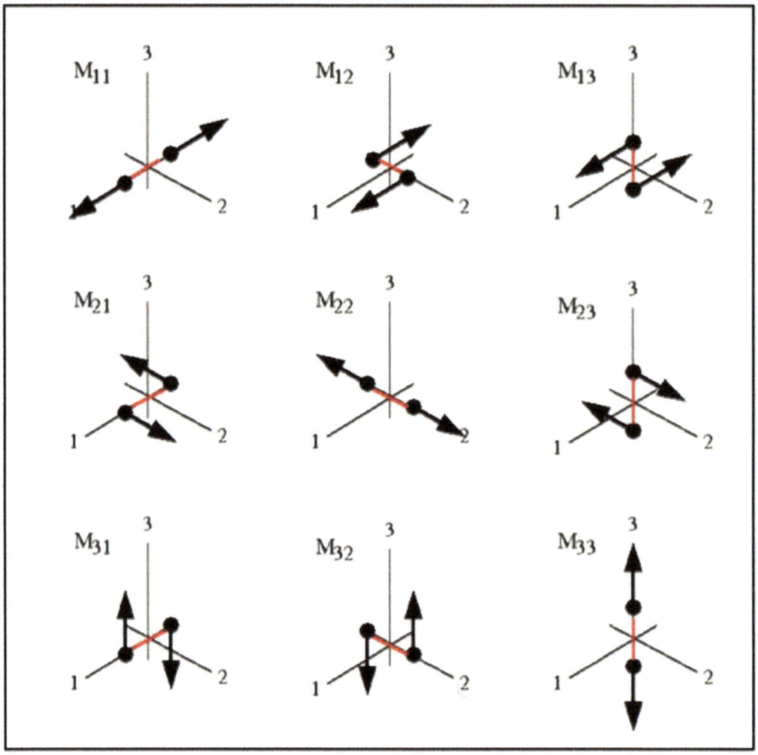

Figure 2.3 Representation of seismic moment tensor.

The solution of equation (2.3) in terms of moment tensor can be expressed as (reference Figure 2.3):

$$
\begin{aligned}
u(x,t) \\
&= \frac{1}{4\pi\rho} A^{N} \frac{1}{r^4} \int_{r/\alpha}^{r/\beta} \tau M_0(t-\tau)\,d\tau + \frac{1}{4\pi\rho\alpha^2} A^{IP} \frac{1}{r^2} M_0\left(t-\frac{r}{\alpha}\right) \\
&+ \frac{1}{4\pi\rho\beta^2} A^{IS} \frac{1}{r^2} M_0\left(t-\frac{r}{\beta}\right) + \frac{1}{4\pi\rho\alpha^3} A^{FP} \frac{1}{r} \dot{M}_0\left(t-\frac{r}{\alpha}\right) \\
&+ \frac{1}{4\pi\rho\beta^3} A^{FS} \frac{1}{r} \dot{M}_0\left(t-\frac{r}{\beta}\right)
\end{aligned}
\tag{2.11}
$$

The radiation term for near-field, intermediate-field P and S, and far-field P and S are given respectively by,

$$
\begin{aligned}
A^{N} &= 9 \sin 2\theta \cos \phi \, \hat{r} - 6\left(\cos 2\theta \cos \phi \, \hat{\theta} - \cos \theta \sin \phi \, \hat{\phi}\right) \\
A^{IP} &= 4 \sin 2\theta \cos \phi \, \hat{r} - 2\left(\cos 2\theta \cos \phi \, \hat{\theta} - \cos \theta \sin \phi \, \hat{\phi}\right) \\
A^{IS} &= -3 \sin 2\theta \cos \phi \, \hat{r} + 3\left(\cos 2\theta \cos \phi \, \hat{\theta} - \cos \theta \sin \phi \, \hat{\phi}\right) \\
A^{FP} &= \sin 2\theta \cos \phi \, \hat{r} \\
A^{FS} &= \cos 2\theta \cos \phi \, \hat{\theta} - \cos \theta \sin \phi \, \hat{\phi}
\end{aligned}
\tag{2.12}
$$

The first term in equation (2.11) is known as near-field term and is proportional to $\frac{1}{r^4}$, similarly, the second and third term of equation (2.11) is referred to as the intermediate P- and S-wave term and the intermediate terms are proportional to $\frac{1}{r^2}$. Further, fourth and fifth term in equation (2.11) referred to as far-field P- and S-wave term, respectively, and proportional to $\frac{1}{r}$. It is important to inform here that far-field terms in equation (2.11) are the only terms that depend on the derivative of the moment term $\dot{M}_0$.

2.4 SEISMIC MOMENT (M_0)

The time-dependent seismic moment M_o released by an active fault can be expressed as (Aki and Richards, 2002):

$$M_0(t) = \mu u(t) A \tag{2.13}$$

where μ is the shear modulus ($3.3*10^{11}$ dyne-cm), u is the displacement and A is the area of rupture. Further, $M_0(t) = 0$ before the rupture begins in an active fault and $M_0(t) = M_0$, when the slip on every point on the fault surface has finished. $M_0(t)$ treats this process as if slip occurs at a point and ignores the fault infiniteness.

Figure 2.4 represents the shape of $M_0(t)$; it is evident from the figure that $M_0(t)$ is zero before the earthquake rupture initiates. As the rupture

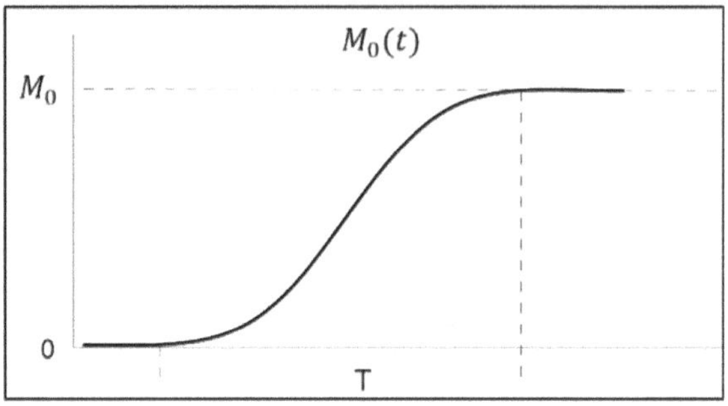

Figure 2.4 Demonstration of rise time for an earthquake source time function.

grows, $M_0(t)$ increases and reaches a value M_0, when the fault plane completely ruptured. Time required to complete the rupture is known as 'rise time' for the earthquake, which is represented by 'T' in Figure 2.4 in the source time function. Several source time functions are proposed for an earthquake rupture by various researchers such as Ramp, boxcar, Brustel-Muller and Wang-Herrman source time functions.

If the source time function of an earthquake is represented by a ramp function (Figure 2.5a), then the far-field displacement is presented by the first order derivative of $M_0(t)$ (Figure 2.5b), similarly, the far-field velocity and acceleration term can be obtained by second and third derivatives of $M_0(t)$ (Figures 2.5c, d). If the simplest source time function is a ramp function, the possible far-field displacement pulse is a one-sided pulse, the simplest velocity pulse is a two-sided pulse and the simplest acceleration pulse is a three-sided pulse (Figure 2.5). Thus, numerically, a synthetic seismogram can be generated using equation (2.11).

It is also possible to construct the *P*, *SV*, or *SH* motion for a moment tensor by summing the moment tensor and weighted Green's functions using:

$$u_n(x,t) = \sum_{i=1}^{5} m_i * G_{in} \tag{2.14}$$

where G_{in} are the Green's functions corresponding to each of the respective moment tensor elements. Thus, if we have an observed

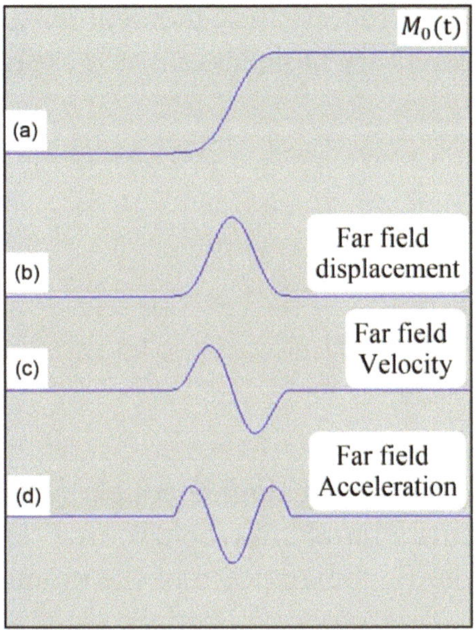

Figure 2.5 (a) shows the source time function of an earthquake, which is represented by a ramp function, (b) far-field displacement as a one-sided pulse (c) far-field velocity by two-sided pulse and (d) far-field acceleration by three-sided pulse.

seismogram, moment tensor information can be extracted using inversion technique.

2.5 MOMENT TENSOR INVERSION

The displacement field on any point of the earth's surface due to an applied force can be expressed by a linear combination of time-dependent moment tensor elements. At large distances from the fault (fault dimension is small w.r.t. the wavelengths of the seismic waves), the displacement $u_n(x,t)$ can be expressed as,

$$u_n(x,t) = M_{pq} * G_{np,q} \tag{2.15}$$

where p, q denotes three Cartesian coordinates, and '*' stands for temporal convolution. The moment tensor can be expressed by a linear combination of six elementary tensors given by,

$$M_{pq} = \sum_{i=1}^{6} a_i M_{pq}^i . \tag{2.16}$$

It is a convenient way of representing moment tensor by six scalar coefficients a_i. The six elementary moment tensor is expressed by,

$$M^1 = \begin{pmatrix} 0 & 1 & 0 \\ 1 & 0 & 0 \\ 0 & 0 & 0 \end{pmatrix} \quad M^2 = \begin{pmatrix} 0 & 0 & 1 \\ 0 & 0 & 0 \\ 1 & 0 & 0 \end{pmatrix} \quad M^3 = \begin{pmatrix} 0 & 0 & 0 \\ 0 & 0 & -1 \\ 0 & -1 & 0 \end{pmatrix}$$
$$M^4 = \begin{pmatrix} -1 & 0 & 0 \\ 0 & 0 & 0 \\ 0 & 0 & 1 \end{pmatrix} \quad M^5 = \begin{pmatrix} 0 & 0 & 0 \\ 0 & -1 & 0 \\ 0 & 0 & 1 \end{pmatrix} \quad M^6 = \begin{pmatrix} 1 & 0 & 0 \\ 0 & 1 & 0 \\ 0 & 0 & 1 \end{pmatrix} \tag{2.17}$$

The M^1–M^5 tensors represent double-couple (DC) focal mechanisms, whereas M^6 represents a purely isotropic source. The scalar seismic moment is defined as the Euclidian norm of the moment tensor (Silver and Jordan, 1982).

$$M_0 = \sqrt{\frac{\sum_{p=1}^{3}\sum_{q=1}^{3}(M_{pq})^2}{2}} \tag{2.18}$$

From equation (2.15), the moment tensor can be expressed as

$$M = G^{-1}u \tag{2.19}$$

In seismology, the moment tensor is decomposed into an isotropic part and a deviatoric part, expressed as:

$$M = M_{iso} + M_{dev} \tag{2.20}$$

where M_{iso} and M_{dev} are the isotropic and deviatoric components of the moment tensor, respectively. Isotropic term represents the volumetric change, whereas M_{dev} is further classified into,

$$M_{dev} = M_{DC} + M_{CLVD} \tag{2.21}$$

where M_{DC} is the double-couple component and M_{CLVD} represents the compensated linear vector dipole (Knopoff and Randall, 1970).

The components of the moment tensor can be expressed using the scalar moment (Mo), strike (ϕ_f), dip (δ) and rake (λ) of a shear dislocation (Aki and Richards, 1980) using the following relations:

$$\begin{aligned}
M_{11} &= -M_o\left(\sin\delta \cos\lambda \sin 2\phi_f + sin\, 2\delta\ \sin\lambda\ sin^2\phi_f\right) \\
M_{22} &= M_o\left(\sin\,\delta \cos\,\lambda \sin 2\phi_f - sin\ 2\delta \sin\,\lambda\ cos^2\,\phi_f\right) \\
M_{33} &= M_0\left(\sin 2\delta\,\sin\,\lambda\right) = -\left(M_{xx} + M_{yy}\right) \\
M_{12} &= M_o\left(\sin\,\delta \cos\,\lambda\, \sin 2\phi_f + \frac{1}{2}\, sin\ 2\delta \sin\,\lambda\ sin\ 2\phi_f\right) \\
M_{13} &= -M_o\left(\cos\delta\ \cos\lambda \cos 2\phi_f + cos\ 2\delta \sin\lambda\ sin\ \phi_f\right) \\
M_{23} &= -M_o\left(\sin\delta\,\cos\lambda\,\sin\phi_f - cos\ 2\delta \sin\lambda\ cos\ \phi_f\right)
\end{aligned} \tag{2.22}$$

Thus, earthquake source parameters (seismic moment, strike, dip, rake, etc.) can be obtained from the observed seismogram by inversion as presented in equation (2.19). It is now a practice to invert the waveforms recorded at various seismic stations round the globe to obtain the moment tensor elements and subsequently derive the best double-couple solution and the principle maximum/most compressive (*P*), intermediate (*B*) and Tension (*T*) axes orientations.

The moment tensor can also be decomposed in terms of its principal axes, the eigen values as well as the directions of the eigen vectors. The axis associated with the largest positive eigen value is the *tension* or *T*-axis, the largest negative value is the *compressional* or *P*-axis, and the intermediate the *null* or *B*-axis. Generally, the quality of moment tensor inversion depends, to a large extent, on the number of data available and the azimuthal distribution of stations about the source. In inversion with body waves, Dufumier (1996) gives a systematic overview of the effects caused by differences in azimuthal coverage and effects caused by the use of only *P* waves, *P* and *SH* waves, or *P*, *SH* and *SV* waves. An overview with respect to the effects caused by an erroneous velocity model for the Green function calculation and effects due to erroneous hypocentre coordinates is made by Kravanja et al. (1999).

Seismograms (three component), from a set of stations, represent the data, *d*, while the moment tensor term is represented by *m*. The linear inverse problem $d = G\,m$ is solved by the least-square method, where G is the Green's function:

$$m = (G^T G)^{-1} G^T d, \tag{2.23}$$

where G^T is G transposed and $()^{-1}$ is the inverse of the system matrix $(G^T G)$. Inversion may be done in time domain or frequency domain. Care must be taken to match the synthetic(s) and observed (u) seismograms.

$$Corr = \frac{\int us}{\sqrt{\int u^2 s^2}} \tag{2.24}$$

where $\int us = \Sigma_i \int u_i(t) s_i(t) dt$ and summation is taken over all components and stations used in the inversion. The misfit (χ) between observed and synthetic seismograms is given by

$$\chi = \int (u - s)^2 \tag{2.25}$$

Also, the misfit can be represented by variance reduction (ξ):

$$\xi = 1 - \frac{\chi}{\int u^2} \tag{2.26}$$

The relation between correlation and variance reduction is given by the formula

$$Corr^2 = \xi \tag{2.27}$$

In most inversion schemes, the focal depth is assumed to be constant (Lay and Wallace, 1995) but there are many variations of the algorithm used to obtain a moment tensor solution from the waveforms of body or surface waves (e.g. Kanamori, 1980; Patton, 1980; Stump and Johnson, 1977; Ward, 1980). All these methods have a priori assumption of the parameters defining the earth's structure, the event location and the origin time. Steps to obtain fault plane solution (FPS) from observed seismogram is summarized next:

1. Location of the event and obtain the source parameters, such as earthquake location, origin time, depth by analyzing seismic waveform records.
2. Select the waveforms of at least a minimum three stations; otherwise, the obtained results may not be stable.
3. Convert the seismic waveforms to displacement by integrating the records
4. The displacement records are to be low pass filtered, as high-frequency components are difficult to model, which requires a detailed knowledge of subsurface crustal velocity model.
5. The displacement spectra are down sampled and stored for inversion.
6. Collect the known velocity model of the study region.
7. Synthetic Green's functions are to be computed in spectral domain using the known velocity structure of the region and at fixed earthquake location. However, earthquake depth can be varied. Green's functions are than convolved with source time function and instrument response to get the synthetic seismogram.
8. Inversion is carried out to get the moment tensor and subsequently FPS, scalar seismic moment, earthquake depth and directions of *P*, *B* and *T*-axes can be obtained.

An example of obtaining FPS by waveform inversion using ISOLA software (Sokos and Zahradnik, 2008) is presented next.

2.6 EXAMPLE OF MOMENT TENSOR INVERSION

A moderate earthquake occurred in Rudraprayag, Uttarakhand, of northwest Himalaya on February 6, 2017. The epicenter and focal depth of the event were reported to be 30.5° N, 79.1° E and 33 km, respectively, by National Centre for Seismology (NCS), New Delhi.

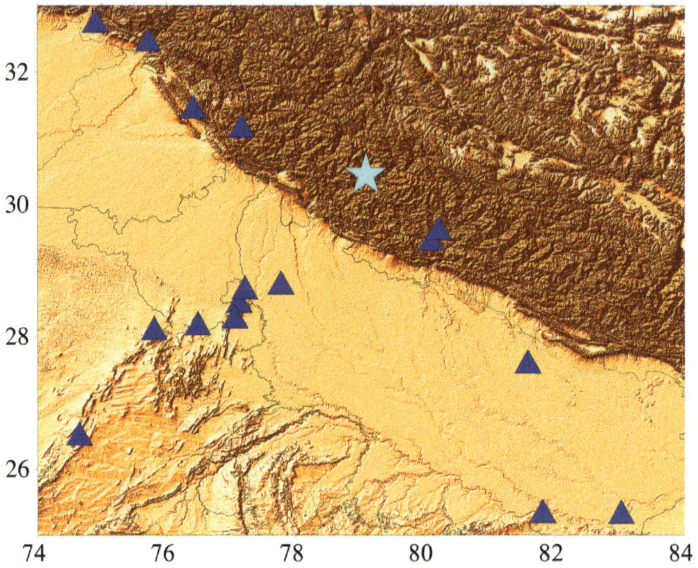

Figure 2.6 Location of Rudraprayag earthquake on February 6, 2017 (Mw 5.5), is plotted by a cyan star and the regional stations recorded the event are shown by blue triangles. (Modified after Sutar et al., 2020.)

The event was recorded by a network of seismic stations, equipped with broadband seismometers (BBS) and strong motion accelerometers (SMA) located within ~700 km distance from the epicenter (Figure 2.1).

The location and the origin time of the event were estimated by minimizing the root mean square (RMS) error between theoretical and observed travel times of various seismic phases recorded at different stations. As the recording stations were spread over a larger area of ~800 × 900 sq km, the global ISPEI91 velocity model (Kennett and Engdahl, 1991) was selected for the earthquake location. The event was reported to be located in 30.467 °N latitude, 79.118 °E longitude with depth 15 km with origin time, dated 2017/02/06 at 17 hour, 03 minutes and 06.2 sec (Sutar et al., 2020) (Figure 2.6).

Full waveform inversion has been carried out for all the available records using moment tensor inversion approach to understand the style of faulting for the event. Recorded waveform data of broadband stations were used for full-waveform inversion to get the FPS using ISOLA software package (Sokos and Zaharadnik, 2008).

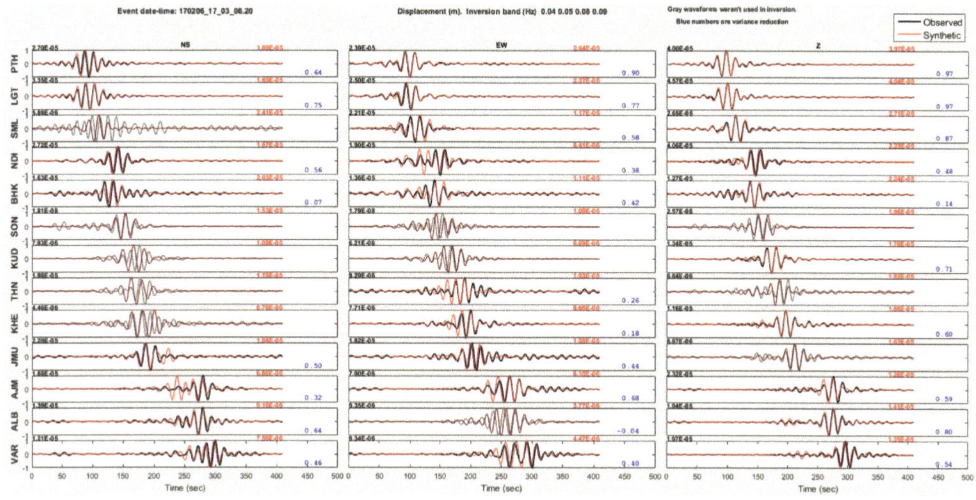

Figure 2.7 Three components (N-S, E-W and Z) normalized correlation plot between the observed and synthetic displacement waveforms of the February 6, 2017, event (Mw 5.5). Black-and-red lines represent observed and synthetic seismograms, respectively. The waveforms data have been band-pass filtered in the frequency range (0.04–0.07) Hz.

At the beginning, recordings having poor signal to noise (SNR) were discarded from the analysis. The recorded data were filtered and integrated to get the displacement waveform and the Green's function (Bouchon, 1981) was computed in the lower-frequency range below 0.5 Hz using the regional velocity model of Kennett and Engdahl (1991). The inversion was carried out in the frequency range between 0.04 and 0.08 Hz to get the FPS. During the inversion, stations whose observed displacement spectra does not match with the synthetic data were neglected in the analysis (grey waveforms in Figure 2.7). The inversion was finalized when the global correlation coefficient and DC percentage values exceed 0.8% and 80%, respectively. It was reported that, the strike, dip and slip vector (rake) of the first nodal plane to be 275°, 47°, 110°, respectively; while, the other nodal plane has strike, dip and slip 68°, 47° and 70°, respectively. Further, the centroid depth of the event was improved by considering multiple sources at different depths. The correlation Vs depth plot (Figure 2.8) shows maximum correlation at 19 km depth. Although correlation at depths between 14 km and 22 km appeared similar but correlation at 19 km depth is higher and hence the centroid depth was selected to be 19 km (Sutar et al., 2020).

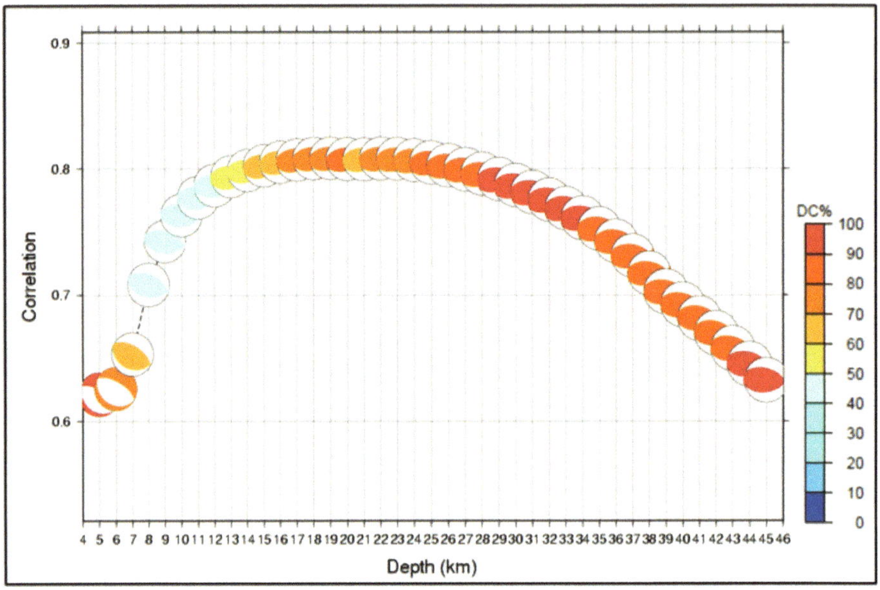

Figure 2.8 Correlation versus DC percentage plot at various depths.

2.7 WAVEFORM MODELLING TO EXTRACT EARTH'S SUBSURFACE PHYSICAL PROPERTIES/WAVEFORM TOMOGRAPHY

Tomography is a technique to extract information on earth's internal structure and numerous techniques are available in the literature, namely, body wave travel time tomography, surface wave tomography, attenuation tomography and ambient noise tomography (Wang and Rao, 2020). Waveform-modelling technique can also be used to extract information on earth's subsurface physical properties such as, seismic velocity structure, attenuation, amplification, also known as seismic waveform tomography. Many seismic tomography methods use the arrival time or amplitude of the waveform as the input data. These methods are generically referred to as travel time or amplitude tomography. In these methods, the input data consists of travel times, amplitudes, or some other secondary attributes of the recorded data, whereas, in waveform tomography the complete waveform data is used giving a better resolution compared to the travel time or amplitude tomography methods. The underlying numerical method for waveform tomography is based on the full-wave equation, as opposed to a ray approximation or a Born approximation for the others. Thus, waveform tomography becomes more accurate than the travel time or amplitude tomography, but it makes the inverse problem more difficult to solve. It uses every wiggle of seismograms to determine the structure of earth's interior, which is constrained by the laws of seismic wave propagation. Therefore, it can image earth's interior even in complex environment that cannot be approximated using simple models.

Waveform tomography was introduced in the late 1970s with an ultimate goal to use all of the information in a seismogram to understand the structure and dynamics of earth. Earth can be regarded as spherically symmetric on a global scale, whereas, it can be considered as flat and horizontally layered on a local scale. In the laterally homogeneous models, numeral computation on expected arrival times and waveforms of seismic waves are easy to calculate. Calculated travel times and waveforms often closely resemble observations, and differences between calculated and observed values are used to constrain earth's lateral variations (Blom et al., 2020; Yuan et al. 2015 and references therein).

Synthetic seismograms are computed by simulating seismic wave propagation using the following equation:

$$\rho\ddot{u} = f(x,t) + \nabla\cdot\sigma\,(x,t) \tag{2.28}$$

where ρ denotes density, u is the displacement field, $\ddot{u}$ is the acceleration, σ is the stress field and f is the forcing term. The equation can be solved in 3D and synthetic waveforms can be generated. The equation can be solved either by finite difference technique, finite element or more advanced available numerical techniques. Using a finite difference method, equation (2.28) may be presented in a matrix form as:

$$Mu = s \tag{2.29}$$

where M is a matrix containing the partial differential operators, u is a vector representing the wavefield at all grids, and s represents the source term, which is zero everywhere except at the source location. To obtain wavefield u, the inverse matrix M^{-1} need to be solved and information on the source term is needed, which is opposite to the moment tensor inversion presented in the previous section. Thus, seismic waveform tomography is also an inverse problem to find the earth structure.

In order to quantify the differences between observed and synthetic waveforms, methodology given by Fichtner et al. (2010) can be used, which is given by:

$$\xi(m) = \iint_{t,\omega} W_p^2(t,w)\left[\varnothing(m,t,\omega) - \varnothing_{obs}(t,\omega)\right]^2 dtd\omega \tag{2.30}$$

$$W_p = \frac{log\left(1 + \lceil u_{obs} \rceil\right)}{max\left(log\left(\lceil 1 + u_{obs} \rceil\right)\right)} \tag{2.31}$$

where W_p is a weighting function that determines the region within that space have sufficient amplitude such that genuine measurements can be done, u_{obs} is the observed seismic signal, and $(\varnothing - \varnothing_{obs})$ is the phase difference between observed and synthetic data.

An illustration of waveform tomography is shown in Figure 2.9 following the results of Blom et al. (2020). In this exercise, seismic waveform tomography of the crust and upper mantel beneath the central and eastern Mediterranean has been estimated using data from 17,000 unique source receiver pairs, gathered from 80 earthquakes to image the crust and upper mantel beneath the study area. This study clearly imaged the high velocity structures across the domain that can be linked to current and past subduction (Figure 2.9c), including the Cyprus subduction zone and high-velocity anomalies beneath the Italian Peninsula and the Dinarides. They have used long period waveform data for the inversion with a windowing technique that optimally separates the small amplitude body waves from large amplitude surface waves. They have imaged several high-velocity structures imaged down to the transition zone, which can be correlated to the current and past tectonic settings of the area. Due to the natural dominance of surface-wave signals in waveform tomography, their model can be constrained for S velocity better compared to P-wave velocity, which is found to be much smoother as a result of its longer wavelength for the same frequency but broadly displays the same structures. They have also reported that density is poorly constrained due to the sensitivity of the measurements and reported that the Hellenic slab is the most prominent

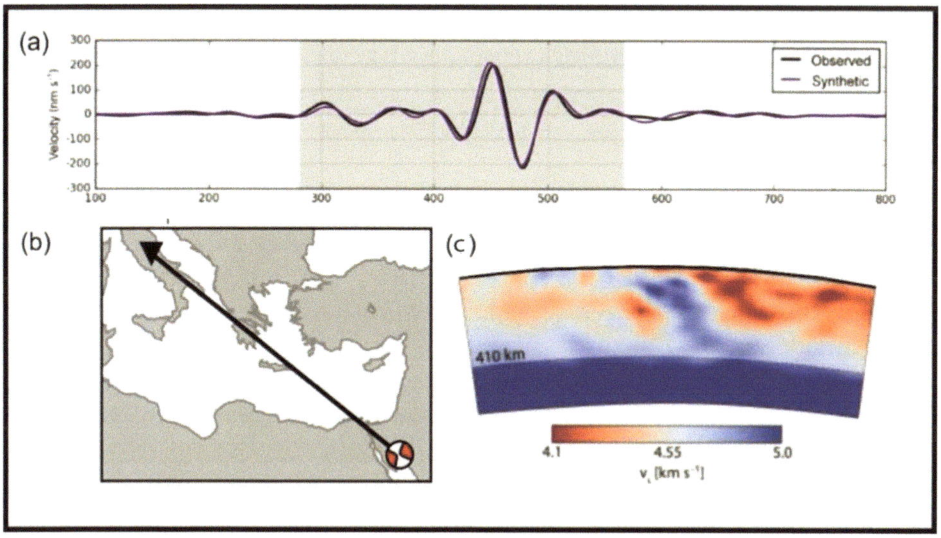

Figure 2.9 (a) Comparison of an observed and synthetic waveform of the vertical component, (b) Station (Triangle) recorded the earthquake of Mw 5.6 in the Sinani region, Egypt. Fault plane solution of the event is overlapped over the map, (c) Model outcome from waveform tomography. (Modified after Blom et al., 2020.)

feature in the model and extends from the surface down to the transition zone in a bent, arcuate shape. Also, it is shown that two more or less parallel anomalies are visible beneath the Italian Peninsula and the Dinarides. The Italian anomaly is visible at depths below ~100 km, whereas the Dinaric anomaly extends towards the surface beneath the southern Dinarides and extends towards Greece in the south. The high-velocity anomaly beneath Anatolia interpreted as the Cyprus slab, which dips towards the north and east beneath the western and central parts of Anatolia.

2.8 CONCLUSION

Waveform modelling is a numerical technique that can extract information on earthquake source parameters and earth's internal structure from the observed seismogram. This can be achieved by minimizing the difference between observed and synthetic seismograms by adjusting the controlling parameters. If the earth structure is known, earthquake source parameters can be extracted using the moment tensor inversion technique, whereas, earth's subsurface information like velocity, density, attenuation, can be extracted from the observed seismic waves with the knowledge of earthquake source parameters using seismic waveform tomography.

REFERENCES

Aki, K., and Richards, P. G. (1980). *Quantitative Seismology. Theory and Methods*, 1 (557), 169.

Aki, K., and Richards, P. G. (2002). *Quantitative Seismology*. Sausalito, CA: University Science Books.

Blom, N., Gokhberg, A., and Fichtner, A. (2020). Seismic waveform tomography of the central and eastern Mediterranean upper mantle. *Solid Earth*, 11, 669–690. https://doi.org/10.5194/se-11-669-2020

Bouchon, M. (1981). A simple method to calculate Green's functions for elastic layered media. *Bull. Seismol. Soc. Am.*, 71, 959–971.

Dufumier, H. (1996). On the limits of linear moment tensor inversions of teleseismic body wave spectra. *Pageoph*, 147, 467–482.

Fichtner, A. (2010). *Full Seismic Waveform Modelling and Inversion*. Heidelberg: Springer. https://doi.org/10.1007/978-3-030-10475-7_211-1.

Kanamori, H. (1980). The size of earthquakes. *Earthquake Information Bulletin (USGS)*, 12 (1), 10–15.

Kennett, B. L. N., and Engdahl, E. R. (1991). Traveltimes for global earthquake location and phase identification. *Geophys. J. Int.*, 105, 429–465.

Knopoff, L., and Randall, M. J. (1970). The compensated linear-vector dipole: A possible mechanism for deep earthquakes. *J. Geophys. Res.*, 75 (26), 4957–4963. doi:10.1029/JB075i026p04957.

Kravanja S., Panza G. F., and Sileny J. (1999). Robust source parameters retrieval from point source time function. *Geophys. J. Int.*, 136 (2), 385–394.

Lay, T., and Wallace, T. (1995). *Modern Global Seismology*. Cambridge, MA: Academic Press, Inc.

Patton, H. (1980). Crust and upper mantle structure of the Eurasian continent from the phase velocity measurements and Q of surface waves. *Rev. Geophys.*, 18, 605–625.

Silver, P. G., and Jordan, T. H. (1982). Optimal estimation of scalar seismic moment. *Geophys. J. Int.*, 70, 755–787. doi:10.1111/j.1365-246X.1982.tb05982.x.

Sokos, E. N., and Zahradnik, J. (2008). ISOLA a Fortran code and a Matlab GUI to perform multiple-point source inversion of seismic data. *Computer Geosci.*, 34, 967–977. doi: 10.1016/j.cageo.2007.07.005.

Stein, S., and Wysession, M. (2003). *An Introduction to Seismology Earthquakes and Earth Structure*. Malden, MA: Blackwell Publishing Ltd.

Stump, B. W., and Johnson, L. R. (1977). The determination of source properties by linear inversion of seismograms. *Bull. Seismol. Soc. Am.*, 67, 1489–1502.

Sutar, A. K., Verma, M., and Bansal, B. K. et al. (2020) Source characterisation of February 06, 2017 Rudraprayag earthquake in Northwest Himalaya and ground motion prediction for a scenario earthquake (M_W 6.8). *J. Geol. Soc. India*, 95, 551–560. https://doi.org/10.1007/s12594-020-1481-5.

Ward, S. N. (1980). A technique for the recovery of seismic moment tensor applied to the Oaxaca, Mexico earthquake of November 1978. *Bull. Seismol. Soc. Am.*, 70 (3): 717–734. https://doi.org/10.1785/BSSA0700030717.

Wang, Y., and Rao, Y. (2020). Seismic, waveform modeling and tomography, H. K. Gupta (ed.), *Encyclopedia of Solid Earth Geophysics, Encyclopedia of Earth Sciences Series*. pp. 1–15. Switzerland AG: Springer Nature. https://doi.org/10.1007/978-3-030-10475-7_211-1.

Yuan, Y. O., Simons, F. J., and Bozdağ, E. (2015). Multiscale adjoint waveform tomography for surface and body waves. *Geophysics*, 80 (5), R281–R302. doi: 10.1190/GEO2014-0461.1.

Chapter 3

Theory and practices involved in depth and source localization of anisotropy

Recent developments in subsurface investigation

Debasis D. Mohanty

3.1 INTRODUCTION

Seismic anisotropy plays a vital role in understanding the deep earth dynamics and structures and gives a true image of current tectonics and deformation patterns (savage, 1999). With a lateral resolution of < 50 km, the shear wave splitting (SWS) measurements are capable of quantifying the deformation patterns in the lithosphere and upper asthenosphere. The two splitting parameters namely fast polarization direction (ϕ, FPD) and delay time (δt) are responsible for characterization of the deformation mechanism and anisotropy of the earth structure. The FPD serves as prolific indicator of external stress pattern due to various mechanisms related to deformations such as subduction, collision, rifting and fossilized anisotropy, which in turn acts as a geological marker for understanding the regional stress and patterns of current tectonics with a possible seismological interpretation of deep earth process (Gao et al., 1997; Gao et al., 2010; Hammond et al., 2014). Another parameter called delay time (δt) is the manifestation of thickness of the anisotropic layer, where it signifies the magnitude of anisotropy of the medium.

The upper mantle is a source of high abundant of dominant minerals like pyroxene and orthopyroxene (Nicolas and Christensen, 1987), which are the major source of vast azimuthal anisotropy in the mantle region. Due to the lattice-preferred orientation (LPO) of these minerals, their fast polarization direction of anisotropy (ϕ) remains parallel to their *a*-axis, which are the prima facie of the induced strain directions in the upper

DOI: 10.1201/9781003177692-3

mantle, helping to know the geodynamics and tectonics of the underneath earth. SWS is simply associated with the shear strain of olivine aggregates caused due to the lithospheric movement. The olivine and orthopyroxene get deformed due to the diffusion creep at a depth of > 200 km, which are responsible for the absolute plate motion (APM) related strain due to LPO-induced shear at the base of the lithosphere (Vinnik et al., 1989; 1992). Again the past significant tectonic events are responsible for the frozen anisotropy in the cool lithospheric mantle regions. The anisotropic direction (FPD) has a significant role in fault system or mountain formation and deformation processes, as the pattern acts as parallel to the strike of these orogenic structures and regional scale faults (Silver et al., 2001). Similarly, in areas like the continental rifts, the fast axis direction predicts the external stress pattern derived by the molten packets subparallel to the rift direction (Gao et al., 1997, 2010; Hammond et al., 2014).

Though, comparatively, many studies on the deformation analysis of Indian subcontinent are available on the basis of the analysis from core refracted phases (SKS, SKKS, PKS and Direct-S phases) (Singh et al., 2006; Mohanty and Mondal, 2020; Mondal and Mohanty, 2021; Mohanty and Singh, 2021; Mohanty et al., 2021; Mohanty and Mondal, 2021 for northeast Himalayan region; Kumar and Singh, 2008; Heintz et al., 2009 for Indian shield; Oreshin et al., 2008 for western Himalaya; Kumar et al., 2010 for Godavari graben; Rao et al., 2013 for northwest Deccan volcanic province; Kumar et al., 2015 for southern India), very limited studies are attributed towards the possible central depth and source localization of anisotropy of a particular region. Though SWS measurements are keys to understand the geodynamics of underneath earth, they are not capable of providing a complete and comprehensive understanding of the exact underneath dynamics and source localization of anisotropy, unless a proper justified analysis for depth of anisotropy is made.

The present chapter aims to characterize the methods and analysis, therefore, to measure the possible depth and source localization of anisotropy, along a suitable example comprising of a case study. A more suitable and vivid description regarding the process of establishment of central depth of anisotropy with a newly developed approach named spatial coherency method (Gao and Liu, 2012) is thoroughly conceptualized in this chapter. The new technique is a grid search-based 'Fresnel zone' concept, which complements the splitting results in a great extent with strengthening our understanding of deformation mechanism beneath a critical geological set up. A well-documented case study is also described to complement the understanding and application of this method to deduce the central source of anisotropy for a better geodynamics model of earth.

3.2 METHODOLOGY

SWS is the most direct way to estimate the azimuthal anisotropy in recent time. For the analysis of mantle anisotropic thickness and its influence, the delay time is generally used but the actual depth anisotropy can't be determined (Alsina and Snieder, 1995). To understand the source localization of anisotropy, it is important to have a good idea about the distribution of anisotropy with respect to the depth.

3.2.1 Concept of Fresnel zone

To estimate the lateral resolution of an unmigrated stacked core refractive wave data, often seismologists use the concept of Fresnel zone (Lindsey, 1989; Sheriff, 1996). Fresnel zone determines the spatial resolving power for unmigrated seismic data, for which important lithological changes may be analysed along with the seismic profile direction. Fresnel zone is an area of the reflector dependent on the frequency and range from which most of the energy of a reflection is returned where the arrival times differ by less than half a period of time from the first break. These waves after reflection from the Fresnel zone (with such arrival times) will interfere constructively and detected as a single arrival. Subsurface features cannot be detected using seismic waves if they have wavelengths smaller than the Fresnel zone. The zone which indicates the part of a reflector (Figure 3.1) from which energy of a reflection can reach a detector where the waves travel with a one-fourth wavelength is called the first Fresnel zone. In the second Fresnel zone, the energy comes delayed one-half to one cycle adding destructive to the energy from the first zone. There is a third Fresnel zone and so on. The size of the Fresnel zone determines the minimum size feature that can be resolved from the seismic data (Dewangan et al., 2007). The radius of the Fresnel zone for zero offset is:

$$R = \frac{1}{2}\sqrt{(\mathrm{V}\ \lambda \mathrm{T})} \tag{3.1}$$

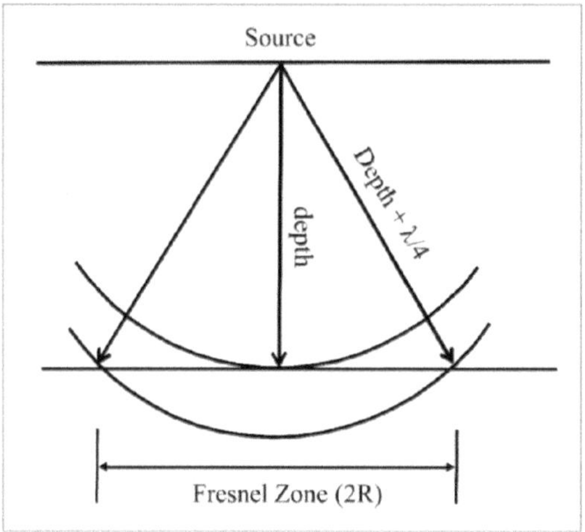

Figure 3.1 Diagram illustrating Fresnel zone (2R), lateral distance over which the reflection signal is averaged. λ is the dominant wavelength of the seismic wave and *R* is the radius of the Fresnel zone.

where, V is the medium velocity, T is the two-way travel time of the reflection (T), and λ is the dominant wavelength of the seismic wavelet.

3.2.2 Fresnel zone concept for central depth of anisotropy

Alsina and Snieder (1995) proposed a different method using the core refracted waves to calculate the depth of anisotropy. The resolution of these waves can be improved using the individual phases recorded at different stations of relatively dense array and by analyzing the lateral variability of the estimated splitting. In the vicinity of the geometrical ray path, the ray arrival is influenced by the material properties. Thus, the ray is treated as a tube, whose diameter is defined as the first Fresnel zone. The size of the Fresnel zone is fully dependent on the length of the ray path. The travel time can be directly converted into the depth for the steep incident waves. As the dimension of the Fresnel zone is a function of depth and period, it is possible to calculate the depth parameter (Alsina and Snieder, 1995). The diameter of the tube increases with the dominant period and depth, and the splitting parameters reflect the combined effect of seismic anisotropy inside the tube. If two events of core refractive phases (PKS, SKS and SKKS, collectively known as XKS) from opposite directions have different fast polarization directions and delay times at the same station, then the layer of anisotropy must reside below the depth at which the two tubes intersect. Similarly, the XKS event which has different splitting parameters at two stations, the depth at which two tubes intersect must be shallower (Alsina and Snieder, 1995).

3.2.2.1 *Drawback of Fresnel zone concept of depth distribution*

Though the Fresnel zone concept was adopted previously to scale out the anisotropic parameters at certain depths, which in turn allowed to compute the central depth of anisotropy, this method had a serious drawback in connection with the localization of events and stations throughout different azimuthal ranges. This particular method needs couple of seismic events with a back azimuthal (BAZ) difference of 180° apart, which sometimes is not feasible and really rare in terms of event distributions, and leads towards difficulties for the prediction of anisotropic depth. Even in case of available events, lesser dense of SWS parameters from less events lead to some unfair measurements. To overcome such difficulties, spatial coherency method (Gao et al., 2010; Liu and Gao, 2011; Gao and Liu, 2012), which is more dependable on grid search process, is advanced for the estimation of depth of anisotropy and source localization.

3.2.3 Spatial coherency approach for central depth determination and source localization

To overcome the drawbacks from Fresnel zone method (as discussed earlier), the new approach called 'spatial coherency' method was adopted, which is based upon the function of Variation factor (F_v), defined as the sum of weighted mean of simple deviation of delay time (*dt*) and mean of circular standard deviation of first polarization direction (F_ϕ) (Fisher, 1993). The minima of variation factor corresponding to a particular depth is termed as the source and depth of anisotropy for that particular region. This particular depth of anisotropy is comprehended by the maximum coherency between the splitting parameters when plotted at their ray piercing points for this particular depth (Gao and Liu, 2012). The variation factor stated as:

$$F_v = \frac{1}{N}\sum_{i=1}^{N}\sqrt{\frac{1}{M_i - 1}}\left(w_\Phi \sqrt{\sum_{j=1}^{M_i}\left(\Phi_{ij} - \bar{\Phi}_i\right)^2} + w_{\delta t}\sqrt{\sum_{j=1}^{M_i}\left(\delta t_{ij} - \bar{\delta t}_i\right)^2}\right) \quad (3.2)$$

where M_i represents the number of samples in the *i*th block out of total *N* number of blocks and *ij* refers to *j*th splitting parameters in *i*th block. $\overline{\Phi}_i$ and $\overline{\delta t}_i$ are average of Φ and δ*t* in the *i*th block. The weighting factors w_Φ and $w_{\delta t}$ are the functions of overall range of splitting parameters (ϕ and δt).

To estimate the depth of anisotropy, the area of research divided into square degree blocks of *dx* dimensions and variation factor F_v is calculated for each and every possible depth of anisotropy. To get the ideal anisotropic depth, we have used a certain series of *dx* values, which lead to the minimum variation factor (F_v) (Liu and Gao, 2011; Gao and Liu, 2012). To examine the central depth of anisotropy, spatial coherency method mainly depends on the maximum coherency between first polarization direction and delay time, which is characterized by minimum coherency curve for a particular depth. At a specific depth, the distribution and configuration of splitting parameters of piercing point also helps to get the maximum coherency. The method computes the central depth with certain assumptions and factors such as: (i) assuming the presence of single layer of anisotropy, (ii) close proximity of seismic stations and (iii) spatial varying splitting parameters form a wider range of BAZ. Alongside, the grid size selection plays a huge role in the examination of depth of anisotropy because, for the higher variability coherency curve, the grid size should be lower/minimum, while the curve gets wider in case of higher value of grid size (Liu and Gao, 2011). The shape of the best coherency curve with ultimate precision should be a smooth 'U' shape.

3.3 CASE STUDY

For better understanding of the processes and logic behind the spatial coherency method, a suitable case study, which has been carried out in the Shillong plateau mass and the Assam foredeep of northeast Indian region (Mohanty et al., 2021), has been thoroughly discussed and represented in the next section. This study attributed for the first time towards the establishment of the possible central depth and source localization of anisotropy in this complex tectonic domain. This finding clearly leads a way to better understand and image the source and evolution of complex tectonics beneath this complex northeast Indian part.

3.3.1 Geological settings of Shillong and Assam foredeep region

The overall geological settings of northeast India is convergent where the Indian, Eurasian, and Sunda plate meet each other. During the late Cretaceous, India became separated from the Gondwana, where it was drifted towards northeast at a speed of ~180 mm/yr and finally subducted at a rate of ~54 mm/yr in the early Eocene to form the E-W trending Himalayan Mountain Range (Molnar and Tapponnier, 1975). In the northeast part of India, the subduction is oblique at a speed of ~50 mm/yr towards N20E and forms the Indo-Burma Range trending in NE-SW (DeMets et al., 1994). At Namcha-Barwa syntaxis, the Himalayan Mountain Range takes a sharp bent and merged with Indo-Burma Range. Comprised with a complex tectonic setup, northeast India is surrounded by many significant faults, namely, Main Central Thrust (MCT), Main Boundary Thrust (MBT) (Angelier and Baruah, 2009), Mishmi Thrust, Lohit Thrust, Naga Thrust, Disang Thrust, Kopili Fault, Dauki Fault (Kayal et al., 2012), Dhubri Fault and Brahmaputra Fault (Kayal and De, 1991; Nandy, 2001).

Shillong plateau, situated in the northeastern part of Indian Shield in between the Indo-Burma subduction zone and Indo-Eurasian subduction zone, is tectonically very active (Figure 3.2). This plateau is surrounded by E-W trending Dauki Fault in the south and the Brahmaputra Fault in the north (Angelier and Baruah, 2009). Dhubri Fault and Kopili Fault demarcates the western and northeastern boundaries of the plateau, respectively (Evans, 1964). Among the other faults, Dapsi Thrust, Dudhnoi Fault, Oldham Fault, Chedrang Fault, Samin Fault and Barapani Shear Zone, are major geological structures to mention. These complex fault systems and up arching of the Shillong plateau were formed during the late Jurassic and early Cretaceous periods, when the Indian plate separated from Gondwana and passed through the Kerguelen Hotspot (Gupta and Sen, 1988). During this time, volcanic eruption took place along the Dauki Fault system and was responsible for the formation of Sylhet Trap (Gupta and Sen, 1988). Bilham and England (2001) suggests that the Shillong plateau shows pop-up geometry because of the surrounding boundary fault (Dauki Fault and Brahmaputra Fault/Oldham Fault) system, complex tectonic settings, and horizontal compressive

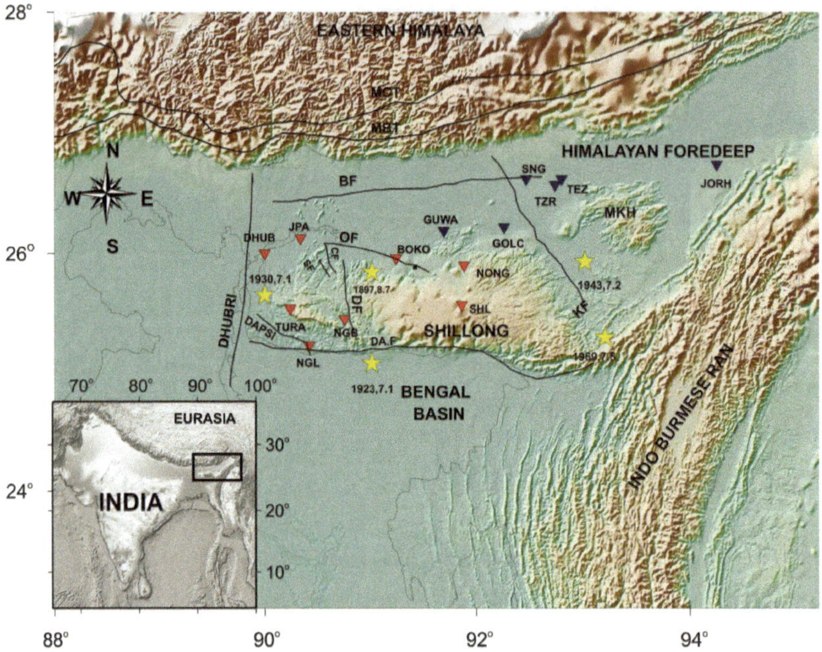

Figure 3.2 Location and geological distribution of 14 broadband seismic stations in 2 tectonically distinct regions of northeast India, namely, Shillong plateau and Himalayan foredeep. The stations in Shillong are marked by red inverted triangles and labelled with bold letters. The stations used for study from HF region are marked with blue inverted triangles and labelled with bold letters. The black rectangle on the map of India (lower left) indicates the location of the study area. The biggest earthquakes occurred in this study region are marked by yellow stars and their year with magnitudes are labelled with bold letters. Important topographic units and faults are abbreviated as BF, Brahmaputra Thrust; CF, Chedrang Fault; D.A.F., Dauki Fault; DAPSI, Dapsi Thrust; DF, Dudhnoi Fault; DHUBRI, Dhubri Fault; IBR, Indo-Burmese Range; KF, Kopili Fault; MBT, Main Boundary Thrust; MCT, Main Central Thrust; MKH, Mikir Hill; OF, Oldham Fault; TT, Tidding Thrust.

force. Though a precise confirmation regarding the unique pop-up tectonics of the Shillong plateau is still a hot debate, many fold research is the need of the hour for postulating the true dynamics and mechanism behind the unique tectonics of this plateau mass. Kailasam (1979), Khattri et al. (1983), Kayal and De (1987) and Nayak (1989) suggest this plateau as a horst structure formed during the Jurassic time due to thermal disturbance. This plateau consists of Archean gneissic rock with sedimentary and metasedimentary deposits of Khasi Group, Jhantia

Group, Garo Group, and Dupitila series of rocks with younger alluvium deposit (G.S.I., 1974). More than 20 large earthquakes have struck this region in the past two centuries. Some of the significant earthquakes are 1869 Cacher earthquake (Mw ~7.5) at the southeast end of the Kopili Fault (Baro and Kumar, 2015), 1897 Assam earthquake (Mw ~8.7) at Oldham Fault (CNDM, 2002), 1923 Meghalaya earthquake (Mw ~7.1) at Dauki Fault (CNDM, 2002) and 1930 Dhubri earthquake (Mw ~7.1) at Dhubri Fault (Kayal, 2008).

Himalayan foredeep is a peripheral foreland basin at the foothill of Himalaya trending East-West. The trend of the northeastern part of Himalayan foredeep is around NNE-SSW, bounded by two younger fold mountain belts, that is, Himalayan Mountain Range and Indo-Burma Range. This region is bounded by Main Boundary Thrust and Naga Thrust in NW and SE, respectively. NE dipping Mishimi Thrust demarcates the northeastern boundary of this region, and Kopili Fault marks the southwestern boundary. It consists of crystalline igneous and metamorphic rock of Archean age overlain by recent terrigenous tertiary sediments of more than 4 km thick, towards the foothill Himalaya sediment thickness that exceeds 7 km (Bhandari et al., 1973).

3.3.2 Source localization and depth determination beneath Shillong plateau

The deformation pattern beneath the Shillong plateau is most likely related to the asthenosphere flow beneath it as the Precambrian basement rocks are free from convergence-related deformation. Mohanty et al. (2020) suggested the reason behind this anisotropy due to lattice-preferred orientation (LPO) of the olivine, which drives the asthenospheric flow. To validate and prove the aforementioned statement, 47 XPS splitting results from 8 broadband stations (Table 3.1 and Figure 3.2) are taken into consideration for modelling purpose to determine the central depth (for details of the splitting parameters refer to Table 3.1, Mohanty et al., 2021). A grid search approach as discussed earlier is applied with the grid size ranging from 0.36° to 0.40° along a central difference of 0.11°. Using the weighting factor of W_Φ = 1/180° and $W_{\Phi t}$ =1/2.30s,

Table 3.1 Details of broadband seismic stations along with their geological distributions

Station	*Lat (deg)*	*Lon (deg)*	*Elevation (m)*
SHILLONG PLATEAU			
DHUB	26.02	90.00	33
SHL	25.57	91.86	1600
TURA	25.54	90.24	560
NGB	25.45	90.75	159
NONG	25.90	91.88	551
BOKO	25.96	91.24	56
JPA	26.13	90.34	46
NGL	25.23	90.42	320
HIMALAYAN FOREDEEP			
FOREDEEP			
GUWA	26.19	91.69	88
JORH	26.74	94.25	75
TZR	26.57	92.74	77
SNG	26.62	92.47	70
GOLC	26.22	92.26	139

Lat: Latitude; Lon: Longitude; deg: degree; m: meter.

possible central depth of anisotropy is established around 100 km by the spatial coherency method and the piercing point plot (Figure 3.3).

In case of the Shillong plateau, the case study satisfied all the conditions required for the central depth calculation and found with the minima of the variation factor (F_v) at a depth of around 100 km. The exact minimum is at 94 km, and 102 km for grid sizes 0.36 and 0.40, respectively, with a fixed central difference of 0.11°. Also, XKS splitting parameters plotted at the ray-piercing point of 100 km depth and positively showed maximum coherency (Figure 3.3) than all other depths. Devi et al. (2011) suggests the approximate depth of the lithosphere-asthenosphere boundary (LAB) at around 100 km from the converted phases receiver function analysis. So the carried study (Mohanty et al., 2021) successfully defends the fact that the LAB must be the source of this high strength anisotropy at this particular depth range of 100 km, which is established from the spatial coherency model. The shear wave splitting (SWS) results of this region show a fast polarization direction (FPD) towards the NE-SW direction coinciding with the absolute plate motion (APM) of the Indian plate. Overall, the source localization of anisotropy at a depth of 100 km is suggestive of the fact that, the asthenospheric drag at the base of the lithosphere (LAB region) is the major source of anisotropy which governs the present-day deformation patterns beneath the plateau mass and is responsible for the present tectonics.

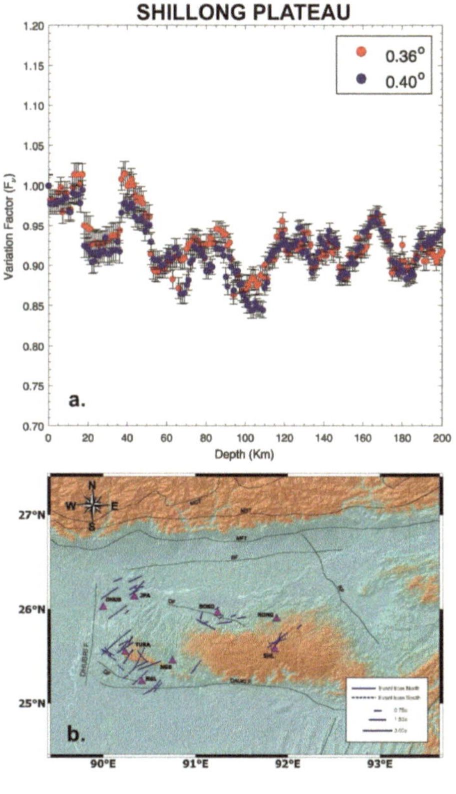

Figure 3.3 (a) Scatter plot of variation factor as a function of depth in Shillong plateau region. SK(K)S and PKS splitting results corresponding to eight broadband seismic stations are incorporated in the spatial coherency model with grid size of 0.30° and 0.40° in order to estimate the probable depth of anisotropic layer beneath this region. The scatter plot of variation factors (Fv) associated with error bars (black vertical lines) represents the minima of the coherency curve around 100 km. (b) SK(K)S and PKS splitting parameters are plotted at the ray-piercing points of 100 km depth, showing the maximum coherency at this particular depth. The length and direction of the blue bars represent delay time (δt) and fast polarization directions (ϕ), respectively, for each station (marked in purple colour). SK(K) S and PKS phases coming from the north of the stations are marked by solid lines and those coming from south are demarcated by dotted lines. Similar abbreviations are used to mark the major faults as in Figure 3.1. *(Continued)*

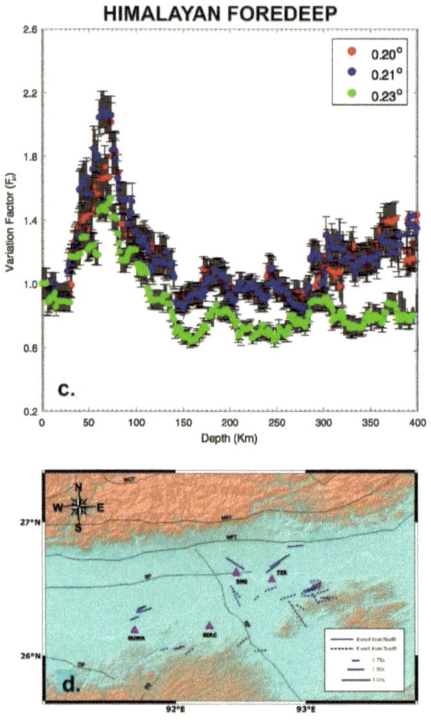

Figure 3.3 (Continued) (c) Scatter plot of variation factor as a function of depth in Himalayan foredeep region. SK(K)S and PKS splitting results corresponding to six broadband seismic stations from Himalayan foredeep region are incorporated in the spatial coherency model with grid size of 0.20°, 0.21° and 0.23° in order to estimate the probable depth of anisotropic layer beneath this region. The scatter plot of variation factors (*Fv*) associated with error bars (black vertical lines) represents the minima of the coherency curve around 150 km. (d) SK(K)S and PKS splitting parameters are plotted at the ray-piercing points of 150 km depth, showing the maximum coherency at this particular depth. The length and direction of the blue bars represent delay time (δt) and fast polarization directions (ϕ) respectively for each station (marked in purple colour). SK(K)S and PKS phases coming from the north of the stations are marked by solid lines and those coming from south are demarcated by dotted lines. Similar abbreviations are used to mark the major faults as in Figure 3.1.

3.3.3 Central depth and source localization of anisotropy beneath Assam foredeep

The study also encompasses the anisotropic behavior of the Himalayan foredeep/Assam foredeep region and established the central depth of anisotropy using 40 XKS splitting results (refer Table 1 of Mohanty et al., 2021) from six broadband stations (Table 3.1). Using the fixed grid sizes of 0.20°, 0.21 and 0.23° with a central difference of 0.16°, the weighting factors are found out as, W_{Φ} = 1/180° and $W_{\delta t}$ =1/2.15s. Using the spatial coherency method with specific grid sizes of 0.20°, 0.21° and 0.23°, the possible depth of the anisotropy is determined for a particular minima of the variation factor (F_v) at a depth of around 150 km, suggests that the possible anisotropy is at 150 km (Mohanty et al., 2021). Also, XKS splitting parameters plotted at the ray-piercing points at 150 km depth showed the maximum coherency (Mohanty et al., 2021) between the parameters at this particular depth. Devi et al. (2011) estimated that the lithosphere-asthenosphere boundary at the Himalayan foredeep region is around 135 km. So, here, the anisotropy is not related to the lithosphere-asthenosphere boundary despite asthenospheric flow beneath this region, which drags the Indian plate towards the northeast direction that plays a major role in deformation pattern beneath this region.

3.4 CONCLUSION

The recent trend and approach in getting the possible central depth and source localization of anisotropy is really productive and qualitative in analyzing the geodynamics and deformation patterns of earth compared to the traditional previous methods. Particularly, the spatial coherency method, which deals in the modified Fresnel zone concept, is so accurate and devoid of event-based difficulties as compared to previous algorithms, where the events and station distributions along any azimuthal ranges are managed to give the exact results. Moreover, the best coherency between the splitting parameters, exact U-shaped smooth attribution of variation factor curve for a minimized value and dense spatial distribution plot for anisotropic parameters for a particular depth, make the coherency method the best and effective one. Though, shear wave splitting measurements allow us to redefine and analyze the deformation patterns for a better geodynamical model establishment, the best picture of underneath dynamics and anisotropic behaviour of the medium goes on missing without the proper justification and information about the source localization and central depth of anisotropy. A much more understanding thought process will lead the way to comprehend the underearth mechanism that is only possible once the depth range of anisotropy is available. Hence, recent techniques and advances are really the way ahead to characterize, understand and decipher the geodynamical models with a high resolution and better logic, after the implication of the knowledge of depth of anisotropy.

ACKNOWLEDGEMENT

D. D. M. is especially thankful to Prof. Rajib Biswas, Editor, for the communication and opportunity given to write this chapter for this esteem book edition. D. D. M. is thankful to Dr. G. N. Sastry, Director, CSIR-NEIST for giving permission towards this invited publication. This study is fully supported by an Early Career Research grant from Science and Engineering Research Board (SERB), India, vide project grant no. ECR/2018/001293.

REFERENCES

Alsina, D., & Snieder, R. (1995). Small-scale sublithospheric continental mantle deformation: Constraints from SKS splitting observations. *Geophysical Journal International*, 123(2), 431–448. https://doi.org/10.1111/j.1365-246X.1995.tb06864.x.

Angelier, J., & Baruah, S. (2009). Seismotectonics in Northeast India: A stress analysis of focal mechanism solutions of earthquakes and its kinematic implications. *Geophysical Journal International*, 178(1), 303–326. https://doi.org/10.1111/j.1365-246X.2009.04107.x.

Baro, O., & Kumar, A. (2015). A review on the tectonic setting and seismic activity of the Shillong plateau in the light of past studies. *Disaster Advances*, 8, 34–45.

Bhandari, L. L., Fuloria, R. C., & Sastri, V. V. (1973). Stratigraphy of Assam valley, India. *American Association of Petroleum Geologists*, 57(4), 642–654.

Bilham, R., & England, P. (2001). Plateau pop-up during the 1897 Assam earthquake. *Nature*, 410, 806–809. https://doi.org/10.1038/35071057.

Centre for Natural Disaster Management. (2002). *Scenario of seismic hazard in Assam, A report by Assam Administrative staff college. CNDM*. Guwahati, Assam: Assam Administrative Staff College.

DeMets, C., Gordon, R. G., Argus, D. F., & Stein, S. (1994). Effect of recent revisions to the geomagnetic reversal time scale on estimates of current plate motions. *Geophysical Research Letters*, 21(20), 2191–2194. https://doi.org/10.1029/94GL02118.

Devi, E. U., Kumar, P., & Kumar, M. R. (2011). Imaging the Indian lithosphere beneath the Eastern Himalayan region. *Geophysical Journal International*, 187(2), 631–64. https://doi.org/10.1111/j.1365-246X.2011.05185.x.

Dewangan, P., Ramprasad, T., & Mangipudi, R. V. (2007). Finite difference modelling of scattered hydrates and its implications in gas-hydrate exploration. *Current Science*, 93(9), 1287–1290.
Evans, P. (1964). The tectonic framework of Assam. *Journal of the Geological Society of India*, 5, 80–96.
Fisher, N. I. (1993). *Statistical Analysis of Circular Data*. London: Cambridge University Press. https://doi.org/10.1017/CBO9780511564345.
Gao, S. S., & Liu, K. H. (2012). AnisDep: A FORTRAN program for the estimation of the depth of anisotropy using spatial coherency of shear-wave splitting parameters. *Computers & Geosciences*, 49, 330–333. https://doi.org/10.1016/j.cageo.2012.01.020.
Gao, S. S., Liu, K. H., & Abdelsalam, M. G. (2010). Seismic anisotropy beneath the Afar Depression and adjacent areas: Implications for mantle flow. *Journal of Geophysical Research: Solid Earth*, 115(B12), 330. https://doi.org/10.1029/2009JB007141.
Gao, S., Davis, P. M., Liu, H., Slack, P. D., Rigor, A. W., Zorin, Y. A., … Logatchev, N. A. (1997). SKS splitting beneath continental rift zones. *Journal of Geophysical Research: Solid Earth*, 102(B10), 22781–22797. https://doi.org/10.1029/97JB01858.
Geological Survey of India. (1974). *Geology and mineral resources of Asam and adjoining states*. GSI.
Gupta, R. P., & Sen, A. K. (1988). Imprints of ninety-east ridge in the Shillong plateau, Indian Shield. *Tectonophysics*, 154, 335–341.
Hammond, J. O. S., Kendall, J.-M., Wookey, J., Stuart, G. W., Keir, D., & Ayele, A. (2014). Differentiating flow, melt, or fossil seismic anisotropy beneath Ethiopia. *Geochemistry, Geophysics, Geosystem*, 15, 1878–1894. https://doi.org/10.1002/2013GC005185.
Heintz, M., Kumar, V. P., Gaur, V. K., Priestley, K., Rai, S. S., & Prakasam, K. S. (2009). Anisotropy of the Indian continental lithospheric mantle. *Geophysical Journal International*, 179(3), 1341–1360. https://doi.org/10.1111/j.1365-246X.2009.04395.x.
Kailasam, L. N. (1979). Plateau uplift in peninsular India. *Tectonophysics*, 61, 243–269.
Kayal, J. R. (2008). *Microearthquake Seismology and Seismotectonics of South Asia*. Springer Science Business Media. https://doi.org/10.1007/978-1-4020-8180-4.
Kayal, J. R., & De, R. (1987). In velocity study using a temporary seismographic network in the Shillong plateau, northeast India. *Bulletin of the Seismological Society of America*, 77(5), 1718–1727. https://doi.org/10.1785/BSSA0770051718.
Kayal, J. R., & De, R. (1991). Microseismicity and tectonics in northeast India. *Bulletin of the Seismological Society of America*, 81(1), 131–138.
Kayal, J. R., Arefiev, S. S., Baruah, S., Hazarika, D., Gogoi, N., Gautam, J. L., … Tatevossian, R. (2012). Large and great earthquakes in the Shillong

plateau–Assam valley area of Northeast India Region: Pop-up and transverse tectonics. *Tectonophysics*, 532–535, 186–192. https://doi.org/10.1016/j.tecto.2012.02.007.

Kelly, L. H., & Gao, S. S. (2011). Estimation of the depth of anisotropy using spatial coherency of shear-wave splitting parameters. *Bulletin of the Seismological Society of America*, 101(5), 2153–2161. https://doi.org/10.1785/0120100258.

Khattri, K., Wyss, M., Gaur, V. N., Saha, S. N., & Bansal, V. K. (1983). Local seismic activity in the region of the Assam Gap, northeast India. *Bulletin of the Seismological Society of America*, 73(2), 459–469. https://doi.org/10.1785/BSSA0730020459.

Kumar, M. R., & Singh, A. (2008). Evidence for plate motion related strain in the Indian shield from shear wave splitting measurements. *Journal of Geophysical Research: Solid Earth*, 113, B08306. https://doi.org/10.1029/2007JB005128.

Kumar, N., Kumar, M. R., Singh, A., Raju, P. S., & Rao, N. P. (2010). Shear wave anisotropy of the Godavari rift in the south Indian shield: Rift signature or APM related strain? *Physics of the Earth and Planetary Interiors*, 181(34), 82–87. https://doi.org/10.1016/j.pepi.2010.05.002.

Kumar, V. P., Prakasam, K. S., Rai, S. S., & Gupta, S. (2015). Upper-mantle anisotropy beneath the south Indian shield: Influenced by ancient and recent Earth processes. *Lithosphere*, 7(2), 108–116. https://doi.org/10.1130/L405.1.

Lindsey, J. p. (1989). The Fresnel zone and its interpretive significance. *The Leading Edge*, 8(10), 33–39. https://doi.org/10.1190/1.1439575.

Liu, K. H., & Gao, S. S. (2011). Estimation of the depth of anisotropy using spatial coherency of shear-wave splitting parameters. *Bulletin of the Seismological Society of America*, 101, 2153–2161. https://doi.org/10.1785/0120100258.

Mohanty, D. D., & Mondal, P. (2020). Disparate behaviour of deformation patterns beneath northeast Indian lithosphere inferred from shear wave splitting analysis. *Physics of the Earth and Planetary Interiors*, 298, 106315. https://doi.org/10.1016/j.pepi.2019.106315.

Mohanty, D. D., & Singh, A. (2022). Shear wave birefringence and current configuration of active tectonics of Shillong plateau: An appraisal of Indian plate motion and regional structures. *International Journal of Earth Sciences*, 111, 269–286. https://doi.org/10.1007/s00531-021-02114-9.

Mohanty, D. D., & Mondal, P. (2021). Evidence of heterogeneous and unstable anisotropic settings beneath the Northeast Indian lithosphere from characterization of null splitting measurements. *Acta Geophysica*, 69, 2035–2050. https://doi.org/10.1007/s11600-021-00676-5.

Mohanty, D. D., Biswal, S., Phukan, M. K., & Ayodeji, E. A. (2021). Possible depth and source localization of seismic anisotropy beneath Shillong plateau and Himalayan foredeep region: An implication towards deformation mechanisms. *Geological Journal*, 1–12. https://doi.org/10.1002/gj.4334.

Mohanty, D. D., Singh, A., O'Driscoll, L. J., Kumar, M. R., Srinagesh, D., & Humphreys, E. D. (2016). P wave velocity structure below India and Tibet incorporating anisotropic delay time effects. *Geochemistry, Geophysics, Geosystems*, 17(3), 725–738. https://doi.org/10.1002/2015GC006064.

Molnar, P., & Tapponnier, P. (1975). Cenozoic tectonics of Asia: Effects of a continental collision. *Science*, 189(4201), 419–426. https://doi: 10.1126/science.189.4201.419.

Mondal, P., & Mohanty, D. D. (2021). Mantle deformation and seismic anisotropy beneath Northeast India inferred from SKKS birefringence. *Studia Geophysica et Geodaetica*, 65, 36–52. https://doi.org/10.1007/s11200-020-1121-y.

Nandy, D. R. (2001). *Geodynamics of North Eastern India and the Adjoining Region*. Kolkata: ACB Publications.

Nayak, P. N. (1989). Seismicity and Geodynamics of Meghalaya Plateau, India. Abstracts of 28th *Int. Geol. Congress* (IGC), Washington, DC, 2–497.

Nicolas, A., & Christensen, N. I. (1987). Formation of anisotropy in the upper mantle peridotites – A review. *Composition, Structure and Dynamics of the Lithosphere-Asthenosphere System*, 16, 137–154. https://doi.org/10.1029/GD016p0111.

Oreshin, S., Kiselev, S., Vinnik, L., Prakasam, K. S., Rai, S. S., Makeyeva, L., & Savvin, Y. (2008). Crust and mantle beneath western Himalaya, Ladakh and western Tibet from integrated seismic data. *Earth and Planetary Science Letters*, 271, 75–87.

Rao, K. M., Kumar, M. R., Singh, A., & Rastogi, B. K. (2013). Two distinct shear wave splitting directions in the northwestern Deccan volcanic province. *Journal of Geophysical Research: Solid Earth*, 118, 5487–5499. https://doi.org/10.1002/2013JB010644.

Rao, N. P., & Kumar, M. R. (1997). Uplift and Tectonics of the Shillong Plateau, Northeast India. *Journal of Physics of the Earth*, 45, 167–176.

Roy, S. K., Kumar, M. R., & Srinagesh, D. (2014). Upper and lower mantle anisotropy inferred from comprehensive SKS and SKKS splitting measurements from India. *Earth and Planetary Science Letters*, 392, 192–206. https://doi.org/10.1016/j.epsl.2014.02.012.

Savage, M. K. (1999). Seismic anisotropy and mantle deformation: What have we learned from shear wave splitting? *Reviews of Geophysics*, 37(1), 65–106. https://doi.org/10.1029/98RG02075.

Sheriff, R. E. (1996). Understanding the Fresnel zone. *AAPG Explorer*, October, 18–19.

Silver, P. G., Gao, S., & Liu, K. H. (2001). Mantle deformation beneath southern Africa. *Geophysical Research Letters*, 28(13), 2493–2496. https://doi.org/10.1029/2000GL012696.

Singh, A., Eken, T., Mohanty, D. D., Saikia, D., Singh, C., & Kumar, M. R. (2016). Significant seismic anisotropy beneath southern Tibet inferred from splitting of direct S-waves. *Physics of the Earth and Planetary Interiors*, 250, 1–11. https://doi.org/10.1016/j.pepi.2015.11.001.

Singh, A., Kumar, M. R., Raju, P. S., & Ramesh, D. S. (2006). Shear wave anisotropy of the northeast Indian lithosphere. *Geophysical Research Letters*, 33(16). https://doi.org/10.1029/2006GL026106.

Vinnik, L. P., Makeyeva, L. I., Milev, A., & Usenko, A. Y. (1992). Global patterns of azimuthal anisotropy and deformations in the continental mantle. *Geophysical Journal International*, 111(3), 433–447. https://doi.org/10.1111/j.1365-246X.1992.tb02102.x.

Vinnik, L., Farra, V., & Romanowicz, B. (1989). Azimuthal anisotropy in the earth from observations of SKS at GEOSCOPE and NARS broadband stations. *Bulletin of the Seismological Society of America*, 79(5), 1542–1558.

Chapter 4

Sedimentary basin configuration of different basin structures of Northeast India by teleseismic body wave imaging

Sowrav Saikia and Santanu Baruah

4.1 INTRODUCTION

The study of crustal structure is a primary and important subject in seismology because it is often required as a priori information for various geological and geophysical researches. The topmost layer of crust or the upper crust, which is the main contributing structure of seismic hazard as well as petroleum and mineral exploration, is a very fascinating aspect of crustal structure study. The crustal structure can be studied by various geophysical data acquired on the surface of the earth, like magnetic, electromagnetic, gravity, passive, and active seismology. The receiver function (RF) technique is a passive seismological study using teleseismic body waves of natural source earthquakes. This technique is nowadays a routine technique used by various researchers to determine major discontinuities, velocity structure of the underlying crust and upper mantle using a three-component seismic station. The technique was first introduced by Phinney et al. (1964) and widely developed by Langston (1979) and later improved by various researchers on their different processing steps and a new way to finding crustal parameters (Owens, 1988; Ammon et al., 1990, Cassidy, 1992; Shibutani et al., 1996; Ligorria and Ammon, 1999; Sambridge, 1999; Zhu and Kanamori, 2000). Briefly, RFs primarily depend on teleseismic earthquakes which sample the earth along the propagation path by removing common features with the help of deconvolving processes such as source time response, propagation path effect, and instrument response. The RF technique can be used to map shallow structures and up to upper mantle global discontinuity and hence this technique is used for earthquake location, geodynamics of the region, and seismic hazard study.

DOI: 10.1201/9781003177692-4

The northeastern part of India has a complex geological and seismotectonic setup comprising Himalayan collision zone, deep river basin, plateau, river-oceanic delta, fold belt, and subduction tectonics (Figure 4.1). Therefore, the study of shallow to the deeper crustal structure of each tectonic can yield geodynamics of each region and will be very useful to earthquake locations and earthquake hazard studies. In this chapter, we make an attempt to find sedimentary basin configurations of all the sedimentary basins of Northeast India.

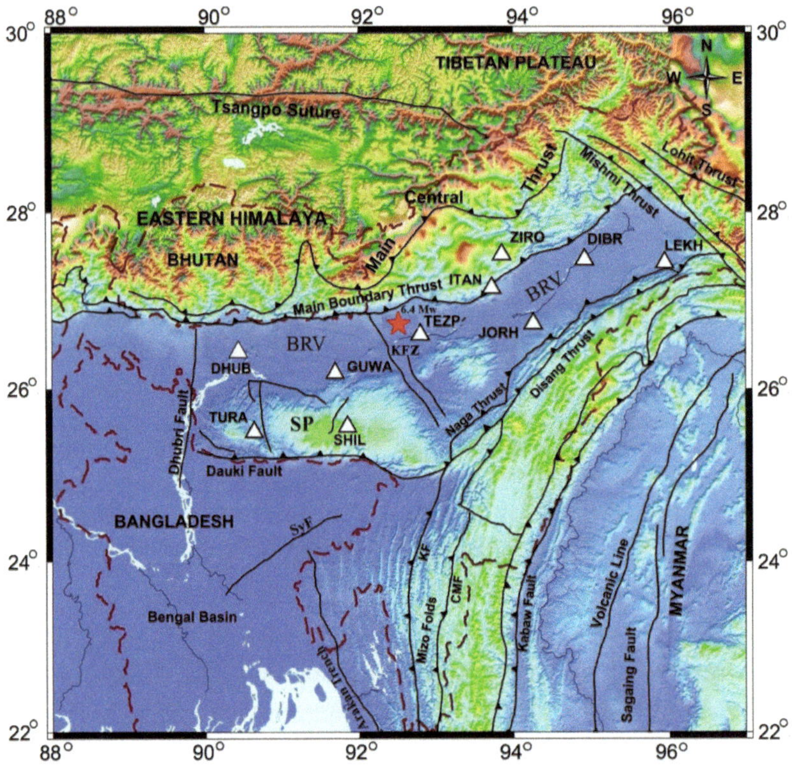

Figure 4.1 Tectonic map of Northeast India and broadband seismograph station along Brahmaputra valley and nearby region in Northeast India. In this figure, major seismotectonic structures like Main Central Thrust, Main Boiunadry Thrust, Dauki Fault, Dhubri Fault, Bengal Basin, Naga Thrust, Kabaw Fault, Arakan Trench, Disang Thrust, Mishmi Thrustet are also shown. Abbreviated names like BRV: Brahmaputra Valley SP: Shillong Plateau, KFZ: Kopili Fault Zone, SyF: Sylhet Fault, CMF: Churachandpur Mao Fault, KF: Kaladan Fault are also shown. Seismic stations around the Brahmaputra valley basin are shown in white-filled triangles. The 28 April, Mw 6.4 Dhekiajuli earthquake, near Kopili Fault is shown in a red star.

4.2 PREVIOUS STUDIES

A good number of geophysical studies have been carried out to map the crustal structure of Northeast India comprises mainly Shillong Plateau, Indo-Myanmar, and its nearby region (Verma and Mukhopadhyay, 1977), using RF analysis (Kumar et al., 2004; Mitra et al., 2005; Singh et al., 2016; Mitra et al., 2018), seismic tomography study (Kayal and Zhao, 1998; Bhattacharya et al., 2008), magnetotelluric study (Gokarn et al., 2008), and deep borehole drilling study for shallow subsurface study (Handique and Dutta, 1980; Kumar et al., 2012). But very few studies emphasized sedimentary basin configuration on different basin structures in Northeast India, except upper Assam Brahmaputra basin. In Northeast India, there are mainly three prominent basin structures; the Brahmaputra River valley basin, the Northeast margin of Bengal delta, and the Imphal basin in the core part of the Indo-Myanmar subduction region.

The basin structures of Brahmaputra valley and Bengal deltaic part of NE India was first reported by Verma and Mukhopadhyay (1977) from an extensive gravity study throughout part of Northeast India. The isostatic Bouguer gravity anomaly map by Verma and Mukhopadhyay (1977) illustrates the gravity anomaly value is high (–50 mGal to 100 mGal) among the Bengal basin and Shillong Plateau. They also observed low gravity values in the Brahmaputra basin up to a value of –255 mGal in upper Assam Brahmaputra basin and nearly –70 mGal in the Northeast margin of Bengal basin. They explained that the high positive isostatic anomaly value is due to the intrusion of incorporation of basic material from the mantle into the crust underneath the plateau. The predominantly negative values of Bouguer anomaly over the Northeast margin of Bengal basin and Brahmaputra valley observed up to a minimum value of –250 mGal, which suggests the burden of low-density sediments overlying the Brahmaputra and Bengal basin.

The upper Assam Brahmaputra basin, which is boasted of oldest oil fields of Asia, the basinal structure has been studied in some pocket regions for explorations of petroleum and minerals (Handique and Dutta, 1980; Pahari et al., 2008; Kumar et al., 2012; Lahiri and Sinha, 2014). From Oil India Limited (OIL) exploration of petroleum in upper Brahmaputra valley, Kumar et al. (2012) reported a borehole and seismic study for a 100 km long profile in the upper Assam Brahmaputra basin using 230 drilled well data comprising of upper Assam shelf. They explained that the basement depth in the upper Assam Brahmaputra valley is more than 4 km and the depth increases to 6 km towards the Naga hill and sub-Himalaya region. Lahiri and Sinha (2014) explained the basement configuration seismic and geomorphologic studies across the Brahmaputra River in the North-South direction. From their study, they inferred that the depth of the basement near the Dibrugarh (DIBR site) region is around 3.2 km, while their profile across the Brahmaputra River Island has a basement depth of around 3.1 km which is nearby to Jorhat (JOR site). Handique and Dutta (1980) had performed extensive drilling operations in 700 wells in the upper Assam Brahmaputra valley for a long profile of 300 km across the Brahmaputra River from Northwest to Southeast up to the margin of the Brahmaputra basin. From their measurement of 11 wells which were penetrated well to the tertiary succession down to the basement rock. They concluded that the depth of the basement undulated between 3 and 6 km depth range across the Brahmaputra valley in which higher thickness of sedimentations is observed in near Naga belt and Northern Sub or Siwalik Himalaya region. From gravity study, Mukhopadhyay and Dasgupta (1988) calculated basement undulation of the Southwest margin of Bengal basin or Surma basin and they found sedimentary thickness is around 7 km thick, whereas in some places, basement is mapped in a deeper depth of 12 km. Although no active seismology or borehole studies have been reported on the Southwest margin of the Bengal basin and Imphal basin to date.

4.3 DATA AND METHODOLOGY

Since 2011, the National Center for Seismology (previously Seismology Division, India Meteorological Department) has been operating the permanent Northeast Telemetry Network (NET), which consists of 19 broadband seismograph stations throughout the region. The NET stations are equipped with Reftek-151 B sensors having a natural period of 120 sec. Here, we utilized the archived seismic dataset and then select teleseismic events of some of the stations located in the basin part. Since the modern NET broadband seismic array began operating only in 2011, we collected 189 good teleseismic events with a high signal-to-noise ratio (SNR) for the period of the subsequent two years. The recorded events are selected with preliminary condition with epicentral distances in between 30° and 90°, magnitude mb > 5.5 with no depth limitations. In general, more events coming from major source locations are from the subduction zones of the western Pacific and Eurasian plates and some from the mid-Atlantic ridge zone. The dataset also contains events from the Northern Eurasia region, Middle East, and Southern Hemisphere events, for which in total cover all back-azimuth quadrants to calculate RFs.

The RF technique is a well-established, popular, and nowadays routine technique to image the crust and upper mantle structure in terms of layers thickness and seismic wave velocity often used to find seismogenic structure underneath a three-component seismic station (Langston, 1979; Ammon, 1991). Seismic body waves pass through a discontinuity between two media of difference impedance, along with reflected and refracted but also reverberated with the discontinuity and free surface. The RFs, which are constructed by deconvolution of the radial or transverse component of the seismic event from the vertical component, are composed of *P*-to-*S* converted waves and its reverberations from different layers beneath the seismic station. Except for the direct *Pp* arrival, the strongest converted pulses of the RF waveform are generally due to the first-order reverberation in the crust-mantle boundary or the Moho discontinuity. But there are some converted phases originated within the crust are also observed before the Moho converted *Pms* phase, having relatively less amplitude than the *Pm* phase. The more complex structures with multiple seismic discontinuities cause complications to the receiver function waveform by adding further *Ps* conversions and reverberations due to the interaction of the incoming *P*-wave front with each discontinuity before Moho *Pms* phase. Before Moho converted *Pms* pulse, the *Ps* pulse converted from the sedimentary basin transition layer is prominent in RF traces. The signature of shallow depth sedimentary basin transition discontinuity is well observed in RFs and can be modeled by considering the first few seconds before the Moho *Pms* phase. The in-depth modelling of the amplitudes and timings of the first few second converted *Ps* arrivals and other immediate reverberated phases can provide valuable constraints of the layered structure within the crust (Owens et al., 1984).

The dataset was processed in SAC (Seismic Analytic Code) software. We cut the seismic waveforms for a 70 sec window; 10 sec before and 60 sec after the teleseismic *P*-wave arrival. Many other operations like removing the mean, de-trend, taper with a 5% cosine window, and filtered the processed band-pass filter the seismograms between 0.05 and 5 Hz to remove low- and high-frequency noise. Rotation of seismograms has been performed on the great-circle path to obtain the radial and tangential component seismograms. After that seismograms are then down-sampled to 10 samples per second. Furthermore, we compute the RFs at two overlapping frequency ranges with fc < 1.25 Hz and fc < 0.5 Hz, using a Gaussian filter with width factors of a = 2.5 and a = 1.0, which facilitates discriminating rapid velocity variations from gradual velocity transitions. In our present study, we took six stations in Brahmaputra valley (DHUB, GUWA, TEZ, JORH, DIBR, MDP), two stations located in outer Himalaya (ITAN, ZIRO), two stations in Shillong Plateau (SHL, TURA), four stations in Northwest margin of Bengal basin and associated Tripura fold belt region (SILC, AGT, BELO, AIZW, SAIH) and one station (IMPH) located at Imphal basin. The SHL (Shillong) and TURA (Tura) stations are equipped with the CMG-40T sensors having a natural period of 40 s. The quality of deconvolved RFs having more than 80% waveform fit in the deconvolution process is selected for final processing and analysis. The corresponding tangential RF for each radial RF is critically visually inspected, and those tangential RFs showing similar amplitudes comparable with the radial RFs are eliminated from further analysis. Finally, the remaining radial RFs are overlaid to visually identify and exclude possible outliers. After all these steps, there were 610 RFs that finally passed the quality control and were used to find the upper crustal structure. In many parts around the globe, the neighbourhood algorithm inversion has been used to obtain the *S*-wave velocity structure. The neighbourhood algorithm inversion proposed by Sambridge et al. (1999), which independently optimizes the searching of synthetic RF model parameters to produce a group of best-fitting *S*-wave velocity structure models, is used to find upper crust structure underneath the station.

4.4 INVERSION

The stacked RF data in which converted phases are well observed translate to depth domain in order to model the velocity structure underneath a station. Generalized inversion plays a major role in this transformation by constraining many parameters. However, there is a nonlinear dependency of the RF. This dependency has a profound effect on the inversion scheme to be carried out. Shibutani et al. (1996) applied a nonlinear genetic algorithm type of inversion and able to find crustal structure by using stacked RFs data. This non-linear inversion scheme was improved by Sambridge (1999) with a fully nonlinear, derivative-free direct search algorithm termed as "neighbourhood algorithm". This inversion starts with a given number of cells, which are entirely distributed in model space. This model space is partitioned into a number of subregions, each containing one sample. The nearest neighbour region of around a sample region is termed as L2 norm or Voronoi cell. Each model space is again partitioned into a number of model spaces by an iterative process and the new samples are placed each time in a number of Voronoi cells up to the last process in which models have the lowest misfit. The misfit function gives the quantitative analysis in comparing the observed and synthetic stack RFs and also correlates the individual traces with the background noise. To calculate the misfit in our analysis, we keep the time window for a short interval in each stacked RF data that begins –5.0 sec before the direct-*P* arrival and keep a maximum of 15.0 sec long after the *P* arrival. For most of the locations of our stations, the stacked, as well as individual RFs, show prominent converted phases and their reverberations at short times after the direct-*P* arrival. These suggest that at least one or two very shallow layers are required to model the initial phases of the RFs. For this reason, we assume primarily five layers means that at least three layers are used for the crust and the upper mantle. Each layer is also split into a number of layers randomly. Finally, the input-stacked RF resample to 10 Hz and normalized misfit value is computed for a very small-time window from –2.5 sec to 5.0 sec to recover all the reverberations of the upper crustal discontinuities before the *Pms* pulse.

The many basins structure regions of Northeast India are of different tectonic regimes. Therefore, the reverberated phases after direct-*P* are quite different from those entire sites and visible in RF traces. The *P*-to-*S* converted wave field emerging from the analysed RFs is rather complicated for all the sites; the radial RFs are somewhat variable with back azimuth variation. But the main observations in our analysis are to extract first-order information about the vertical variation of the seismic velocities in the upper crust, and hence we model the stack RF to get 1-D models of underlying geology. The stack RFs are effective in reducing the noise and by suppressing the arrivals arising from the lateral heterogeneities, and thus improving the signal to produce the 1-D properties by the inversion procedure. Here, we analyse the RF data in different filters with Gaussian widths of 1.0 Hz, 2.5 Hz, and 5.0 Hz (Figure 4.2). The RFs computed using the Gaussian width of 5.0 Hz is associated with the high-frequency noise of the subsurface and thus difficult to model. On the other side, a low Gaussian width filter like 1.0 Hz tends to exclude minor reverberated phases after the direct *P* phase and gives the image of the major discontinuities after the inversion, although not sufficient to image some prominent intra-crustal discontinuity. For the inversion procedure, we used stacked RFs of Gaussian width 2.5 Hz to invert the data.

4.4.1 Scenario 1: Sedimentary basin configuration of Brahmaputra valley

The Brahmaputra valley plain is a long NEE-SWW trending stretched basin located at the heart of the northeastern part of India in between the Eastern Himalaya and the Shillong-Mikir massif. The Mesozoic to early tertiary sediments in this region are the source of petroleum resources but also increase the seismic hazard of the region due to largely populated cities with high-rise buildings, schools, colleges, bridges, and so on. The valley is stretched 800 km in length with an average of 110 km width along Southwest to Northeast direction. Some major seismic active regions of the world encircle the region which is the habitat of millions of people of the thickly populated valley with fast-growing urbanization. Recently, a major earthquake of Mw 6.4 occurred on 28 April 2021 at the vicinity near Kopili fault zone and shook the whole Brahmaputra valley where sand dykes, liquefaction, ground cracks were common in the epicentral region. This earthquake caused widespread damage in the epicentral track and nearby cities of Guwahati, Jorhat, and Nagaon. Some geological studies delineated sedimentary basin configuration in some pocket regions based on surface geological investigations (Handique and Dutta, 1980; Pahari et al., 2008; Lahiri and Sinha, 2014).

Here we partition the upper crust into the sediment, basement layer in modelling procedure such that it can accurately find the low-velocity sedimentary column in different layers. Except for clays, alluvial, and

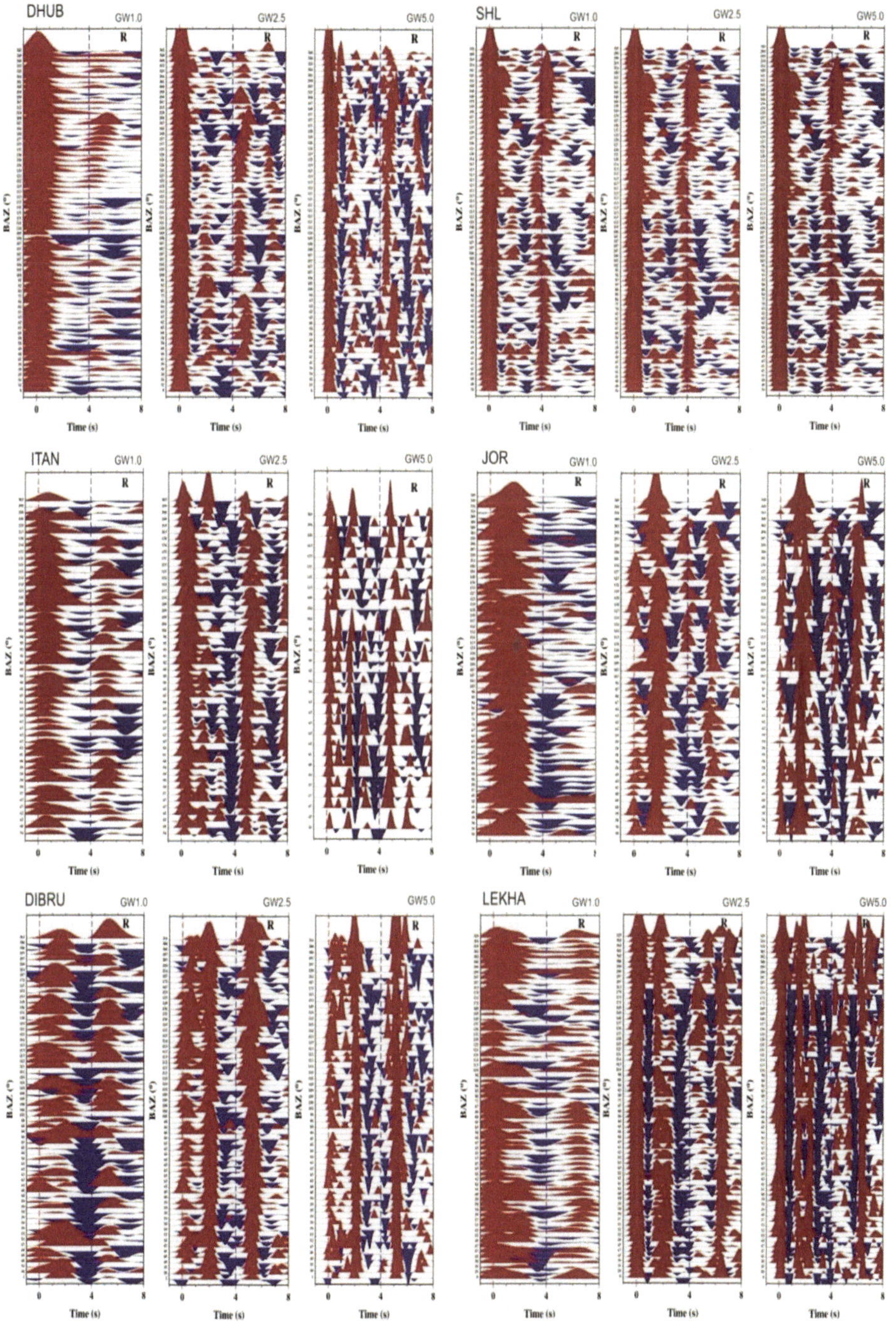

Figure 4.2 The receiver function traces of some stations located at Brahmaputra valley (DHUB, JOR, DIBRU LEKHA), Shillong Plateau (SHL), sub-Himalaya (ITAN) plotted with respect to event back azimuth for different Gaussian widths (Gw) 1.0 Hz, 2.5 Hz and 5.0 Hz. (After Saikia et al., 2017.)

water-filled sediments, here it is referred to as a surface material that has been compacted and possibly metamorphosed during burial and then re-exposed in erosion.

The ten stations are located at three different tectonic regimes; Brahmaputra valley region, Shillong Plateau region, and sub-Himalayan region (Figure 4.1). Hence, we can expect RFs will be influenced by the site-specific heterogeneity due to the underlying crust and upper mantle. For the hard rock-located stations, the RFs consist of a direct *P* arrival and only large amplitude secondary phases. Some other phases were also observed before Moho-converted *Pms* phase, indicating the presence of intra crustal structure. The secondary phase is a distinct *P*-to-*S* converted phase from Moho. In contrast, for the stations inside the basin with soft sediments, a large secondary arrival occurs about ~1-2 sec after an insignificant appearance of the direct *P* arrival, caused by strong reverberations in the sedimentary layers (Figure 4.2). This phenomenon is attributed to the lower edge of the sedimentary strata or the topmost surface crystalline basement of the site. The *P*-to-*S* converted phase from the Moho discontinuity could not be directly identified due to heavy interweavement of the reverberation with the Moho converted phase. Sharp shear velocity contrasts across the transition of sediment basement is commonly observed in sedimentary basin in RF data, which is seen as a strong pulse in the RF at some few seconds delay time (Zheng et al., 2005). This is due to the presence of sedimentary layer (low-velocity layer), the incoming ray turns sub-vertical and the horizontal component of the *P* phase gets diminished. Teleseismic RFs technique is also successfully used to map sedimentary basin structure around different parts of the world, for example, Mediterranean region (Meijde et al., 2003), Bohai Bay basin of China (Li et al., 2007), Rocky Mountains of the United States (Sheehan et al., 1995), Gujarat region (Chopra et al., 2014, Gangetic basin (Srinivas et al., 2011). By searching in a suitable range of velocities and layered depth in a genetic algorithm (Shibutani et al., 1996; Clitheroe et al., 2000; Chopra et al., 2014) and neighbourhood algorithm (Srinivas et al., 2011) has been performed to map sedimentary basin configuration map.

The first peak represents the Ps-converted phase generated at the bottom layer of the sedimentary layer and dominates the direct *P* arrival. The time shift is smaller for the stations located on the western part and hence the depth of sedimentary is also relatively shallow. Here we apply the neighbourhood algorithm for all those ten stations located at Brahmaputra valley and its adjoining Shillong Plateau and sub-Himalaya region. Similar findings were observed in some sites located at upper Assam Brahmaputra valley. Figure 4.2 represents the RFs at JORH, DIBRU station, in which we observed converted phases from the sediment-basement boundary at ~1.5 sec delay time are visible by a positive pulse. This delay pulse indicates a negative velocity contrasts for an incident wavefield from below.

Figure 4.3 shows the results of velocity modelling and V_P/V_S ratio for the DIBRU, JOR, and ITAN sites. The first two stations are located in the upper Assam Brahmaputra valley and the other in sub-Himalaya (ITAN

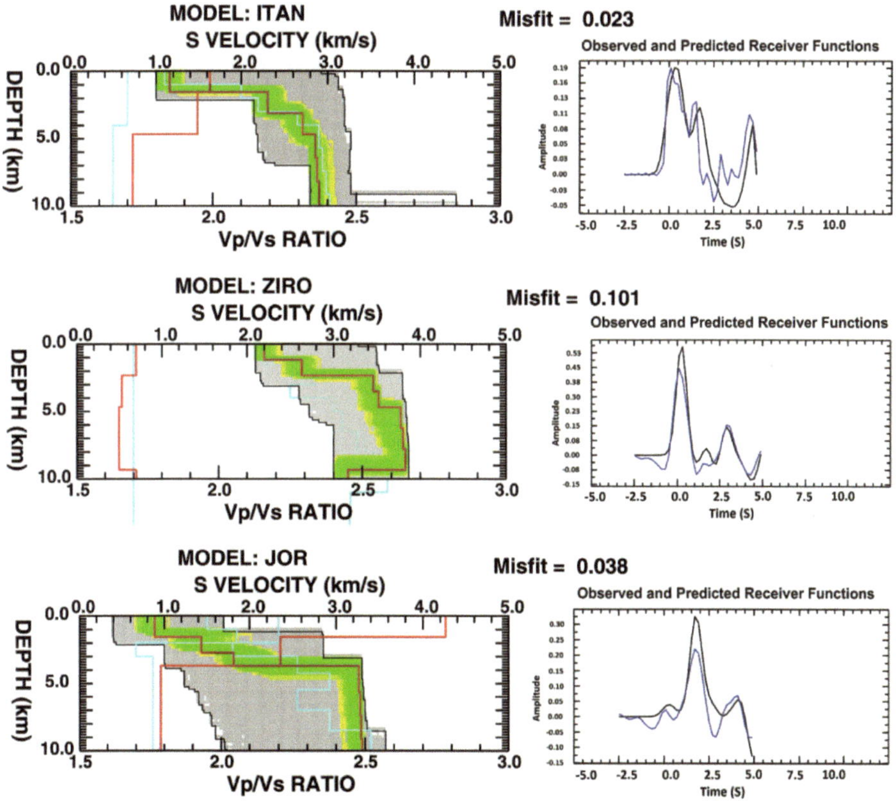

Figure 4.3 Upper crustal shear-wave velocity structure of some individual stations; ITAN, ZIRO, JOR, DIBRU, LEKHA are obtained by inversion of stacked RFs. The corresponding depth-V_P/V_S ratio model for each station is represented in red-coloured trace on the left-hand side. The grey-shaded region shows the entire sample model space, the yellow- to green-coloured areas shows the best 1,000 models to shows *(Continued)*

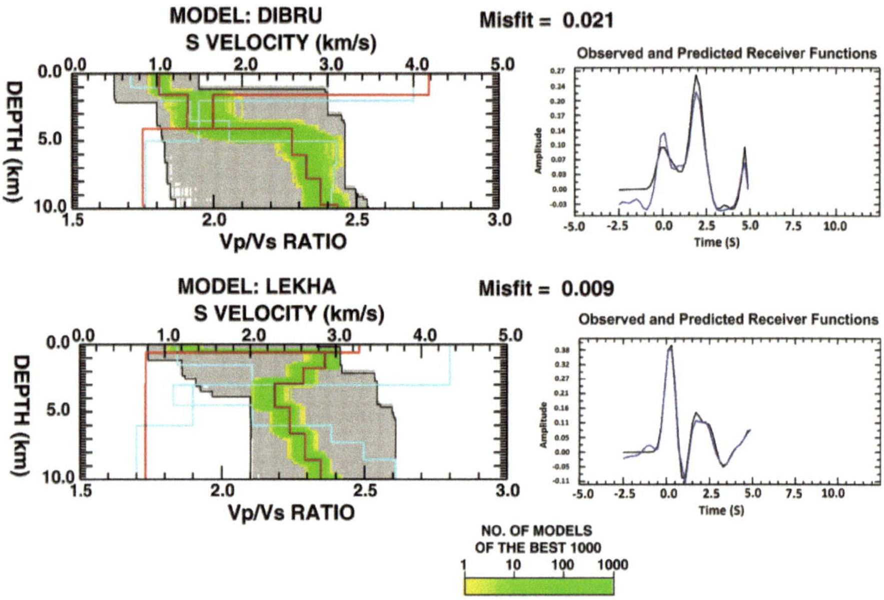

Figure 4.3 (Continued) increment of a minimum misfit by yellow to green colour, and the best model with the lowest misfit is represented by red-coloured trace. The blue trace model represents the model obtained from IASPEI91, while blue trace is V_P/V_S modeled from using the Iaspei91 model. The pink-coloured V_P/V_S is the best fit V_P/V_S approximation which is corresponding to the final model shown in red colour. On the right-hand side, the corresponding matching of observed RF is shown in black and synthetic RF is shown in blue-coloured RFs having lowest misfit. (After Saikia et al., 2017.)

station). The thick sedimentary column at these stations is seen by low shear-wave velocity of the top surface, with low S-wave velocity ranges (0.8–1.3 km/sec) and high values of V_P/V_S ratio around 1.8–2.3. The shear-wave velocity sharply increases to 2.8–3.3 km/sec at some depths and this can be inferred as reaching the basement layer. For the stations, DIBRU and JORH, the thickness of this sedimentary column varies between 4.6 and 3.7 km, respectively. The SHL station, which is located in Shillong Plateau, has a subsurface S-wave velocity of around 2.6 km/sec, which signifies the subsurface is lacking loose sediments. The shear-wave velocity sharply increases in velocity within 1 km depth of basement. The high shear velocity in the subsurface signifies that the Shillong site has meta-sediment cover on its surface. For JOR and DIBRU stations stacked RFs, the amplitudes of P-to-S converted phase and reverberated phases are seen in between 1.0 sec and 3.0 sec and have much larger amplitude than the direct *P* phase. A common observation from these two stations is that the RFs waveforms are highly correlated for the majority of the event throughout all back-azimuth groups and thus it can be inferred that the artifacts in the mapping of the unmodelled structure will be lowest.

The station ITAN is located at sub-Himalaya south to the Main Boundary Thrust, on the boundary between the sub-Himalayan orogeny and the Brahmaputra valley. In this station, the near-surface shear-wave velocity is around 1.2 km/sec with high V_P/V_S ratio of 1.9. The *S* velocity is quite low up to 2 km and gradationally increases to 2.2 km/sec. The low velocity signifies thick sedimentary cover and increases to 2.8 km/sec at about 4.6 km.

The stations JOR (Jorhat) and DIBRU (Dibrugarh) are located in the northeastern part of the Brahmaputra valley exhibit more or less similar configuration of the first 10 km of upper crust. The JOR station has a low subsurface *S* velocity of around 0.90 km/sec. The *S* velocity gradationally increases stepwise in two layers and reaches *S* velocity around 3.25 km/sec at 3.7 km depth, which can be interpreted as the basement. The subsurface layer is 1.5 km thick overlying by a layer of 1.0 km thickness with the *S*-wave velocity of 1.4 km/sec. After that shear-wave velocity slowly increases to 1.8 km/sec and extended to 2.5 km depth. The shear-wave velocity has a sharp velocity increment on 3.7 km depth and attains a velocity 3.2 km/sec, which can be referred to as basement depth. The near-surface velocity for DIBRU station is around 1.05 km/sec and extended to 1.5 km, having a high V_P/V_S of 1.80. This layer is followed by another low-velocity layer gradationally increases from *S* velocity around 1.4 km/sec and continues up to the basement at 4.1 km with a shear-wave velocity 2.8 km/sec with V_P/V_S ratio of around 1.91.

We have observed that the thickness of sedimentary increases from the southwest part of Brahmaputra valley towards the Northeast part of Brahmaputra valley. The thickness of sediments goes on increasing parallels to the Moho phase undulation towards the Northeast corner of Brahmaputra valley.

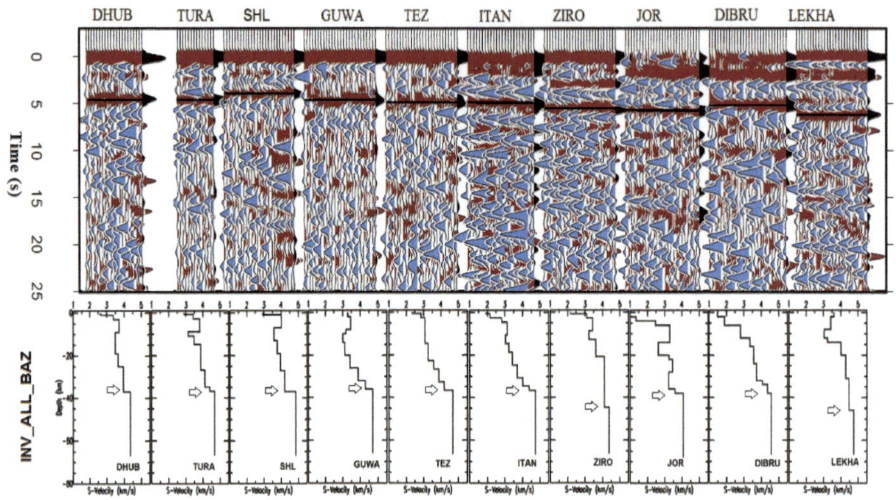

Figure 4.4 Receiver functions and their corresponding S-velocity profiles along Southwest-Northeast profile along the Brahmaputra valley are arranged by combining all the results. The variation in the basement depth (1.0–5.5 km) along the Brahmaputra valley is well observed throughout the region. A small variation of sedimentary thickness from 0.2 km to 1 km is observed in the Shillong Plateau and considerable variation from 4 km to 5.5 km is observed in the sites located at the Brahmaputra valley and Himalayan foredeep region.

4.4.2 Scenario 2: Crustal structure of Southwest of Northeast India region

The southwestern part of Northeast India (Figure 4.5) is also seismically very active and experienced two big earthquakes: the Cachar earthquake (10 January 1869, M 7.5) and Srimangal earthquake (8 July 1916; M 7.5). This region is the Northeast corner of the Bengal basin delta, world's biggest river delta region. During the earthquake, the casualty in human lives and property loss is very high due to the amplification of ground motion of its loose and thick sediments. This part of the Northeast India region has a long history of big to great earthquakes occurring from time to time and this highly populated region can be severely affected by the seismic hazard. The Manipur earthquake of Mw 6.8, which occurred on 2 January 2016, was centred in between the Bengal basin and the Indo-Burmese ranges and caused severe damage near epicentral region (Gahalaut et al., 2016). Very recently, high seismic activity experienced between June and July 2020, in the Champhai region in the Tripura fold belt region caused widespread panic among the localities (Figure 4.4). Some 23 earthquakes of more than

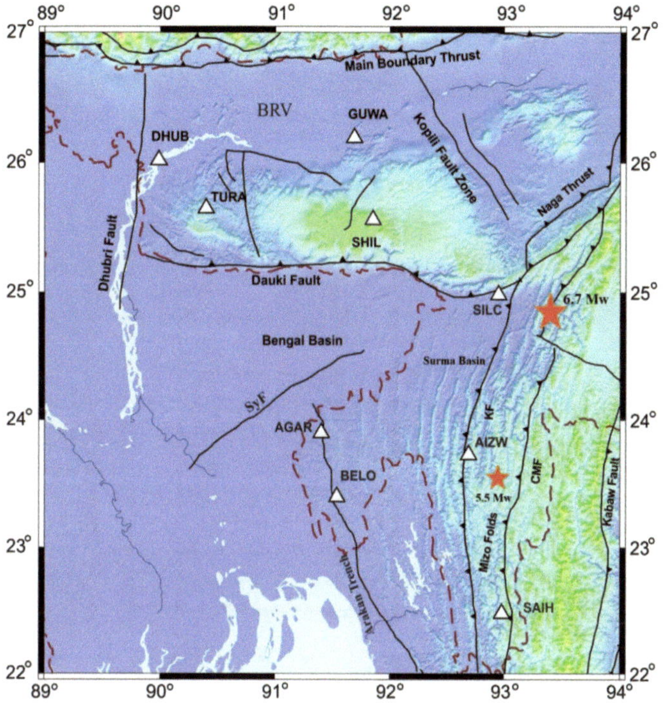

Figure 4.5 Regional tectonic map showing major tectonic features of southwestern part of Northeast India along with the BBS station shown in a black triangle (modified after Saikia et al., 2018); CMF: Churachandpur Mao Fault, SP: Shillong Plateau, DhF: Dhuburi Fault, DuF: Dudhnoi Fault, DrT: Drapsi Thrust. Two prominent earthquakes: the Tamenglong earthquake (Mw 6.7, 4 January 2016) and the Mizoram earthquake (Mw 5.5, 22 June 2020) are shown in red-coloured star.

3.0 magnitude with shallow focus (depth < 30 km) nature occurred within two months. The maximum magnitude earthquake of Mw 5.5 occurred on 22 June 2020 caused widespread damages in terms of property losses and loss of livestock in the Champhai region. The sedimentary structure of the Surma basin region in the Tripura fold belt causes amplification of seismic waves for which the ground vibration was severely felt in the epicentral region (Bharali et al., 2021; Malsawma et al., 2021).

Figure 4.6 shows the RFs for AGAR station located at Bengal basin characterize the converted phases from the sediment-basement transition layer arrived at around 1.0 sec delay for all azimuth arriving events. This delay positive phase signifies a negative velocity or impedance contrast for an incident wavefield arriving from below the basement. We observe

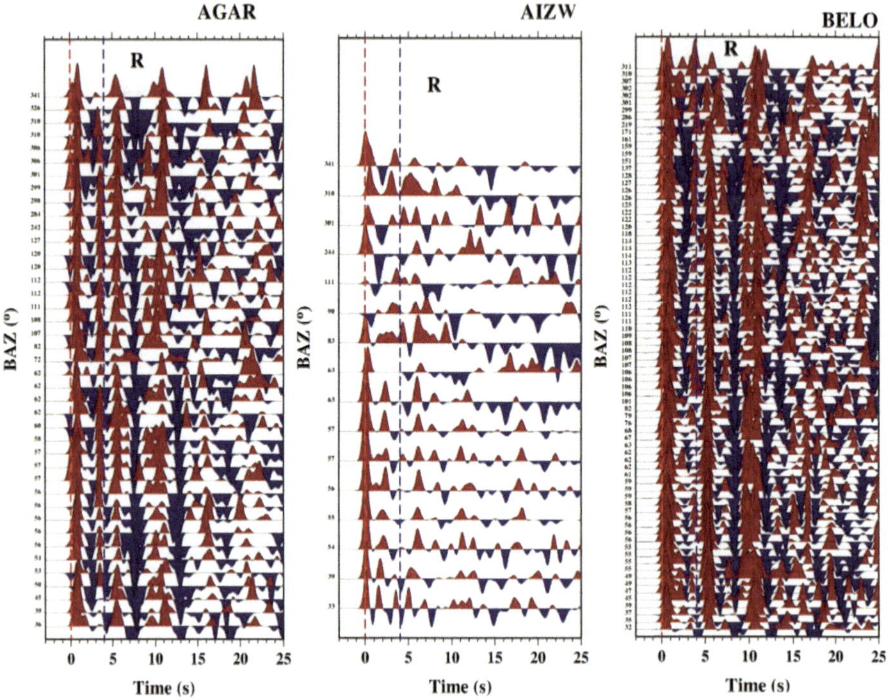

Figure 4.6 Receiver function traces of some stations located at Northwest margin of NE India region.

some negative polarity pulse in between 2.0 sec and 4.0 sec. It can be seen from the synthetic test that the addition of sedimentary layer obscures all the phases up to the Moho multiples and the shape of the first ~5 sec of RFs is mainly due to the sedimentary layer. In the RF traces of all other stations, we observed that variation in upper crustal structure is well signified by the polarity of pulses and associated minor pulses after a few seconds of direct *P* phase before the Moho converted *Ps* phase.

RF analysis indicates large lateral changes in the velocity structure and physio-chemical composition beneath lower Brahmaputra, SP, Bengal basin, and Tripura fold belt region reflecting the nature of transition zone of basin, fold belts, and plateau tectonics. We observed thick sediments of 8–9 km thickness in the eastern flank of the Bengal in Northeast India. The AIZW station located at Tripura fold belt region has some sedimentary rock, not very less *S* velocity and velocity of metasediments of 2.8 km/sec is extended to the first 2 km rather than thick loose sedimentation of Bengal basin sites (Figure 4.7).

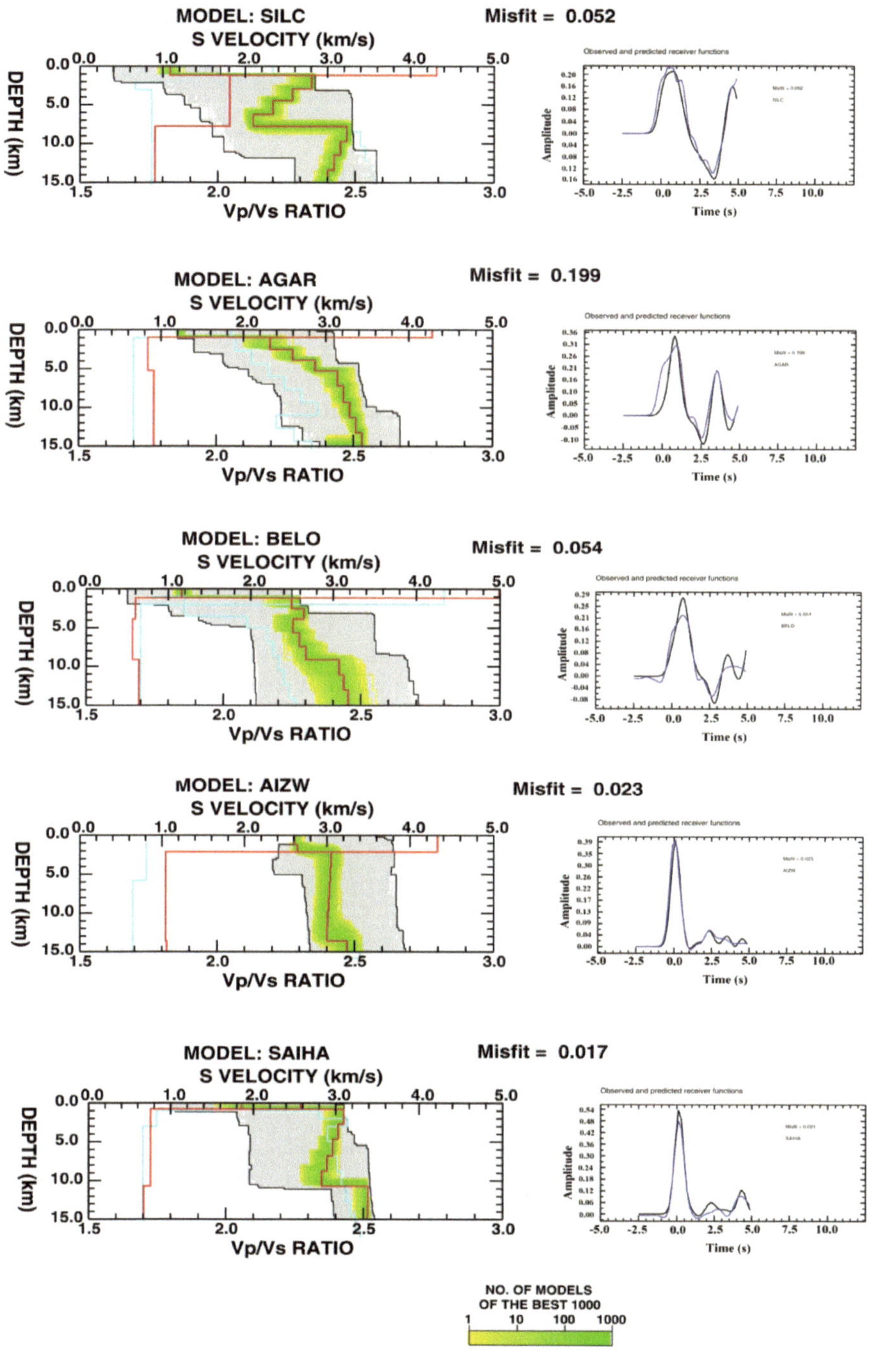

Figure 4.7 The shear velocity structure of upper crust of some stations shows sedimentary basin configuration of Southwest part of Northeast India comprising region eastern flank of Bengal basin (SILC, AGAR, BELO) and Tripura fold belt (AIZ, SAIHA). (After Saikia et al., 2018.)

4.4.3 Scenario 3: The Imphal basin in the Indo-Myanmar subduction region

The Indo-Myanmar subduction region located at the easternmost part of Northeast India is one of the well-known active subduction zones in the entire Indian continent, where the Indian plate is flexed beneath the Burmese microplate (Figure 4.8). This ~700 km long arc is seismically very active and observed around 150 earthquakes of M > 2.5 per year and is a continuous part of the Andaman-Sundra subduction arc. The entire north-south trending and convex westward Indo-Burmese region structurally have different tectonic regimes: Tripura fold belt (TFB), Arakon-Yoma ranges, Myanmar Central Basin, and Shan Plateau highland region. The Imphal valley of the tertiary period is a part of the Assam-Arakan basin (Evans, 1932).

The station IMPH is located at the heart of the Imphal valley of the Cenozoic age, which is also situated at the core part of the nearly North-South trending Indo-Burmese subduction arc. Imphal is the state capital of Manipur state with fast-growing urbanization and thick population of

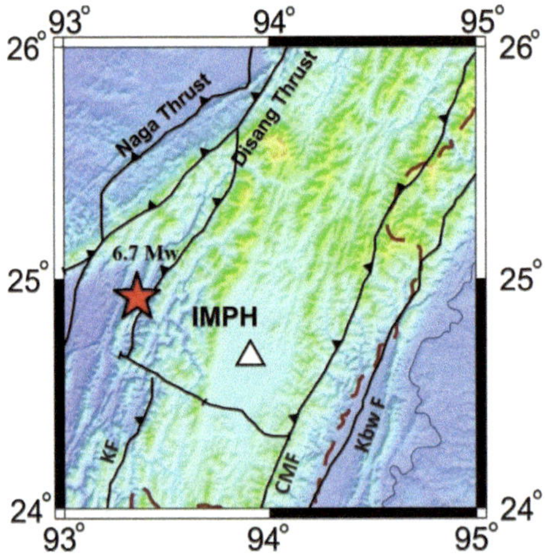

Figure 4.8 Seismotectonic map of the Indo-Myanmar region shows the location of the Imphal station (in a black triangle) within the Imphal valley. The major features of the region and its surrounding regions are Naga Thrust, Disang Thrust, Kaladan Fault (KF), Churachandpur Mao Fault (CMF), Kabaw Fault (KbW F). The Imphal (IMPH) seismic station is shown in a white-filled triangle. The 4 January 2016, Mw 6.7, Tamenglong earthquake is shown in red-coloured star.

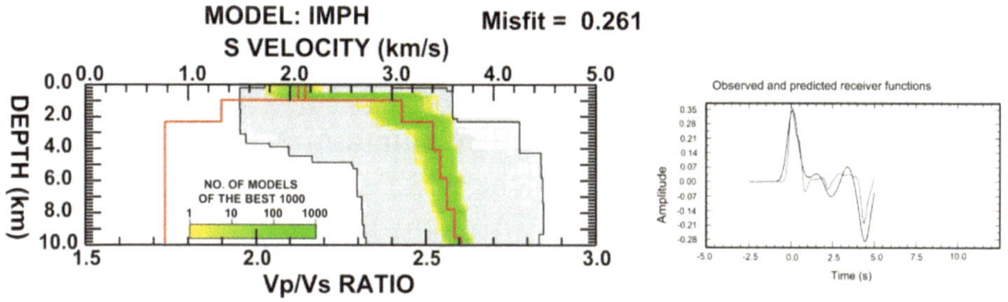

Figure 4.9 The shear-wave velocity structure of the first 10 km of the upper crust to map the sedimentary basin configuration of the Imphal station located at the Indo-Myanmar region.

the city with multi-stories building, bridges, hospitals, academic buildings makes this region seismically more vulnerable. So finding shallow sedimentary basin structures can help seismic risk analysis of this region. Here, we see the upper crustal velocity structure of the Imphal basin up to a depth of 10 km (Figure 4.9).

Figure 4.9 reveals low *S* velocities near the surface and in two layers it reaches 3.4 km/sec with a composite layer thickness of 2.3 km. The subsurface velocity is around 2.1 km/sec and increases to 3.1 km/sec for 1 km depth and further increases to 3.4 km/sec at 2.3 km depth and reaching the basement. The shear velocity is nearly constant to more than 10 km depth after basement depth of 2.3 km.

4.5 CONCLUSION

Our results show that the thick sedimentary layers are widely distributed Cenozoic sediments in the Brahmaputra valley, eastern flank of Bengal basin delta, and Imphal basin, which give prominent modification of broadband waveforms owing to the large velocity contrast at the bottom of the sedimentary layer. This phenomenon is absent in soft sediments in hard rock-situated station like in Shillong Plateau of Archean age correlates with its topography as basement rock is well exposed. In most cases, the layered sedimentary basin configuration consists of low shear-wave velocities and high V_P/V_S ratio or proportional value Poisson ratio values. These examples suggest sedimentary basement configuration of a simple homogeneous horizontal-layered medium can be successfully determined by using the teleseismic RF technique.

ACKNOWLEDGMENT

The first author gratefully and sincerely appreciates the technical staff of the National Center for Seismology in maintaining and data archival of all seismic stations in this study. The GMT software package by Wessel and Smith (1995) was used for drafting the figures. The first author also acknowledged the unanimous reviewer and editor for their constructive comments on the chapter.

REFERENCES

Ammon, C. J., Randall, G. E., Zandt, G. 1990. On the non-uniqueness of receiver function inversions. *J. Geophys. Res.*, 95, 15303–15318.

Ammon, C. J. 1991. The isolation of receiver effects from the teleseismic P waveforms. *Bull. Seismol. Soc. Am.*, 81, 2504–2510.

Bharali, B., Rakshit, R., Dinpuia, L., Saikia, S., Baruah, S. 2021. The 2020 Mw 5.5 Mizoram earthquake and associated swarm activity in the junction of the Surma Basin and Indo-Myanmar Subduction Region. *Nat. Hazards*, 109, 2381–2398. doi:10.1007/s11069-021-04924-1.

Bhattacharya, P. M., Mukhopadhyay, S., Majumdar, R. K., Kayal, J. R. 2008. 3-D seismic structure of the northeast India and its implication for local and regional tectonics. *J. Asia. Ear. Sci.*, 33, 25–41.

Cassidy, J. F. 1992. Numerical experiments in broadband receiver function analysis. *Bull. Seismol. Soc. Am.*, 82 (3), 1453–1474.

Chopra, S., Chang, T. M., Saikia, S., Yadav, R. B. S., Choudhary, P., Roy, K. S. 2014. Crustal structure of the Gujarat region, India: new constraints from the analysis of teleseismic receiver functions. *J. Asian Earth Sci.*, 96, 237–254.

Clitheroe, G., Gudmundsson, O., Kennett, B. L. N. 2000. Sedimentary and upper crustal structure of Australia from receiver functions. *Aust. J. Earth Sci.*, 47, 209–216.

Evans, P. 1932. Tertiary succession in Assam. *Trans. Min. Geol. Inst.*, 27:155.

Gahalaut, V., Martin, S., Srinagesh, D., Kapil, S. L., Suresh, G., Saikia, S., Kumar, V., Dadhich, H., Patel, A., Prajapati, S., Shukla, H. P., Gautam, J. L., Baidya, P. R., Mandal, S., Jain, A. 2016. Seismological, geodetic, macro-seismic and historical context of the 2016 Mw 6.7 Tamenglong (Manipur) India earthquake. *Tectonophysics*, 688, 36–48. doi: 10.1016/j.tecto.2016.09.017.

Gokarn, S. G., Gupta, G., Walia, D., Sanabam, S. S., Hazarika, N., 2008. Deep geoelectric structure over the lower Brahmaputra valley and Shillong Plateau, NE India using magnetotellurics. *Geophys. J. Int.*, 173, 92–104.

Handique, G. K., Dutta, S. K. 1980. A study of the Surma-Tipam groups of Assam valley, south of Brahmaputra. *Geo. Survey of India*, 25, 42–52.

Kayal, J. R., Zhao, L. 1998. Three dimensional seismic structure beneath Shillong Plateau and Assam Valley, northeast India. *Bull. Seismol. Soc. Am.*, 88, 667–686.

Kumar, M. R., Raju, P. S., Devi, E. U., Saul, J., Ramesh, D. S. 2004. Crustal structure variations in northeast India from converted phases. *Geophys. Res. Lett.*, 31, L17605.

Kumar, A., Satish, T., Bharali, B. R., Verma, A. K. 2012. Basement configuration and structural style in OIL's operational areas of Upper Assam. AAPG, #50739.

Lahiri, S. K., Sinha, R. 2014. Morphotectonic evolution of the Majuli Island in the Brahmaputra valley of Assam, India inferred from geomorphic and geophysical analysis. *Geomorphology*, 227, 101–111.

Langston, C. A. 1979. Structure under Mount Rainer, Washington, inferred from teleseismic body waves. *J. Geophys. Res.*, 84, 4749–4762.

Li, J., Tian, B., Wang, W., Zhao, L., Yao, Z. 2007. Short note lateral variation in the sedimentary structure of West Bohai Bay Basin inferred from P-multiple receiver functions. *Bull. Seismol. Soc. Am.*, 97, 1355–1363.

Ligorria, J. P., Ammon, C. J. 1999. Iterative deconvolution and receiver function estimation. *Bull. Seismol. Soc. Am.*, 89, 1395–1400.

Malsawma, J., Lalnuntluanga, P., Sailo, S., Vanthangliana, V., Tiwari, R., Gahalaut, V. 2021. The 22 June 2020 Mizoram, India earthquake (Mw 5.5): An unusual intra-wedge shallow earthquake in the Indo-Burmese Wedge. *Curr. Sci.*, 9, 120, 1514–1516. doi: 10.18520/cs/v120/i9/1514-1516.

Meijde, M. V. D., Van der Lee, S., Giardini, D. 2003. Crustal structure beneath broad-band seismic stations in the Mediterranean region. *Geophys. J. Int.*, 152, 729–739.

Mitra, S., Priestly K., Bhattacharya, A. K., Gaur, V. K. 2005. Crustal structure and earthquake focal depths beneath the northeastern India and Southern Tibet. *Geophys. J. Int.*, 160, 227–248. doi:10.1111/j.1365-246X.2004.02470.x.

Mitra, S., Priestley, K., Borah, K., Gaur, V. K. 2018. Crustal structure and evolution 1068 of the Eastern Himalayan plate boundary system, Northeast India. *J. Geophys. Res.*, 123, 621–640. https://doi.org/10.1002/2017JB014714.

Mukhopadhyay, M., Dasgupta, S. 1988. Deep structure and tectonics of the Burmee arc: Constraints from earthquake and gravity data. *Teconophysics*, 149, 299–322.

Owens, T. J., Zandt, G., Taylor, S. R. 1984. Seismic evidence for an ancient rift beneath the Cumberland Plateau, Tennessee. *J. Geophys. Res.*, 116, 618–636.

Owens, T. J., Crosson, R. S. 1988. Shallow structure effects on broadband teleseismic P-waveform. *Bull. Seismol. Soc. Am.*, 78, 98–108.

Phinney, R. A. 1964. Structure of the Earth's crust from spectral behaviour of long period body waves. *J. Geophys. Res.*, 69, 1069–1074.

Saikia, S., Baruah, S., Chopra, S., Singh, U. K., Gogoi, B., Gohain H. B. 2018. Study of crustal structure and geological implications of south western margin of Northeast India. *J. Seismol.*, 22, 229–249. doi: 10.1007/s10950-017-9701-1.

Saikia, S., Baruah, S., Chopra. S., Gogoi, B., Singh, U. K., Bharali, B. 2019. An appraisal of crustal structure of the Indo-Burmese subduction region. *J. Geodyn.*, 127 (1), 16–30. doi: 10.1016/j.jog.2019.05.002.

Saikia, S., Chopra, S., Baruah, S., Singh, U. K. 2017. Shallow sedimentary structure of the Brahmaputra valley constraint from receiver function analysis. *Pure Appl. Geophys.*, 174 (1), 229–247. doi: 10.1007/s00024-016-1371-3.

Sambridge, M. 1999. Geophysical inversion with a neighborhood algorithm-I. Search a parameter space. *Geophys. J. Int.*, 138, 479–494. doi: 10.1046/j.1365-246X.1999.00876.x.

Sambridge, M. 1999a. Geophysical inversion with a neighborhood algorithm-I. Search a parameter space. *Geophys. J. Int.*, 138, 479–494.

Sambridge, M. 1999b. Geophysical inversion with a neighbourhood algorithm-II. Appraising the ensemble. *Geophys. J. Int.*, 138, 727–746.

Sheehan, A. F., Abers, G. A., Jones, C. H., Lerner-Lam, L. 1995. Crustal thickness variations across the Colorado Rocky Mountains from teleseismic receiver functions. *J. Geophys. Res.*, 100 (B10), 20391–20404.

Shibutani, T., Sambridge, M., Kennett, B. L. N. 1996. Genetic algorithm inversion for receiver functions with application to crust and uppermost mantle structure beneath eastern Australia. *Geophys. Res. Lett.*, 23, 1829–1832.

Singh, A., Bhushan, K., Singh, C., Stecler, M., Akhtar, H., Seeber, L., Kim, W., Tiwari, A. Biswas, R. 2016. Crustal structure and tectonics of Bangladesh: New constraints from inversion of receiver functions. *Tectonophysics*, 680, 99–112. http://dx.doi.org/10.1016/j.tecto.2016.04.046.

Srinivas, D., Srinagesh, D., Chadha, R. K., Kumar, M. R. 2011. Sedimentary thickness variations in the Indo-Gangetic foredeep from inversion of receiver functions. *Bull. Seismol. Soc. Am.*, 103, 2257–2265.

Verma, R. K., Mukhopadhyay, M. 1977. An analysis of the gravity field in northeastern India. *Tectonophysics*, 42, 283–317.

Wessel, P., Smith, W. H. F. 1995. New version of the generic mapping tools released. *EOS Trans. AGU*, 76, 329.

Zheng T., Zhao, L., Chen, L. 2005. A detailed receiver function image of the sedimentary structure in the Bohai Bay Basin. *Phys. Earth Planet. Int.*, 152, 129–143.

Zhu, L., Kanamori, H. 2000. Moho depth variations in southern California from receiver functions. *J. Geophys. Res.*, 105, 2969–2980.

Chapter 5

Seismic wave amplification and localized ground effects

A geological tool for future land use code and land use planning in the low seismic area

Sambit P. Naik

5.1 INTRODUCTION

Earthquake-induced ground effects play a significant role in characterizing the seismic ground motions. The seismic wave may strongly amplify (or deamplify) depending on the local geology and presence of geological structures such as faults, cavities, and sedimentary basins (Spudich and Olsen, 2001; Li et al., 2004; Panzera et al., 2013). The role of local geology in seismic wave amplification was widely observed during the 1985 Mexico City earthquake, 1994 Northridge earthquake, 1999 Izmit earthquake, and the 2015 Nepal earthquake (Kawase and Aki, 1989; Pitarka and Irikura, 1996; Ergin et al., 2004; Gautam, 2017). The damage associated with the earthquakes mentioned earlier, as well as earthquakes taking place recently around the world, indicated that most of the damage was caused by the coseismic geological effects or by the ground shaking (Porfido et al., 2020). It is well understood that tectonic crustal structures are capable of generating moderate to strong earthquakes (Mw>5) and caused extensive ground effects such as surface faulting, ground cracks, upliftment and subsidence slope failure, drainage changes, and liquefaction (Ambraseys and Sarma, 1969; Caputo, 2005; Daniell et al., 2017). These earthquake-induced ground effects are known as earthquake environmental effects (EEE). The EEEs are mainly controlled by the local geological structure and

DOI: 10.1201/9781003177692-5

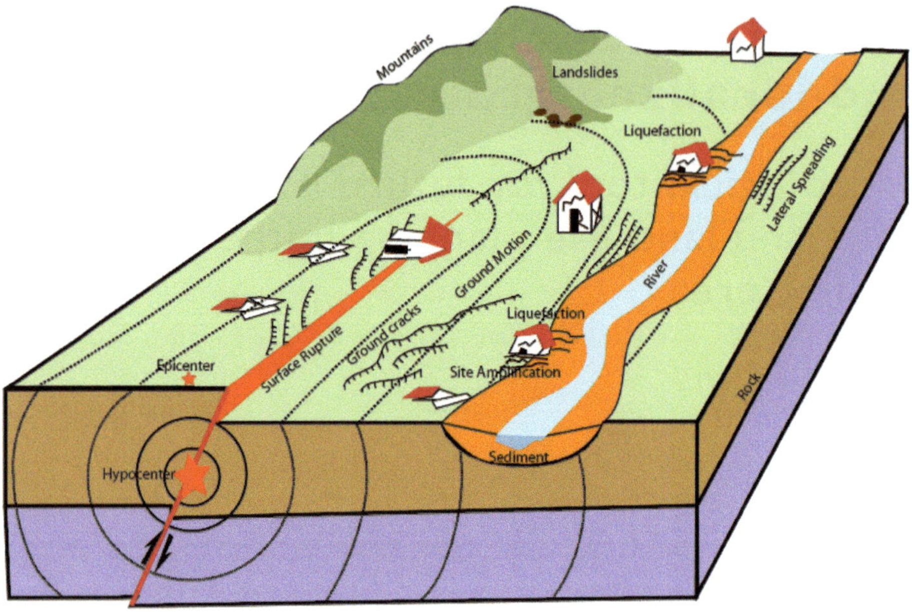

Figure 5.1 Schematic diagram showing the typical ground effects and phenomenon observed around epicentral area during an earthquake.

geotechnical site conditions. The localized ground effects are governed by the soil response to the seismic waves at the ground surface due to an earthquake (Figure 5.1). Local site effects are considered the most critical factors in ground motion micro zonation, evident during several recent moderate to large magnitude (Serva, 1994; Somerville and Moriwaki, 2003; Daniell et al., 2017). From those earthquakes and damage associated, it can be inferred that significant damage occurred within sedimentary basins or areas having soft sediments induced by amplification of seismic waves. Such effects can be minimized by using the comprehensive database of recorded strong motions around the sedimentary basins. With a proper post-earthquake damage survey, documentation of the ground effects and its analysis may act as a validation tool for the experimental and numerical study to understand the role of local geology in seismic wave amplification. This may help the future seismic risk mitigation and loss. Ideally, such a database is not available for areas of minimal records of damaging earthquakes, for developing countries like India, or areas sitting far from the plate boundary with large interval of recurring earthquake regions like South Korea.

Due to very small earthquakes between interseismic times in those areas, the earthquake hazard studies are in a very primitive stage with minimal recording stations, making the site amplification studies more difficult. Also, the number of recorded signals is relatively low. The installation of local arrays in each zone with a high level of seismicity is too expensive and requires a longtime interval to gather statistically significant data sets. Despite decades of observational studies in both laboratory and simulation, the processes leading to extensive damage within sedimentary basins remain enigmatic. So, one must find an alternative approach to understand the critical factors controlling the seismic wave propagation and excitation due to soft sedimentary layers in areas with minimal historical earthquake records or limited seismic station arrays. In this case, proper documentation of such ground effects known as EEEs can lead us to derive the role of local geology in seismic wave amplification and understand the subsurface geology for seismic hazard delineation.

For documented earthquakes from specific tectonic environments such as intraplate regions, a detailed characterization of seismic-induced ground effects can be used as a fundamental tool to define the seismic landscape expected for a given area and level of seismicity in the future. This also can be used as a useful back analysis tool for assessing the actual vulnerability of the environment and for predicting its future response to seismic wave amplification in that area (Serva, 1994; Profido et al., 2002; Liu-Zheng et al., 2017; Naik et al., 2020; Shimony et al., 2020), which is much needed for a proper definition of land use codes and land planning in low seismic areas. Also, this approach may guide locating the suitable sites for installing seismic stations, understanding the subsurface geology, which may be helpful to understand the site amplification studies for future seismic hazard assessment.

In this chapter, proper documentation of ground effects can guide us to understand the linkage between the presence of discontinuities such as faults and cavities and topographic irregularities with local site amplification in an area of low seismicity and regions with limited seismic station arrays. Case studies of three sites located in low seismic intraplate areas illustrating the role of local geology in seismic wave tapping and its importance in land-use planning are described.

5.2 METHODS

The earthquakes in the intraplate region pose significant seismic threats to the nearby urban center as the cause of these earthquakes is often uncertain (Stein and Mazzotti, 2007). In many cases, the causative fault is profoundly buried or blind. Because of their rare nature and absence of accompanying surface ruptures, the mechanisms involved in generating the intraplate earthquake, their associated ground effects are not well understood (Wesnousky et al., 1984; Bolt, 1996; Gangopadhyay and Talwani, 2003; Stein and Mazzotti, 2007). In these cases, it is always easy to neglect the role of local geological structures in seismic hazard assessment (Wang, 2011). Due to its rare occurrence, the earthquake hazard studies in these areas are in a very primitive stage with very limited recorded stations. With this, it is difficult to understand the role of local geological structures in seismic wave amplification. But the geological evidence of coseismic damage associated with such rare events may help understand the possible mechanism involved. They may contribute to the proper seismic hazard assessment for these areas. To test the hypothesis as the geological effect may guide us to understand the subsurface geology and its role in seismic damage, we have re-analyzed the aerial distribution of the ground effects and local geology of the affected area for the 2001 Bhuj earthquake, India; 2008 Wenchuan earthquake, China; 2017 Pohang earthquake, Republic of Korea. We have considered these earthquakes intraplate earthquakes and have occurred in areas with very limited records of historical earthquakes and few seismic stations.

5.3 GROUND EFFECT DISTRIBUTION AND SEISMIC WAVE AMPLIFICATION DURING DAMAGING EARTHQUAKES

5.3.1 2001 Bhuj earthquake

The 2001 Bhuj earthquake (Mw 7.7), known as Republic Day, occurred on January 26, 2001, in the Kachchh region, in Gujarat, India (Figure 5.2).

The earthquake was one of the unique earthquakes in the world as the quake occurred many hundreds of km away from the nearest plate boundary. Despite its shallow hypocenter and large magnitude, it produced no

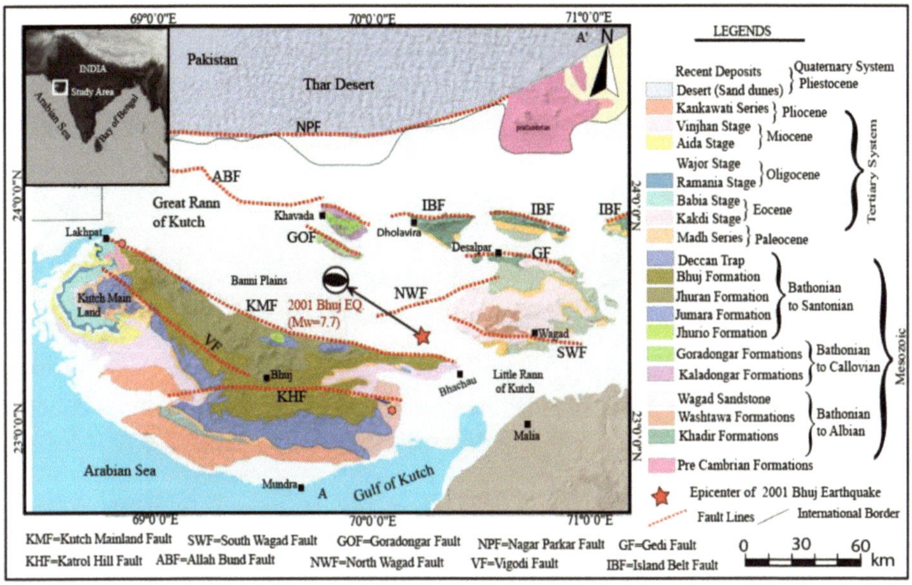

Figure 5.2 Geological map distribution of major fault systems around the epicentral area of 2001 Bhuj earthquake along with the epicenter of the earthquake. The geological map suggests the area is covered by recent deposits which are bounded by the major active faults. (Modified from Naik et al., 2020.)

surface rupture. The earthquake was felt all across India, even in Nepal, covering an area 16 times that of the 1906 San Francisco earthquake. The earthquake caused extensive ground effects, which includes secondary surface rupture, ground cracks, liquefaction, lateral spreading, rockfall (Figures 5.2 and 5.3; McCalpin and Thakkar, 2003; Rydelek and Tuttle, 2004; Pande, 2013; Naik et al., 2020). We have systematically analyzed the ground effects (mostly liquefaction-related phenomena) caused by the 2001 Bhuj earthquake, which suggests the local clustering of liquefaction-related phenomena within the Kachchh basin.

This indicates the amplification of seismic waves within the basin. These observations allow us to speculate the linkage between the observed ground effects and geological structures. For instance, the liquefaction phenomenon was clustered within the Kachchh basin between the Kachchh Mainland fault and Allah Bund fault-Island Belt fault (Figures 5.3 and 5.4). This broad spatial pattern of ground effects could be related to the effect of rupture directivity, trapping of seismic waves, and seismic wave amplification within the fault zones.

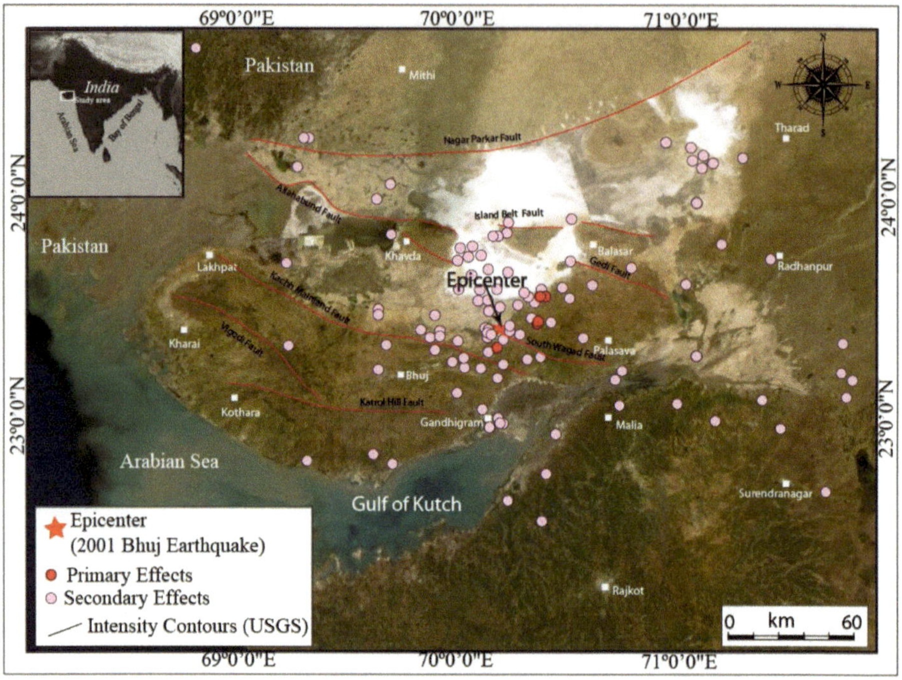

Figure 5.3 Ground effects of 2001 Bhuj earthquake which suggests the clustering of the secondary effects within the Kachchh basins.

The ground effects such as liquefaction, lateral spreading, ground cracks, automatic water ejection, and densification of soft sediments are primarily associated with the intensity of ground shaking (Ambrasey and Sarma, 1969; Obermeier and Dickenson, 2000; Holzer et al., 2010; Matsuoka et al., 2015; Liu-Zheng et al., 2017). The fault zone contains fractured rock, fault gouge, fluids that may act as a barrier for seismic wave transmission and amplify the seismic waves within the fault zone (Chester and Logan, 1986; Chester and Chester, 1998; Marra et al., 2000; Donati et al., 2001; Spudich and Olsen, 2001; Rovelli et al., 2002; Ben-Zion and Sammis, 2003) or by trapped waves within fault zones (Li et al., 1990; Li et al., 2004). The seismic wave amplification by trapping the seismic waves within fault zones is vital, primarily when soft-sediment deposits cover the fault zones (Donati et al., 2001; Liu-Zheng et al., 2017). The trapping of seismic waves within the fault zone is possible in both active and inactive faults (Rovelli et al., 2002). The seismic energy trapping and amplification of seismic waves could explain why soil liquefaction was clustered within the Kachchh basin.

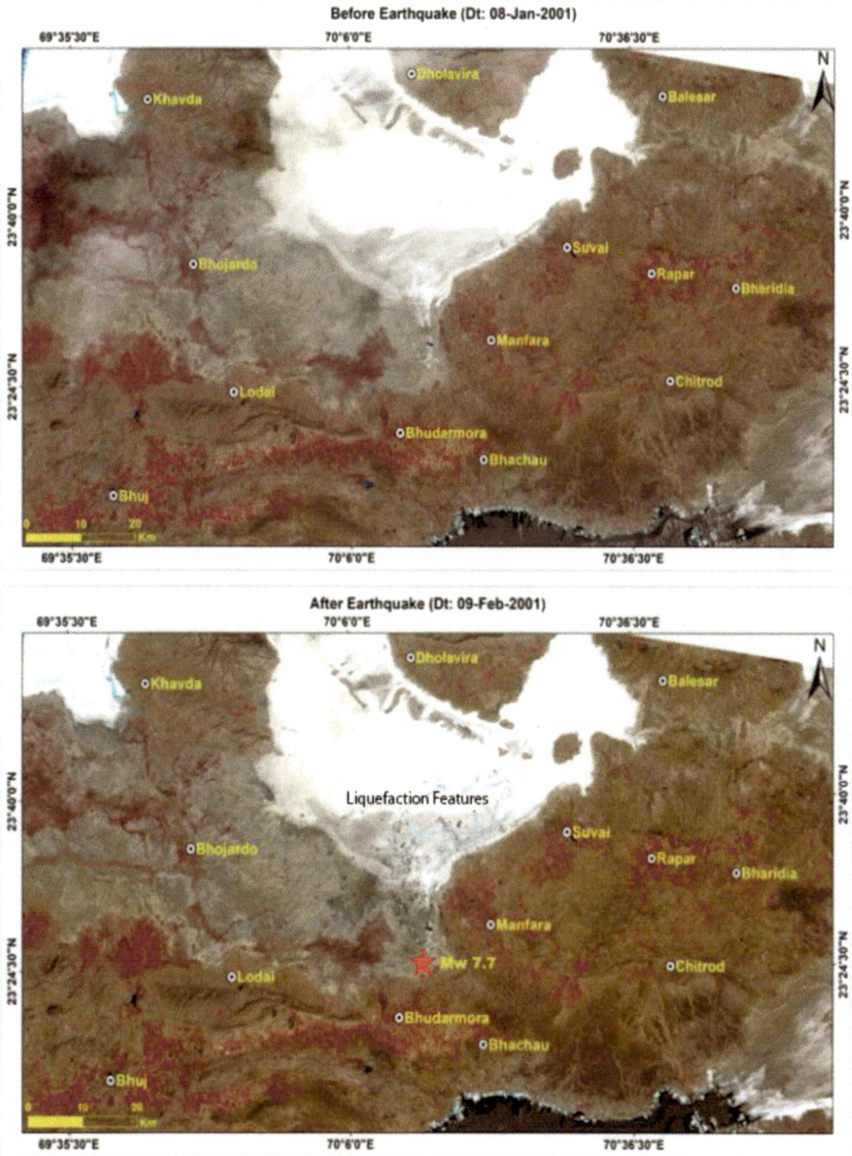

Figure 5.4 Satellite image showing the pre- and post-earthquake image for clustered liquefaction phenomena for the 2001 Bhuj earthquake.

5.3.2 2008 Wenchuan earthquake, China

The May 12, 2008 (Mw 8.0), Wenchuan earthquake, also known as the Great Sichuan earthquake or Wenchuan earthquake, occurred (N31.0N, 103.4E) in Wenchuan County at 75 km to the provincial capital Chengdu.

The focal depth of the earthquake was 19 km, and it caused a surface rupture of 240 km along NE-striking, west-dipping Beichuan and Pengguan faults (Liu-Zeng et al., 2009; Shao et al., 2010; Zhang et al., 2010; Liu-Zeng et al., 2012, Figure 5.5). The earthquake also caused significant ground shaking, which was felt over a vast region of China, up to ~1500 km away from the epicenter. The peak ground acceleration

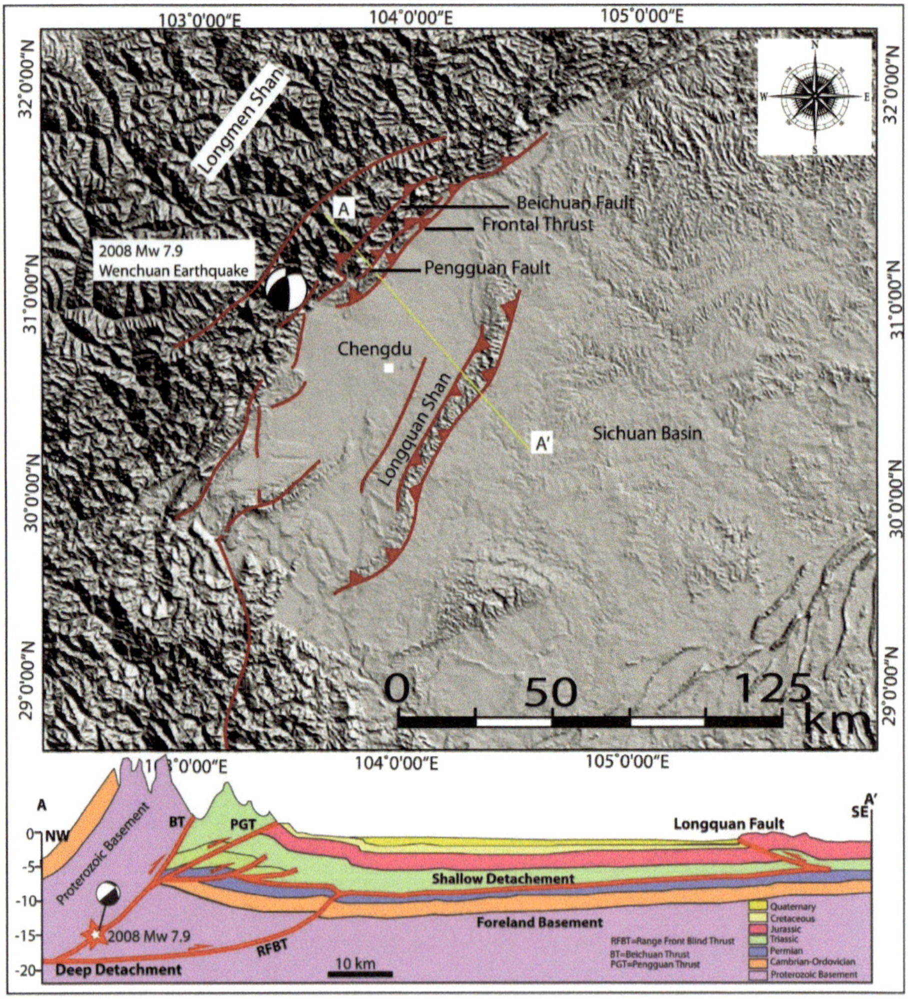

Figure 5.5 Tectonic map showing the major faults and epicentral location of the 2008 Mw 7.9 Wenchuan earthquake, China. The geological cross section shows the epicentral location is a basin bounded by several thrust faults. (Modified from Wang and Lin, 2017.)

(PGA) distribution contour for the earthquake shows the ground motion was attenuated with slight increase away from the fault. The epicentral region was located in the Sichuan basin (Figure 5.5), showing high PGA values varying between 0.20 and 0.8 g (Liu-Zheng et al., 2017).

Apart from the extensive surface rupture, other ground effects were observed on a large scale, including landslides, soil liquefaction, and water ejection (Cao et al., 2011; Li et al., 2014; Liu-Zheng et al., 2017). Several researchers carry forward the post-earthquake survey to map those effects (Liu-Zeng et al., 2009; Lekkas, 2010; Zhang et al., 2010; Hu et al., 2011; Liu Zheng et al., 2017). Most of the studies were based on regional damage assessment due to the earthquake. Liu-Zheng et al. (2017) considered a local-scale approach in understanding the role of local geology in the lustering of ground effects. The systematic analysis of liquefaction, water ejection, and so on, covered more than 200 liquefaction features (Figure 5.6). From the distribution of secondary ground effects, it can be inferred that local-scale seismic wave amplification is controlled by the reverberation of seismic waves within the basin and trapping of seismic waves within the fault zone, which increases the ground amplification within the basin.

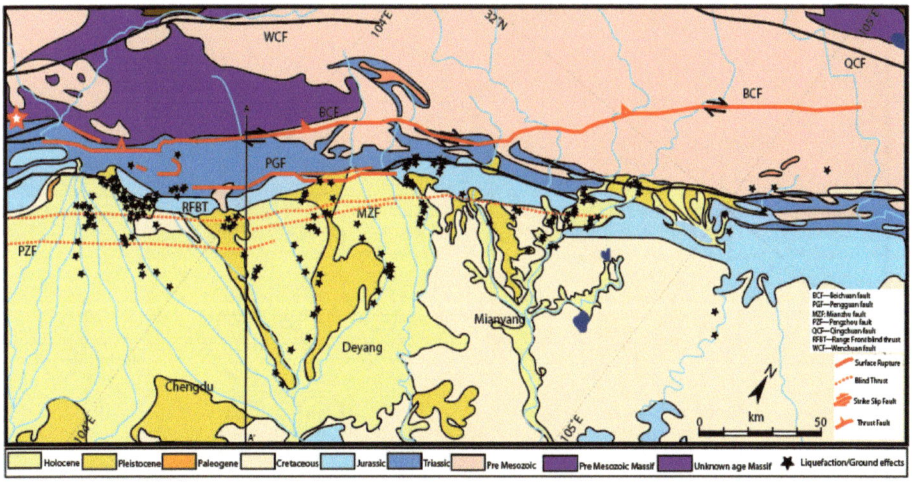

Figure 5.6 Geological map of Wenchaun earthquake epicentral area showing distribution of ground effects bonded by major fault systems. For A-A' refer to Figure 5.7. (Modified from Liu-Zheng et al., 2017.)

5.3.3 Pohang earthquake

On November 15, 2017, a moderate magnitude earthquake of MW 5.4 occurred in Pohang (36.065° N, 129.269° E) with a focal depth of about 4–6 km along SE Korean Peninsula. The epicenter was within the Heunghae basin of the industrial city Pohang. The earthquake was caused by a NE-SW striking, reverse dominant oblique-slip fault (Figure 5.7; Gihm et al., 2018; Choi et al., 2019). The earthquake was the second largest instrumentally recorded earthquake in Korea. There is also no

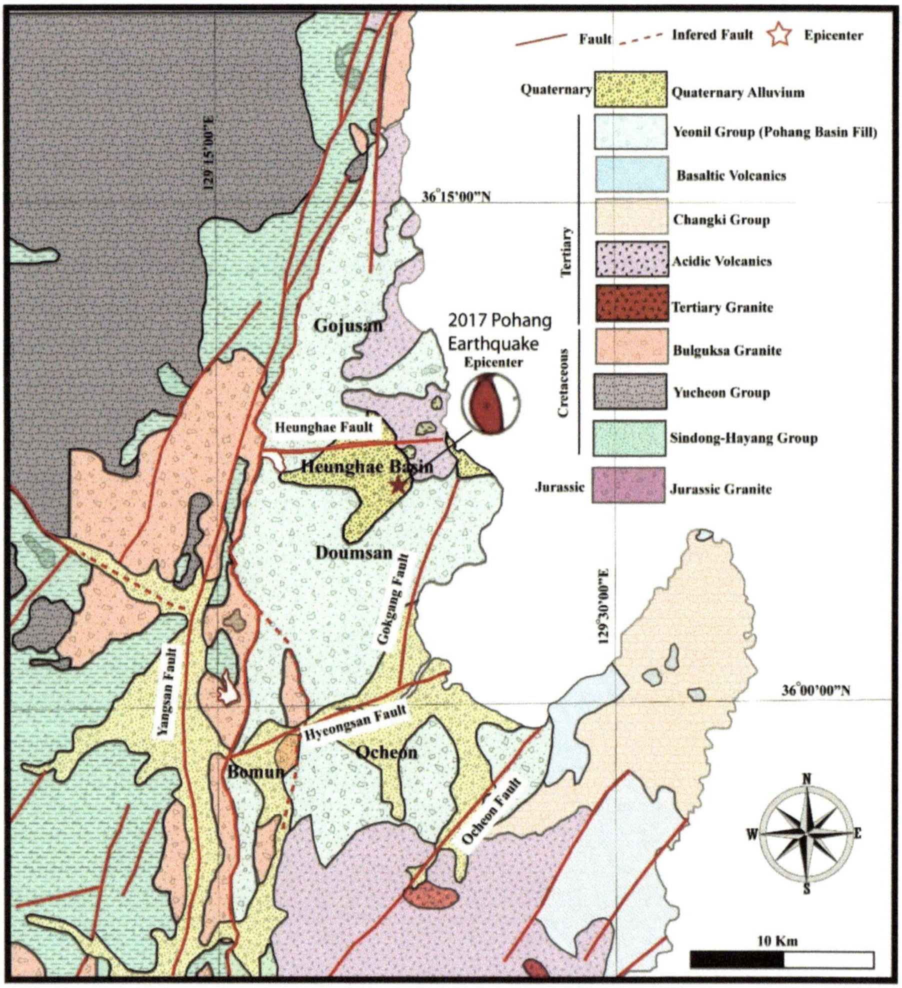

Figure 5.7 Geological map for the epicentral area of the 2017 Pohang earthquake, South Korea, with major structural features around the epicentral area. (Modified from Naik et al., 2020.)

permanent seismic station located within the Heunghae basin. The closest seismic stations located at Pohang show the PGA value of 0.29 g. Due to its epicenter within a small basin covered by thick alluvial deposits and shallow focal depth (5 km; Naik et al., 2019), the earthquake caused extensive, severe seismic wave amplification within the basin, which resulted in the development of extensive ground effects such as liquefaction, lateral spreading, ground cracks. Several researchers performed the post-earthquake field survey to map the ground effects (Gihm et al., 2018; Choi et al., 2019; Naik et al., 2019) and to understand the controlling factors for such widely distributed ground effects within the basin.

We have compiled all the published ground effects, which suggests that more than 600 liquefaction feautres were observed in the basin despite the small earthquake magnitude. The systematic analysis of those effects suggests clustering of the ground effects within the southern part of the basin than the northern part (Figure 5.8). This observation speculates the role of local geological structures such as fault zone at depth in seismic wave amplification within the basin (Donati et al., 2001; Liu-Zheng et al., 2017; Naik et al., 2019).

By analyzing the source of earthquake and clustering of ground effects within the Heunghae basin area, it can be argued that the clustering of ground effects (Figure 5.8) is the combination of mountain basin effect and trapping of seismic waves within the fault zone. The Heuanghae fault bounds the basin in the north.

The causative antithetic blind fault for the 2017 Pohang earthquake in NE. The presence of these two faults led to the differential amplification due to the trapping of seismic waves within the basin and fault zone, which delays the passing of seismic waves and caused extensive liquefaction in the southern part of the basin than the northern part.

5.4 DISCUSSIONS

Seismic wave amplification within the sedimentary basins during strong to moderate earthquakes has been widely observed. This was evident during the 1985 Mw 8.0 earthquake in Mexico City (Kawase and Aki, 1989), the 1994 Mw 6.7 Northridge earthquake in the San Fernando

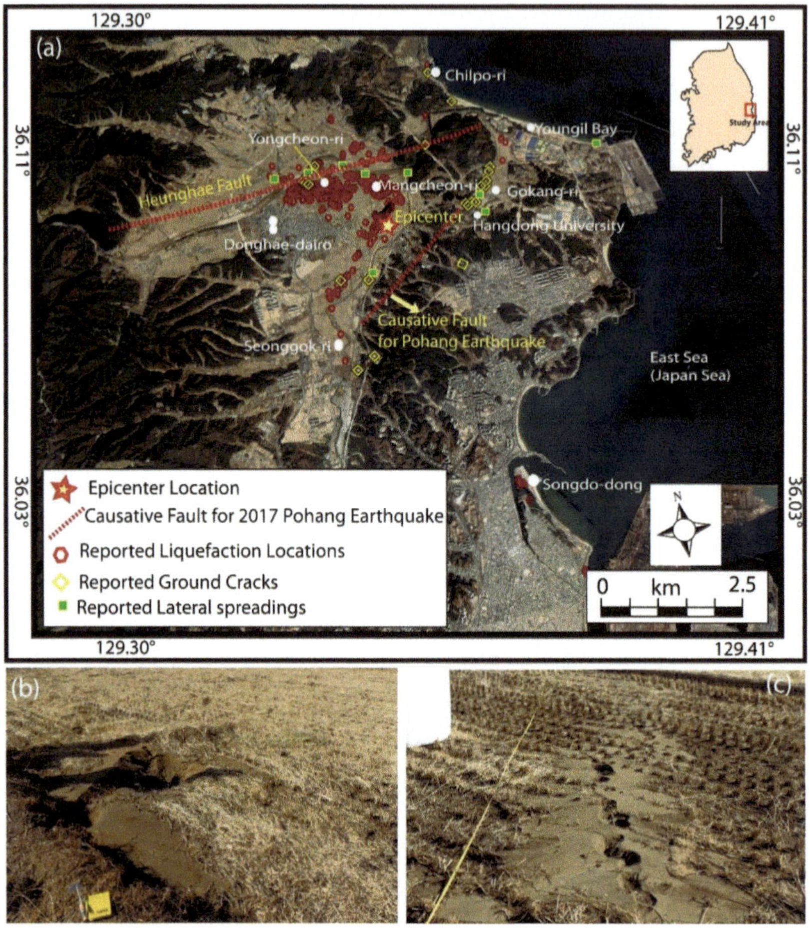

Figure 5.8 (a) Google earth image showing the epicenter, major structural features and clustering of ground effects for 2017 Pohang earthquake, South Korea; (b) and (c) showing the field photos of the typical liquefaction features observed within the epicentral area of the 2017 Pohang earthquake.

valley and Los Angeles basin (Pitarka and Irikura, 1996), the 1999 Izmit Mw 7.4 earthquake in Avcilar and Istanbul (Ergin et al., 2004), the 2015 Gorkha earthquake in Nepal (Dhakal et al., 2016; Gautam, 2017), and the 2017 Pohang earthquake, South Korea (Naik et al., 2019). Some recent studies reported the seismic wave amplification within the basin is related to the seismic wave frequency, rupture processes, and structural and geological setting of the basin (Pitarka and Irikura, 1996; Pratt et al., 2003; Ergin et al., 2004; Lee et al., 2009, Liu-Zheng et al., 2017).

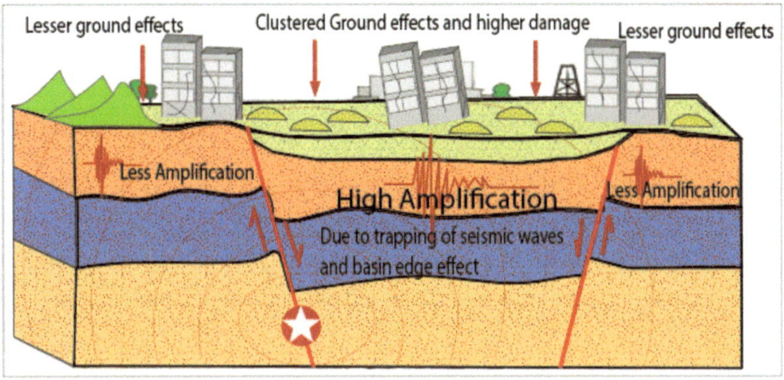

Figure 5.9 Schematic diagram showing possible mechanism involved for localized/clustered ground effects observed during all the above earthquakes.

When the seismic wave enters the basin, the surface wave slows down. It amplifies within the basin for a long time because of contrasting lithology between the basin (soft sediments) and basin boundary (hard rock). The reverberation of seismic waves could amplify the ground motion, which could cause a high amount of damage in some regions with clustered earthquake environmental effects (Graves, 1998; Olsen, 2000; Lee et al., 2009; Liu-Zheng et al., 2017; Naik et al., 2019). This kind of damage pattern can be explained by single factors such as the mountain basin effect, trapping of seismic waves by a fault zone, or a combination of factors of the basin is bounded by active or inactive fault zones (Figure 5.9).

Even if the basin looks homogeneous from a geological point of view, other local geological factors control the distribution of ground effects and damage associated with it. The factors such as the wedge-shaped basement-to-sediment basin interface, which behave as an acoustic lens, causing localized seismic wave amplification and extensive damage within the basin, which was observed in the 2012 Emilia earthquake (Mw 5.9) in Italy, clustering of liquefaction features was observed within the Po-Plain (LombardI et al., 2014; Naik et al., 2019).

Similarly, the trapping of seismic waves by the major fault zones within the basin controls the severity and clustering of the ground effects. Large faults within the sedimentary basins with fault gouges and fractured basement rock can trap the seismic wave, amplifying the upper bound within the basin due to soft sediments.

The systematic analysis of the ground effects from the previously mentioned three earthquakes, that is, the 2001 Bhuj earthquake, the 2008 Wenchuan earthquake, and the 2017 Pohang earthquake, where clustered ground effects were observed within the surrounding basins located close to the epicenter depicts the role of local geology in such effects. During the 2001 Bhuj earthquake (Mw 7.7) in India, most of the ground liquefaction-related ground effects were located within the Bani plains, part of the Kachchh basin, and close to the fault zones (Thakkar and Goyal, 2004). A similar observation was noticed during the 2008 Wenchuan earthquake (Mw 8.0), where most of the liquefaction-related ground effects were clustered to the Range Front Blind fault (Liu-Zheng et al., 2017). During the 2017 Pohang earthquake, most of the liquefaction features were clustered to the south of the Heunghae fault and antithetic causative fault for the generation of the Pohang earthquake. The Heunghae fault and the antithetic blind fault caused the differential amplification due to trapping seismic waves within the same basin and the clustering liquefaction features (Naik et al., 2019).

This kind of detailed study of ground effects for rare seismic events within the intraplate tectonic setting where large or damaging seismic events occurred with many thousand years of recurrence interval. Due to the large recurrence interval of such destructive earthquakes, the seismic hazard in this area is usually underestimated with little focus on understanding the geological factors controlling the seismic hazard.

It can be inferred from the just discussed analysis that coseismic ground effects are a major seismic hazard and can cause significant damage to lifeline structures, increasing both economic losses and loss of life. The clustered ground effects close to fault zones, as documented here, highlight the effect of geologic structures within sedimentary basins on seismic hazards, which has not been included in the seismic hazard assessment studies in most cases. Though in need of further testing with more case studies from the slow tectonic region, this conclusion would infer seismic hazard in areas with similar geological and depositional environments to the present study. This can also be used as a valid back analysis tool for assessing the actual vulnerability of the environment and predicting its future response to seismic wave amplification in that area, which is much needed for a proper definition of land use codes and land planning in seismic regions. Also, this approach may guide locating the suitable sites for installing seismic stations, which may be helpful to understand the site amplification studies for future seismic hazard assessment.

5.5 CONCLUSIONS

We present systematic post-earthquake field investigation data for the 2001 Bhuj earthquake, 2008 Wenchuan earthquake, and 2017 Pohang earthquake, which suggest clustering ground effects within the sedimentary basins close to the fault zones. This indicates the amplification of the seismic waves within the sedimentary basins by combining the mountain basin effect and seismic wave trapping within the fault zones. This study suggests that immediate and careful coseismic geological investigation can be effective for proper earthquake parameter estimation and seismic hazard evaluation on the vulnerability of the particular area. Moreover, this study can significantly contribute to engineering implications for a realistic seismic hazard assessment, particularly in liquefaction zonation. Also, it will be helpful to understand the possible mechanism involved in seismic wave amplification in areas of the lower seismic network or lower seismicity. These aspects need to be considered in liquefaction hazard mapping for similar geological settings on a local and regional scale. This type of study has essential engineering implications for evaluating potential damage in sedimentary basins during strong earthquakes.

REFERENCES

Ambraseys, N. and Sarma, S. 1969. Liquefaction of soils induced by earthquakes. *Bulletin of the Seismological Society of America*, 59(2), 651–664.

Ben-Zion, Y. and Sammis, C.G. 2003. Characterization of fault zones. *Pure and Applied Geophysics*, 160(3), 677–715.

Bolt, B.A. 1996. Intraplate seismicity and zonation. *Bulletin of the New Zealand Society for Earthquake Engineering*, 29(4), 221–228.

Caputo, Riccardo. 2005. Ground effects of large morphogenic earthquakes. *Journal of Geodynamics*, 40, 113–118.

Cao, Z., Youd, T.L. and Yuan, X. 2011. Gravelly soils that liquefied during 2008 Wenchuan, China earthquake, Ms = 8.0. *Soil Dynamics and Earthquake Engineering*, 31(8), 1132–1143.

Chester, F.M. and Logan, J.M. 1986. Implications for mechanical properties of brittle faults from observations of the Punchbowl fault zone, California. *Pure and Applied Geophysics*, 124(1), 79–106.

Chester, F.M. and Chester, J.S. 1998. Ultracataclasite structure and friction processes of the Punchbowl fault, San Andreas system, California. *Tectonophysics*, 295(1–2), 199–221.

Choi, J.H., Ko, K., Gihm, Y.S., Cho, C.S., Lee, H., Song, S.G., Bang, E.S., Lee, H.J., Bae, H.K., Kim, S.W. and Choi, S.J. 2019. Surface deformations and rupture processes associated with the 2017 Mw 5.4 Pohang, Korea, earthquake. *Bulletin of the Seismological Society of America*, 109(2), 756–769.

Daniell, J.E., Schaefer, A.M. and Wenzel, F. 2017. Losses associated with secondary effects in earthquakes. *Frontiers in Built Environment*, 3, 30.

Dhakal, Y.P., Kubo, H., Suzuki, W., Kunugi, T., Aoi, S. and Fujiwara, H. 2016. Analysis of strong ground motions and site effects at Kantipath, Kathmandu, from 2015 Mw 7.8 Gorkha, Nepal, earthquake and its aftershocks. *Earth, Planets and Space*, 68(1), 1–12.

Donati, S., Marra, F. and Rovelli, A. 2001. Damage and ground shaking in the town of Nocera Umbra during Umbria-Marche, central Italy, earthquakes: The special effect of a fault zone. *Bulletin of the Seismological Society of America*, 91(3), 511–519.

Ergin, M., Özalaybey, S., Aktar, M. and Yalcin, M.N. 2004. Site amplification at Avcılar, Istanbul. *Tectonophysics*, 391(1–4), 335–346.

Gangopadhyay, A. and Talwani, P. 2003. Symptomatic features of intraplate earthquakes. *Seismological Research Letters*, 74(6), 863–883.

Gautam, D. 2017. Unearthed lessons of April 25 2015 Gorkha earthquake (MW 7.8): Geotechnical earthquake engineering perspectives. *Geomatics Natural Hazards and Risk*, 8(2), 1358–1382.

Gihm, Y.S., Kim, S.W., Ko, K., Choi, J.H., Bae, H., Hong, P.S., Lee, Y., Lee, H., Jin, K., Choi, S.J. and Kim, J.C. 2018. Paleoseismological implications of liquefaction-induced structures caused by the 2017 Pohang earthquake. *Geosciences Journal*, 22(6), 871–880.

Graves, R.W. 1998. Three-dimensional finite-difference modeling of the San Andreas fault: Source parameterization and ground-motion levels. *Bulletin of the Seismological Society of America*, 88(4), 881–897.

Holzer, T.L., Jayko, A.S., Hauksson, E., Fletcher, J.P., Noce, T.E., Bennett, M.J., Dietel, C.M. and Hudnut, K.W. 2010. Liquefaction caused by the 2009 Olancha, California (USA), M5. 2 earthquake. *Engineering Geology*, 116(1–2), 184–188.

Hu, Y., Wang, J., Li, X., Ren, D. and Zhu, J. 2011. Geographical detector-based risk assessment of the under-five mortality in the 2008 Wenchuan earthquake. *China. PloS One*, 6(6), e21427.

Kawase, H. and Aki, K. 1989. A study on the response of a soft basin for incident S, P, and Rayleigh waves with special reference to the long duration observed in Mexico City. *Bulletin of the Seismological Society of America*, 79(5), 1361–1382.

Kawase, H., Lee, W.H.K., Kanamori, H., Jennings, P.C. and Kisslinger, C. 2003. Site effects on strong ground motions. *International Geophysics Series*, 81(B), 1013–1030.

Lee, S.J., Komatitsch, D., Huang, B.S. and Tromp, J. 2009. Effects of topography on seismic-wave propagation: An example from northern Taiwan. *Bulletin of the Seismological Society of America*, 99(1), 314–325.

Lekkas, E.L. 2010. The May 12 2008 Mw 7.9 Wenchuan, China, Earthquake: Macroseismic intensity assessment using the EMS-98 and ESI 2007 scales and their correlation with the geological structure. *Bulletin of the Seismological Society of America*, 100(5B), 2791–2804.

Li, Y.G., Leary, P., Aki, K. and Malin, P. 1990. Seismic trapped modes in the Oroville and San Andreas fault zones. *Science*, 249(4970), 763–766.

Li, Y.G., Vidale, J.E. and Cochran, E.S. 2004. Low-velocity damaged structure of the San Andreas Fault at Parkfield from fault zone trapped waves. *Geophysical Research Letters*, 31(12).

Li, G., West, A.J., Densmore, A.L., Jin, Z., Parker, R.N. and Hilton, R.G. 2014. Seismic mountain building: Landslides associated with the 2008 Wenchuan earthquake in the context of a generalized model for earthquake volume balance. *Geochemistry, Geophysics, Geosystems*, 15(4), 833–844.

Liu-Zheng, J., Wang, P., Zhang, Z., Li, Z., Cao, Z., Zhang, J., Yuan, X., Wang, W., Xing, X. 2017. Liquefaction in western Sichuan Basin during the 2008 Mw 7.9 Wenchuan earthquake, China. *Tectonophysics*, 694, 214–238.

Liu-Zeng, J., Zhang, Z., Wen, L., Tapponnier, P., Sun, J., Xing, X., Hu, G., Xu, Q., Zeng, L., Ding, L., Ji, C., Hudnut, K.W., van der Woerd, J. 2009. Co-seismic ruptures of the 12 May 2008, Ms 8.0 Wenchuan earthquake, Sichuan: East–west crustal shortening on oblique, parallel thrusts along the eastern edge of Tibet. *Earth and Planetary Science Letters*, 286(3–4), 355–370.

Liu-Zeng, J., Sun, J., Wang, P., Hudnut, K.W., Ji, C., Zhang, Z., Xu, Q. and Wen, L. 2012. Surface ruptures on the transverse Xiaoyudong fault: A significant segment boundary breached during the 2008 Wenchuan earthquake, China. *Tectonophysics*, 580, 218–241.

Lombardi, D. and Bhattacharya, S. 2014. Liquefaction of soil in the Emilia-Romagna region after the 2012 Northern Italy earthquake sequence. *Natural Hazards*, 73(3), 1749–1770.

Marra, F., Azzara, R., Bellucci, F., Caserta, A., Cultrera, G., Mele, G., Palombo, B., Rovelli, A. and Boschi, E. 2000. Large amplification of ground motion at rock sites within a fault zone in Nocera Umbra (central Italy). *Journal of Seismology*, 4(4), 543–554.

Matsuoka, M., Wakamatsu, K., Hashimoto, M., Senna, S. and Midorikawa, S. 2015. Evaluation of liquefaction potential for large areas based on geomorphologic classification. *Earthquake Spectra*, 31(4), 2375–2395.

McCalpin, J.P. and Thakkar, M.G. 2003. 2001 Bhuj-Kachchh earthquake: Surface faulting and its relation with neotectonics and regional structures, Gujarat, Western India. *Annals of Geophysics*, 46(5), 937–956.

Naik, S.P., Kim, Y.S., Kim, T. and Su-Ho, J. 2019. Geological and structural control on localized ground effects within the Heunghae Basin during the Pohang Earthquake (MW 5.4, November 15 2017), South Korea. *Geosciences*, 9(4), 173.

Naik, S.P., Mohanty, A., Porfido, S., Tuttle, M., Gwon, O. and Kim, Y.S. 2020. Intensity estimation for the 2001 Bhuj earthquake, India on ESI-07 scale and comparison with historical June 16 1819 Allah Bund earthquake: A test of ESI-07 application for intraplate earthquakes. *Quaternary International*, 536, 127–143.

Obermeier, S.F. and Dickenson, S.E. 2000. Liquefaction evidence for the strength of ground motions resulting from late Holocene Cascadia subduction earthquakes, with emphasis on the event of 1700 AD. *Bulletin of the Seismological Society of America*, 90(4), 876–896.

Olsen, K.B. 2000. Site amplification in the Los Angeles basin from three-dimensional modeling of ground motion. *Bulletin of the Seismological Society of America*, 90(6B), S77–S94.

Pande, P., 2013. Geoseismological investigation of the January 26, 2001 Bhuj earthquake in western peninsular India. *Natural Hazards*, 65(2), 1045–1062.

Panzera, F., Lombardo, G., D'Amico, S. and Galea, P., 2013. Speedy techniques to evaluate seismic site effects in particular geomorphologic conditions: faults, cavities, landslides and topographic irregularities. *Engineering Seismology, Geotechnical and Structural Earthquake Engineering*, 1, 101–145.

Pitarka, A. and Irikura, K. 1996. Basin structure effects on long-period strong motions in the San Fernando Valley and the Los Angeles basin from the 1994 Northridge earthquake and an aftershock. *Bulletin of the Seismological Society of America*, 86(1B), S126–S137.

Pratt, T.L., Brocher, T.M., Weaver, C.S., Creager, K.C., Snelson, C.M., Crosson, R.S., Miller, K.C. and Tréhu, A.M., 2003. Amplification of seismic waves by the Seattle basin, Washington State. *Bulletin of the Seismological Society of America*, 93(2), 533–545.

Porfido, S., Esposito, E., Vittori, E., Tranfaglia, G., Michetti, A.M., Blumetti, M., Ferreli, L., Guerrieri, L. and Serva, L., 2002. Areal distribution of ground effects induced by strong earthquakes in the Southern Apennines (Italy). *Surveys in Geophysics*, 23(6), 529–562.

Porfido, S., Alessio, G., Gaudiosi, G. and Nappi, R. 2020. New perspectives in the definition/evaluation of seismic hazard through analysis of the environmental effects induced by earthquakes. *Geosciences*, 10(2), 58.

Rovelli, A., Caserta, A., Marra, F. and Ruggiero, V. 2002. Can seismic waves be trapped inside an inactive fault zone? The case study of Nocera Umbra, central Italy. *Bulletin of the Seismological Society of America*, 92(6), 2217–2232.

Rydelek, P.A. and Tuttle, M. 2004. Explosive craters and soil liquefaction. *Nature*, 427(6970), 115–116.
Somerville, P. and Moriwaki, Y. 2003. 65 Seismic hazards and risk assessment in engineering practice. *International Geophysics*, 81(PART B), 1065–1080.
Serva, L. 1994. Ground effects in intensity scales. *Terra Nova*, 6(4), 414–416.
Shao, G., Ji, C., Lu, Z., Hudnut, K., Liu, J., Zhang, W. and Wang, Q. 2010, December. Slip history of the 2008 Mw 7.9 Wenchuan earthquake constrained by jointly inverting seismic and geodetic observations. In AGU Fall Meeting Abstracts (100[5B], pp. S52B–04).
Shimony, R., Gvirtzman, Z. and Tsesarsky, M. 2020. Seismic energy release from Intra-Basin sources along the Dead Sea transform and its influence on regional ground motions. *Bulletin of the Seismological Society of America*, 111(1), 295–308. doi: https://doi.org/10.1785/0120200215
Shimony, R., Gvirtzman, Z. and Tsesarsky, M. 2021. Seismic energy release from Intra-Basin sources along the Dead Sea transform and its influence on regional ground motions. *Bulletin of the Seismological Society of America*, 111(1), 295–308.
Spudich, P. and Olsen, K.B. 2001. Fault zone amplified waves as a possible seismic hazard along the Calaveras fault in central California. *Geophysical Research Letters*, 28(13), 2533–2536.
Stein, S. and Mazzotti, S. 2007. Approaches to continental intraplate earthquake issues. *Special Papers-Geological Society of America*, 425, 1.
Thakkar, M.G. and Goyal, B. 2004. On the relation between magnitude and liquefaction dimension at the epicentral zone of 2001 Bhuj earthquake. *Current Science*, 87, 811–817.
Wang, Z. 2011. Seismic hazard assessment: Issues and alternatives. *Pure and Applied Geophysics*, 168(1), 11–25.
Wang, M. and Lin, A. 2017. Active thrusting of the Longquan Fault in the central Sichuan basin, China, and the seismotectonic behavior in the Longmen Shan fold-and-thrust belt. *Journal of Geophysical Research: Solid Earth*, 122(7), 5639–5662.
Wesnousky, S.G., Scholz, C.H., Shimazaki, K. and Matsuda, T. 1984. Integration of geological and seismological data for the analysis of seismic hazard: A case study of Japan. *Bulletin of the Seismological Society of America*, 74(2), 687–708.
Zhang, P.Z., Wen, X.Z., Shen, Z.K. and Chen, J.H. 2010. Oblique, high-angle, listric-reverse faulting and associated development of strain: The Wenchuan earthquake of May 12, 2008, Sichuan, China. *Annual Review of Earth and Planetary Sciences*, 38, 353–382.

Chapter 6

HVSR as a versatile tool for subsurface investigation

Rajib Biswas, Vickey Sarma, and Nilutpal Borah

6.1 INTRODUCTION

Seismic waves while getting transmitted through the guiding medium may undergo either amplification or suppression. This is in general determined by the underlying medium. Accordingly, the intrinsic properties of the underlying medium have direct impact on the amplitudes of propagating seismic waves which can be referred to as site effect. Site effect influences ground motion. This amplification may vary over short distances. There are several methods by virtue of which this site effect can be estimated. They include passive as well as active. Active methods are often time-consuming as well as cumbersome. For example, earthquake recordings are considered the best to deal with subsurface investigation. However, when and where the event is going to occur remains a mystery. Instead, if we rely on passive methods, these obstacles can be lessened. Ambient noise or microseisms provides valuable information regarding the subsurface structure which has of late become a popular source for researchers. Quite contrary to earthquakes as well as synthetic blasts, the omnipresence of ambient noise makes them as the cost-effective as well rapid method for evaluating the structure. Towards exploiting the ambient noise or microseisms, implementation of horizontal-to-vertical spectral ratio (HVSR) has been widely established as a standard and cost-effective geophysical technique dedicated solely to decipher the underlying structure. The same technique can be applied for earthquake recordings too. This ambient noise mechanism is basically a nondestructive method, and it does not require a longer duration. It has been observed that while studying site characteristics from local and regional earthquakes, it usually takes lot of time towards recording of an event and subsequent interpretation of the event. Relatively, ambient noise analysis is a very useful tool to establish site amplification studies. As a result, ambient noise analysis is gaining popularity.

DOI: 10.1201/9781003177692-6

If we want to characterize a site in the context of peak resonant frequency and amplification factor, horizontal-to-vertical spectral ratio (H/V) can lead to rapid and easy assessment of site effects' evaluation in urban areas. Nogoshi and Igarashi (1971) were the pioneers in formulating this versatile technique. They used earthquake seismograms for evaluating the spectral ratio. However, Nakamura was the researcher who made revolutionary steps in adapting this technique using ambient noise as a prime input (Nakamura, 1989). Since then, there have been tremendous developments related to the practical and theoretical aspects suiting different adaptability of H/V ratio by the scientific community. Of late, it is widely used to evaluate the site characteristics basically in urban environments taking ambient noise data as its input.

In this chapter, the methodologies related to HVSR is outlined along with future recommendations. The implementation of HVSR as a tool for investigating subsurface of the site characteristics is overviewed. In addition, two case studies of evaluating site characteristics are also presented so as to provide the real practices carried in delineating subsurface structure. The study area, that is, Shillong City, is estimated with the amplitude of H/V ratios.

6.2 DATA

In order to estimate the site characteristics, ambient noise data play an important role pertinent to the specific site where noise data are collected. There exist several parameters/criteria before embarking on collection of microseisms/ambient noises. It is often advised that adherence to equispaced grid formulation results in good convergence as well as better control of results. However, in real practice, sticking to an ideal formulated gridding is often not possible which may be ascribed to undulating formations as well as heterogeneous topography.

Pertaining to this data collection, the guidelines of SESAME (2004) (Site Effects Assessment Using Ambient Excitations) have been followed, some of which are listed next.

a. The proper installation of the sensor is the prime requisite, that, there should be proper soil-sensor coupling.

b. Windy conditions should be avoided as far as practicable.
c. Removal of grasses beneath the sensor should be done to ensure that it properly rests on the soil.
d. Avoiding anthropogenic sources such as quarries, industrial machines and so on.
e. Allowing stability of sensor before starting to record ambient noise.
f. Proper insurance of a quiet environment.

Each afore-listed point possesses its own influence on H/V estimations because they somehow prevent the reliable estimation of H/V ratio, thereby aiding spurious results.

6.3 DATA PROCESSING

With the objective of determining site response through H/V ratio, Nakamura (1989) initiated his own algorithm on the basis of varying windows of recorded microtremors. Later, Field and Jacob (1993), Lermo and Chavez-Garcia (1993) put forward their own formulation related to the estimation of H/V ratio. Bard (1998) slightly modified the Nakamura technique by incorporating the ellipticity of Rayleigh waves and designed the modified algorithm towards estimation of H/V ratio. Afterwards, a developed algorithm emerged as a SESARRAY package, built up in LGIT, Grenoble. After its inception, it is widely recognized by the geoscientists, micro-zonation experts as well. The advantage of this package is the utilization of same type of seismometer allowing one to preprocess the data without any instrumental correction. The ambient noise data collected from 70 sites were processed with SESARRAY package. In addition to this, the following steps are adopted as per preprocessing criteria.

a. Determination of stable windows in the 30–40s range using an anti-trigger algorithm with 1s STA; a 30s LTA and STA/LTA ratio thresholds of 0.3–2.0.
b. For each window, a cosine taper with a length of 5% is applied on both sides of the window signal of the vertical (V), North-South (NS) and East-West (EW) component.

c. For each window, a fast fourier transform (FFT) is applied to the signal of the three components in order to obtain the spectral amplitude to which smoothing factor of 40 is applied, followed by arithmetical average (Kono and Ohmachi, 1998).

d. Spectral ratios of different components (NS/V, EW/V) and average (NS, EW)/V are computed.

The adoption of the STA/LTA algorithm is implemented so as to remove the spurious signals (caused by human activity and pedestrians, etc.). Its presence in the data contributes to the erroneous H/V peak estimation. Further, some other important features of H/V ratio have to be taken into account so as to infer the best possible estimates, such as reliability of H/V curves as well as peaks.

6.4 METHODOLOGY

Various researchers (Nakamura et al., 1989, 2008; Bard, 1998; Seht and Wohlenberg 1999; Parolai et al., 2001) tried to estimate site amplification behavior pertaining to a region utilizing H/V ratio from ambient noise since the late 1980s. They all agreed at one point that it is basically caused by shear-wave resonance in the sedimentary layer. Field and Jacob (1995), Lachet et al. (1996), Lermo and Chavez-Garcia (1993, 1994) carried out site estimations using this technique. Additionally, it is assumed that H/V ratio yields a sharp peak when there is a impedance contrast existing between the top layer and the underlying formations. Later, it is supported by many researchers (Wathelet et al., 2004; Mundepi et al., 2009, 2010; Raptakis et al., 2010). Considering all these different views, the theoretical formulation following Seth and Wohlenberg (1999) adopted here seems to be plausible, but it is not considered as the ultimate one so far the theoretical aspect is concerned. They described the existence of resonance peak as exhibited by H/V ratio to the λ/4 resonance hypothesis, that is, the site effect is expressed in terms of a transfer function.

6.4.1 Theory of H/V ratio

Suppose we have a soft soil with thickness m which is on top of a hard rock basement. We associate a shear-wave velocity, v_S. In this situation, there will arise resonant frequencies, f_R of this system when there originate integer multiples of quarter of wavelength. This makes the following transfer function:

$$f_R = nv_s / 4m(\mathrm{n} = 1,2,3.........) \tag{6.1}$$

The dependence of shear-wave velocity with depth in layers can be formulated as

$$v_s(\mathrm{z}) = v_0(1+z)^{\mathrm{x}} \tag{6.2}$$

In this equation, v_0 stands for surface shear-wave velocity at the surface with $z_0 = 1\mathrm{m}$. Meanwhile, x relates the depth dependence of the velocity. Considering all these, we can compute f_R as $1/4\mathrm{T}_0$ where T_0 refers to the time taken by the shear wave in making passage from the bottom of the layer to the surface. As we know that $v(z) = dz/dt$, this leads to estimation of T_0.

$$\mathrm{T}_0 = \int_z \frac{dz}{v_S(z)} = \frac{1}{v_0}\int_0^m (1+z)^{-1} dz$$

$$= \frac{1}{v_0}\frac{(1+m)^{1-x} - 1}{(1-x)}$$

Accordingly, the thickness and resonant frequency becomes interrelated as

$$f_R = \frac{1}{4T_0} = \frac{v_0(1-x)}{4[(1+m)^{1-x} - 1]}$$

$m = \left[\frac{v_0(1-x)}{4f_R} + 1\right]^{1/1-x} - 1$, where f_R is to be given in Hz and v_0 in m/s.

Consequently, the estimation of resonant frequency can now be performed via the aforementioned equation. The H/V spectral ratio $\left[T^{w}_{(H/V)}\right]$ is attained via division of the averaged components of the sediment site $[S_{NS}(w)]$ and $[S_{EW}(w)]$ by the spectrum of the vertical component $[S_v(w)]$ of the sediment site:

$$\left[T^{w}_{(H/V)}\right] = \frac{\left[S^{(w)}_{NS} + S^{(w)}_{EW}\right]^{1/2}}{S_v(w)} \tag{6.3}$$

6.5 RELIABILITY TESTS

6.5.1 H/V curves

In the beginning, the usual practice is to process the whole data in absence of window selection or quality of data. While inspecting the spectral amplitudes and spectral ratios, it is advisable to adopt the conditions for H/V curve's reliability (SESAME, 2004; Hasancebi et al., 2006). The adoption helps in choosing the best curves. Upon not fulfilling SESAME criteria, it is better to discard spectral amplitudes. We enlist the criteria of the reliability tests which are as follows.

a. The H/V curve below $10/l$ is not to be used where l signifies the window length.
b. Each H/V peak frequency to engage 200 or more significant cycles.
c. The curves should possess standard deviation $< \pm 2.0$.

6.5.2 H/V peaks

Likewise, H/V peaks are checked for reliability following Koller et al. (2004), Guillier et al. (2007), SESAME (2004). Based on fulfilling of five of the following listed criteria, the peak is adjudged reliable.

a. peak amplitudes (A_0) should be ≤ 2.
b. Between f_0 and $f_0/4$, existence of a frequency with amplitude less than half of $f_0/2$.
c. Between f_0 and $4^* f_0$, imperative existence of a frequency with amplitude $A < A_0/2$.
d. The frequency standard deviation (σ_f) for each peak (f_0, f_1 ...) should be less than a frequency-dependent threshold.
e. $\sigma_A(f_0)$ (standard deviation of peak) of H/V < threshold frequency as fixed in SESAME (2004).
f. No alteration of H/V peak position (within $\pm 5\%$).

6.6 CASE STUDY

6.6.1 Shillong city

With strict adherence to all the norms as discussed previously, the resonance frequency estimates were carried out in Shillong City with the help of H/V spectral ratios via ambient noise recording. Shillong possesses a very undulating topography with further accompaniment of heterogenous crustal structure. This restrained us in implementing a perfect grid formulation. As such, the ambient recordings were executed on sites with care so that the recordings did not contain any transients. After accruing the ambient noise data, the HVSR were estimated. Accordingly, two range of estimates in resonance frequencies have been attained—one covering lower range and the other one covering high frequency range, as depicted in Figure 6.1.

The ambient noise recordings were carried out as per SESAME (2004). Around 70 sites were covered with proper soil-senor coupling along with adherence to criteria as described in earlier sections. After accruing ambient noise, preprocessing was applied with the aim of obtaining suitable data as per SESAME (2004). Afterwards, HVSR had been applied to decipher the site response. It was then followed by computation of HVSR which yielded site-specific resonant frequencies.

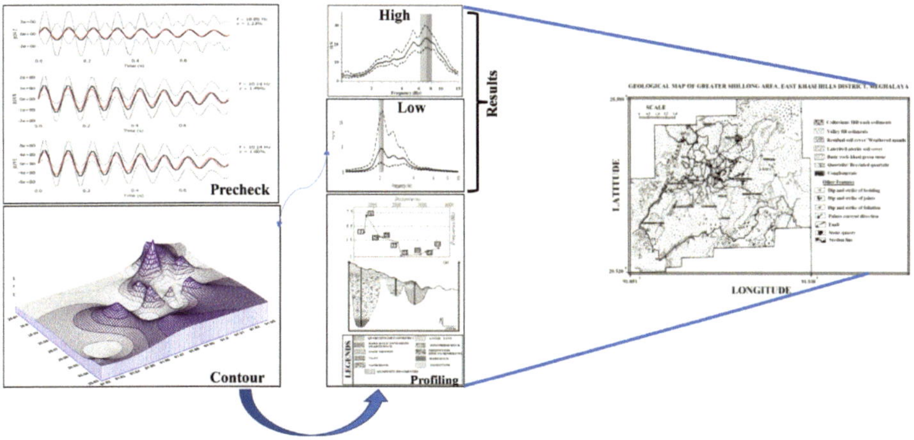

Figure 6.1 HVSR estimates and corresponding subsurface profiling in Shillong city.

However, the testing of H/V curves for reliability were executed before frequency estimates. Additionally, the origin of peak frequency was computed through damping tests. The moto behind the adoption of these criteria was to arrive at the best peak frequency pertinent to the site. Further, they helped in ensuring reliable characterization of site conditions as actuated by resonance frequencies and allied amplifications. While estimating all these, there was always influence from local lithology which could not be trivial at any cost. As described in the theoretical section, it was possible to empirically formulate a relationship frequency estimates and the corresponding overburden thickness. This indirectly paved the way for computation of subsurface velocity with linkage to depth of overburden thickness.

Taking all these preprocessing steps with care, the HVSR estimates for Shillong region were performed. The results of H/V peak frequencies fell in two categories—one with higher side and the other side with lower frequency estimates. Majority of sites yielded higher frequency estimates which engaged in general compact form of strata (Biswas and Boruah, 2011). Simultaneously, smaller thickness-related sedimentary strata could be another reason. This had similarities as observed in many studies, for example, Lombardo and Rigano (2007); Guillier et al. (2007). On the contrary, the sites with lower frequency estimates could be linked to higher overburden thickness. As given in Figure 6.1, the contour distribution of frequency estimates bore a testimony of wide variation. Specially, central part of Shillong showed estimates of 1–2 Hz falling in the lower frequency category (Biswas and Borauh, 2011; Biswas et al., 2013a, b). This tallied the borehole information available at those sites. While certain sites were characterized by possession of higher frequency estimates with good conformity with local lithology which could be well endorsed from the geological map provided in Figure 6.1—revealing existence of hard rock strata in the form of sandstones, quartzites, conglomerates, and so on.

Looking at the wide variability of frequency estimates yielded by HVSR, an attempt had been made to reconstruct the subsurface geology in tune with available borehole information for Shillong City (Biswas et al 2018a, b; Biswas and Boruah, 2019). The reconstructed section was illustrated in Figure 6.1. Notably, the H/V peak frequencies matched well with the site geology. Following Cara et al. (2008), who correlated the variability in H/V peak frequencies in terms of the existing lithology, the reconstructed sections based on borehole information implicated a heterogeneous strata underneath. The H/V peak frequencies were found to be in good resonance with these underlying formations. As for example, lower amplitudes engulfed smaller frequencies while higher frequencies spawned over larger amplitudes. Going by Souriau et al. (2007) and Dubos et al. (2003), these observations led to the fact that topographical undulations might have influence in these variational patterns of HVSR estimates.

It is a usual practice of empirically relating the fundamental frequency with the known thickness towards mapping of sedimentary thickness in a specific region. Towards this, a power law relating estimates of fundamental frequencies with that of thickness of sedimentary column in this region had been derived. Mundepi et al. (2010) mapped the overburden thickness in Jammu City, incorporating the borehole information and the H/V peak frequencies, pertinent to that region. Both these works were based on the pioneering work of Seht and Wohlenberg (1999). Parolai et al. (2001) also developed empirical relationship entailing these two parameters. Interestingly, the geology of the study region resembles to some extent the geology found in the two later works. Thus, a comparison of these two pioneering works relating power law between overburden thickness and resonant frequency was performed (Biswas, 2016; Biswas et al. 2018a, b; Biswas and Boruah, 2019). This led to the fact that the derived relationship between fundamental frequency and thickness of soil/overburden column pertinent to this region was quite consistent with the estimations made by Seht and Wohlenberg (1999) and Parolai et al. (2001), although constrained by shallower depth borehole information. Notably, this attained relationship yield similar estimates in the higher frequency sides (Biswas and Boruah, 2011; Biswas et al., 2018a; Biswas and Boruah, 2019).

Additionally, sedimentary thickness when aligned with H/V estimate of peak frequencies leads to the estimation of average shear-wave velocity of the underlying soil column (Biswas et al., 2015). Exploiting the close connection between peak frequency, shear-wave velocity, and thickness of the soil layer, several studies utilized it for assessing valuable information (e.g., Delgado et al., 2000; Seht and Wohlenberg, 1999; Parolai et al.,

2001). In pursuance of these formulations, a new relationship was built for Shillong City which connected soil thickness and frequencies (Biswas et al., 2014, 2015; Biswas and Boruah, 2016; Biswas et al., 2018a, b; Biswas and Boruah, 2019) The attained relationship agreed well with similar kind of studies. This relationship helped in preliminary assessment of average shear-wave velocity variation corresponding to the varying sedimentary formation.

6.6.2 Tezpur city

Likewise, ambient noise recordings were carried out in urban area of Tezpur City, India. With a grid formulation, microseisms were recorded at 70 sites. In similitude to Shillong, two cases of resonance frequency estimates were obtained. As per findings, there is an emergence of larger values of resonant frequencies (> 2.0 Hz), which spanned the western as well as southern part of the area under study. In the same way, lower frequency estimates less than 1.0 Hz emerged for certain pocket areas of Tezpur City (Bora and Biswas, 2017; Bora et al. 2017; Bora et al. 2020, 2021). On inspection, it was revealed that those areas were characterized by prevalence of low velocity sediments. Going towards the extreme eastern side, the lowest peak frequency of 0.52 Hz had been obtained. Similarly, the HVSR estimates yielded the highest peak frequency of 2.8 Hz towards the east-southern part of the area (Bora et al. 2018; Bora and Biswas, 2018). These lower and higher bound of frequency estimates could be correlated directly with minimum and maximum sediment thickness, respectively. In some cases, there appeared two peaks in the HVSR spectra. The first one f_1 could be interpreted to evolving from the shallower impedance contrast belonging to comparatively younger and older sediments.

Figure 6.2 shows the Tezpur area where ambient noise recordings had been conducted. Around 70 sites were covered. Tezpur has an undulating topography. Some parts of the studied area are overlain by exposed bedrock. Likewise, there are areas which are characterized by overburden thickness. The sediment depositions around the study area varies considerably. As such, the resonance frequency estimates through HVSR technique varied considerably throughout the study area in tune with the overburden thickness. In case of exposed bedrock, the HVSR estimates

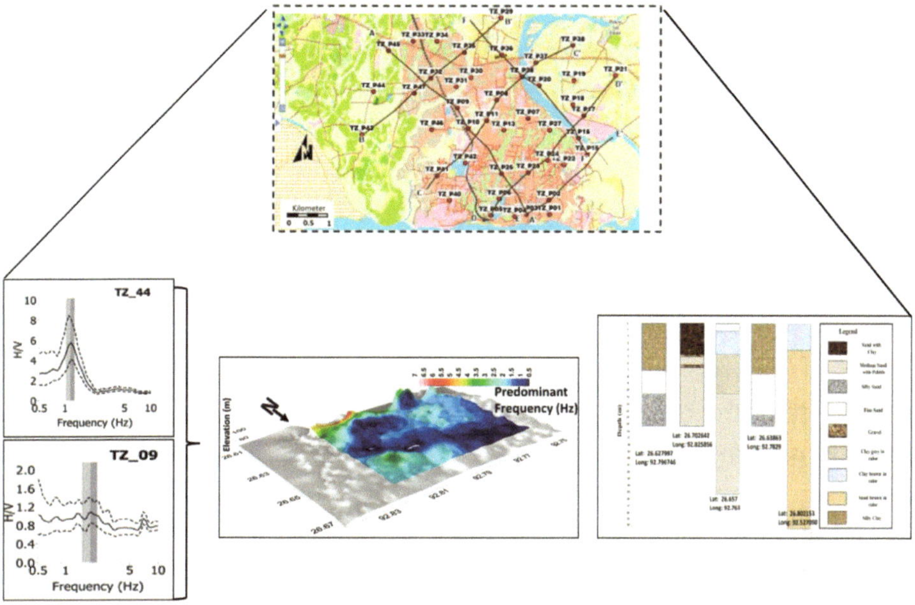

Figure 6.2 Same as Figure 6.1 with case study taken up at Tezpur City. The resonance frequency estimates as example are provided here. The adjacent figure towards right displays the contour distribution of the predominant frequency estimates. Towards the extreme right, reconstructed cross section is shown.

emerged very high. Similarly, for sites having thicker strata of sediments, the HVSR estimates of resonant frequency came out to be very small.

6.7 CONCLUDING REMARKS AND FUTURE OUTLOOK

HVSR is one of the versatile techniques to rapidly assess the site effects of urban area. It helps in evaluation of crucial parameters in the form of resonance frequency, soil amplification, overburden thickness. Exploiting the potential of the robust technique, this chapter presents the basic criteria to be adopted while accruing data in the form of ambient noise. It also highlights the preprocessing steps to be adopted while measuring the HVSR estimates. With strict adherence to these outlined protocols, the chapter also presents two case studies where HVSR was used to evaluate the site effects in Shillong City and Tezpur City of northeast India. The results as well as the analysis have been systematically appraised in this chapter. There is no doubt that HVSR hold lot of potential if used judiciously. As compared to other active methods, this passive geophysical method can provide valuable insights within a very short span of time. It can be easily applied to ambient vibrations as well as seismic recordings. Once the ambient noise is properly processed with all requisite step, it becomes convenient to estimate the resonance frequency of the recorded sites. The resonance frequency estimates help researchers to evaluate other allied parameters of significance. However, HVSR cannot be regarded as a stand-alone procedure to estimate the site effects of the area under investigation. It can lead to ambiguity in results if not properly cross-validated with other techniques. However, it can quickly yield a rapid overview of the area under investigation. Now with assistance from adequate borehole data, soil penetration results, HVSR could reliably infer shear-wave velocity profiles which is considered as one of the vital parameters for subsurface investigation. Once the information related to shear-wave velocity structure is obtained, it helps in building stratigraphical layers. The subsequent validation with other geophysical techniques is essential for better interpretation. It has, therefore, become customary among the researcher community to adopt HVSR with the aim of subsurface investigation with further accompaniment of robust geophysical investigations such as MASW, resistivity method as well as gravity anomaly study. The synergistic amalgamation of results obtained from these with HVSR estimates can lead to prudent subsurface investigation, thereby giving better in-depth analysis.

REFERENCES

Bard, P.-Y. (1998). Microtremor measurements: A tool for site effect estimation? In *Proc. 2nd Intl. Symp. on the Effects of Surface Geology on Seismic Motion*, Irikura, Kudo, Okada and Sasatani (eds.). Rotterdam: Balkema, 1251–1279.

Bhattacharya, P.M., Pujol, J., Mazumdar, R.K. and Kayal, J.R. (2005). Relocation of earthquakes in the northeast India region using joint hypocenter determination method. *Current Science*, 89(8), 1404–1413.

Biswas, R., Bora, N. and Boruah, S. (2018a). Frequency dependent site response inferred from microtremors accompanied by ambient noise. *Physics and Astronomy International Journal*, 2(4), 273–277.

Biswas, R., Baruah, S. and Bora, N. (December 2018b). Assessing shear wave velocity profiles using multiple passive techniques of Shillong region of northeast India. *Natural Hazards*, 94(3), 1024–1041.

Biswas, R. and Baruah, S. (2011). Site response estimation by Nakamura method: Shillong city, northeast India. *Memoir of the Geological Society of India*, 77, 173–183.

Biswas, R., Bora, D., Baruah, S., Kalita, A. and Baruah, S. (2013a). The effects of attenuation and site on the spectra of microearthquakes in the Shillong region of northeast India. *Pure and Applied Geophysics*, 170(11), 1833–1848.

Biswas, R., Bora, D. and Baruah, S. (2013b). Influence of attenuation and site on microearthquakes' spectra in Shillong region, of Northeast, India: A case study. *Acta Geophysica*, 61(4), 886–904.

Biswas, R., Bora, D. and Baruah, S. (2015). Mapping sediment thickness in Shillong city of northeast India through empirical relationship. *Journal of Earthquakes*. Article ID: 572619.

Biswas, R., Baruah, S. and Baruah, S. (2014). *A Brief Overview of Site Response Estimation Techniques*. Italy: Mistral Services.

Biswas, R. and Baruah, S. (2016). Shear wave velocity estimates through combined use of two passive techniques of a tectonically active area. *Acta Geophysica*, 64(6).

Biswas, R. (2016). Non-linear earthquake site response analysis: A case study in Shillong City. *IJEHRM*, 3(4).

Bora, N. and Biswas, R. (2017). Quantifying regional body wave attenuation in a seismic prone zone of northeast India. *Pure and Applied Geophysics*, 2017(174), 1953–1963.

Bora, N., Biswas, R. and Bora, D. (2017). Assessing attenuation characteristics prevailing in a seismic prone area of NER India. *IOP: Journal of Geophysics and Engineering*, 2017(14), 1368–1381.

Bora, N., Biswas, R. and Dobrynina, A. (2018). Regional variation of coda Q along Kopili fault and its implications. *Tectonophysics*, 722, 235–248.

Bora, N. and Biswas, R. (2019). Delineation of subsurface profiles beneath the Kopili fault zone in northeast India utilizing coda portion. *Journal of Asian Earth Sciences*, 173, 237–249. doi: 10.1016/j.jseaes.2019.01.023.

Bora, N. and Biswas, R. (2018). P- and S-wave attenuation in the Kopili region of northeastern India. *Annals of Geophysics*, 61(6), SE668. doi: 10.4401/ag-7917.

Biswas, R. and Boruah, S. (2019). Assessing site response through ambient noise measurements in a seismic prone area. *Physics & Astronomy International Journal*, 3(2), 72–80. doi:10.15406/paij.2019.03.00161.

Bora, N., Biswas, R. and Malischewsky, P. (2020). Imaging subsurface structure of an urban area based on diffuse-field theory concept using seismic ambient noise. *Pure and Applied Geophysics*, 177, 4733–4753. https://doi.org/10.1007/s00024-020-02547-4.

Bora, N., Biswas, R. and Srinivassan, V. (2021). CodaQback—A simplified python code facilitating auto-windowing for estimating seismic coda attenuation parameter. *International Journal of Geotechnical Earthquake Engineering*, 12(1), 1–11. doi: 10.4018/IJGEE.20210101.

Cara, F., Cultera, G., Azzara, R.M., Rubeis, V., Giulio, G., Giammariano, S., Tosi, P., Rovelli, A. (2008). Microtremor measurements in the city of Palermo,

Italy: Analysis of the correlation between local geology and damage. *Bulletin of the Seismological Society of America*, 98, 1354–1372.
Delgado, J., Lopez Casado, C., Giner, J., Estevez, A., Cuenca, A. and Molina, S. (2000). Micro tremors as a geophysical exploration tool: Applications and limitations. *Pure and Applied Geophysics*, 157–9, 1445–1462.
Dubos, N., Souriau, A., Ponsolles, C. and Fels, J.F. (2003). Etude des effets de site dans la ville de Lourdes (Pyrénées, France) par la méthode des rapports spectraux. *Bulletin de la Société Géologique de France*, 174, 33–44.
Field, E. and Jacob, K. (1993). The theoretical response of sedimentary layers to ambient seismic noise. *Geophysical Research Letters*, 20–24, 2925–2928.
Field, E.H. and Jacob, K. (1995). A comparison and test of various site response estimation techniques, including three that are non reference-site dependent. *Bulletin of the Seismological Society of America*, 85, 1127–1143.
Guillier, B., Chatelain, J., Claudet, S.B. and Haghshenas, E. (2007). Use of ambient noise: From spectral amplitude variability to H/V Stability. *Journal of Earthquake Engineering*, 11, 925–942.
Hasancebi, N. and Ulusay, R. (2006). Evaluation of site amplification and site period using different methods for an earthquake-prone settlement in Western Turkey. *Engineering Geology*, 87, 85–104.
Koller, M., Chatelain, J.-L., Guillier, B., Duval, A.-M., Atakan, K., Lacave, C., P.-Y. Bard and SESAME-Team. 2004. Practical user guidelines and software for the implementation of the H/V ratio technique: Measuring conditions, processing method and results interpretation. *Proceedings of the 13th World Conference on Earthquake Engineering Vancouver, Canada*. Paper 3132.
Konno, K. and Ohmachi, T. (1998). Ground-motion characteristics estimated from spectral ratio between horizontal and vertical components of microtremor. *Bulletin of the Seismological Society of America*, 88–1, 228–241.
Lachet, C., Hatzfeld, D., Bard, P.Y., Theodulidis, N., Papaioannou, C. and Savvaidis, A. (1996). Site effects and microzonation in the city of Thessaloniki (Greece): Comparison of different approaches. *Bulletin of the Seismological Society of America*, 86, 1692–1703.
Lermo, J. and Chavez-Garcia, F.J. (1993). Site effects evaluation using spectral ratios with only one station. *Bulletin of the Seismological Society of America*, 83–5, 1574–1594.
Lermo, J. and Chavez-Garcia, F.J. (1994). Are microtremors useful in site response evaluation? *Bulletin of the Seismological Society of America*, 84, 1350–1364.
Lombardo, G. and Rigano, R. (2007) Local seismic response in Catania (Italy): A test area in the northern part of the town. *Engineering Geology*, 94, 38–49.
Mundepi, A.K. and Lindholm, C., Kamal (2009). Site response evolution and sediment mapping using horizontal to vertical spectral ratios (HVSR) of ground ambient noise in Jammu City, NW India. *Journal of the Geological Society of India*, 75–5, 551–558.

Mundepi, A.K. and Mahajan, A.K. (2010). Soft soil mapping using horizontal to vertical ratio (HVSR) for seismic hazard assessment of Chandigarh city in Himalayan foothills, North India. *Journal of the Geological Society of India*, 75, 799–806.

Nogoshi, M. and Igarashi, T. (1971). On the amplitude characteristics of microtremor (part 2) (in Japanese with English abstract). *Journal of the Seismological Society of Japan*, 24, 26–40.

Nakamura, Y. (1989). A method for dynamic characteristics estimation of subsurface using microtremor on the ground surface. *Quarterly Report of Railway Technical Research*, 30, 25–30.

Nakamura, Y. (2008). On the H/V Spectrum, 14th World Conference on Earthquake Engineering.

Parolai, S., Bormann, P. and Milkereit, C. (2001). Assessment of the natural frequency of the sedimentary cover in the Cologne area (Germany) using noise measurements. *Journal of Earthquake Engineering*, 5, 541–564.

Raptakis, D. and Makra, K. (2010). Shear wave velocity structure in western Thessaloniki (Greece) using mainly alternative SPAC method. *Soil Dynamics and Earthquake Engineering*, 20, 202–214.

Seht, I. M. and Wohlenberg, J. (1999). Microtremor measurements used to map thickness of soft sediments. *Bulletin of the Seismological Society of America*, 89–1, 250–259.

SESAME. (2004). Guidelines for the implementation of the H/V spectral ratio technique on ambient vibrations measurements, processing and interpretation. SESAME European research project WP12-Deliverable D23.12.

Wathelet, M., Jongmans, D. and Ohrnberger M. (2004). Surface-wave inversion using a direct search algorithm and its application to ambient vibration measurements. *Near Surface Geophysics*, 2, 211–221.

Souriau, A., Roulle, A. and Ponsolles, C. (2007). Site effects in the city of Lourdes, France, from H/V measurements: Implications for seismic-risk evaluation. *Bulletin of the Seismological Society of America*, 97(6), 2118–2136.

Chapter 7

Application of teleseismic receiver function in investigation of crustal thickness and Poisson's ratio

Monika Wadhawan, Devajit Hazarika, and Sowrav Saikia

7.1 INTRODUCTION

The knowledge of the crustal structure and rock proportion of the continental crust is of fundamental importance in understanding crustal generation and evolution. The uppermost part of the crust being the most accessible part of the Earth has been the subject of several geological investigations. However, the composition and depth distribution of different types of rocks in the crust remains uncertain. The most basic parameter regarding the continental crust is the total thickness of the crust. The study of seismic phases is one of the robust and powerful ways to decipher the Earth's layered structure and compositions. The study of P- and S waves and their reverberated phases enrooted through different discontinuities can be a very useful contributing factor in investigating the crustal composition and petrologic behaviour of Earth's crust. Within the lithospheric depth, the Moho discontinuity which separates the Earth's crust from the mantle represents a major change in seismic velocities, chemical composition, and rheology. The depth of the Moho discontinuity and its variation region to region is an important parameter to describe the evolutionary history and associated geotectonic deformation. By the utility of broadband, seismological data are useful in determining the elastic properties of the crust. Seismic measurements probe the elastic properties, like Poisson's ratio (σ) of the crust at depth, from which physio-chemical composition can be inferred. Estimation of the bulk composition of the Earth's crust is limited by the uncertainty in the composition of the lower crust.

DOI: 10.1201/9781003177692-7

Seismic measurements are perhaps the most direct method for probing crustal composition in the upper and lower part of the crust. The analysis of both compressional and shear-wave velocities is a powerful tool to compliment petrological and geochemical studies in the investigation of the mineralogical composition of the crust as well as its formation and evolution process. The ratio of P-wave (V_P) and S-wave (V_S) velocities, known as V_P/V_S ratio, is closely related to Poisson's ratio which enables to discover the nature of continental crust (Zandt and Ammon, 1995). Poisson's ratio is related to the elastic parameter of rock that defines the material composition and physical conditions of the Earth's crust (Christensen, 1996). For common rock types, σ varies from 0.20 to 0.35, and is particularly sensitive to rock composition; increases the silica contents by lower 'σ' value and high mafic content has a high 'σ' value (Kern, 1982). For lower crustal rocks, low σ (< 0.26), intermediate σ (0.26–0.28), and high σ (> 0.28) are characteristic of felsic, intermediate, and mafic compositions, respectively(Christensen, 1996). Thus, the measurement of Poisson's ratio offers the means of distinguishing between felsic and mafic rocks (Christensen and Mooney, 1995). The elastic parameter Poisson's ratio 'σ' provides the ability to discriminate among possible bulk compositions (Zandt and Ammon, 1995).

Passive seismological experiments are widely accepted for the investigation of the crustal composition and layered structure of the crust. The receiver functions (RFs) technique, a handy and robust technique, is widely used to probe primarily the crustal thickness, seismogenic depth, shear-wave velocity structure, and Poisson's ratio (or equivalently V_P/V_S). The RFs are time series computed from three-component seismograms that contain the relative response of the Earth near the receiver site. The incoming teleseismic wave bears the information of the earthquake source, the path travelled by the seismic wave, and the response of the local Earth's structure beneath the receiver site. It is obtained by removing the effects of source, the travel path, and instrument response by deconvolving vertical component seismogram with the radial and tangential components resulting in radial receiver function (RRF) and tangential receiver function (TRF), respectively. Whenever any teleseismic P wave travels a longer distance, it encounters with discontinuity, a part of this P wave gets reflected and converted into P-to-S phase (PmS) that

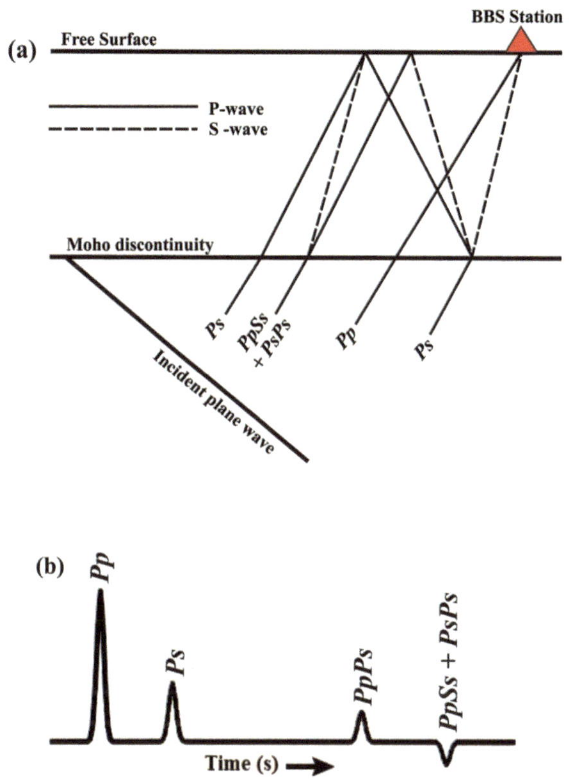

Figure 7.1 (a) Simplified ray diagram of the P-to-S (Ps) converted phases and its crustal multiples (PpPs, PpSs + PsPs) for a simple one-layered Earth model, (b) the corresponding theoretical receiver function.

causes reverberated phases or crustal multiples (e.g. PpPs, PsSs + PsPs) (Figures 7.1a, b).

7.2 RECEIVER FUNCTION COMPUTATION

7.2.1 Rotation

Three-component seismometers are used to record the particle motions of vertical and two horizontal components in vertical (Z), North-South (N-S), and East-West (E-W) directions, respectively. This geographical coordinate system is referred to as the ZNE coordinates system. In this ZNE coordinate system, the two horizontal components rarely correspond exactly to the SH and SV polarizations of the S wave. As a result, rotation is essential for the horizontal components. Rotation converts

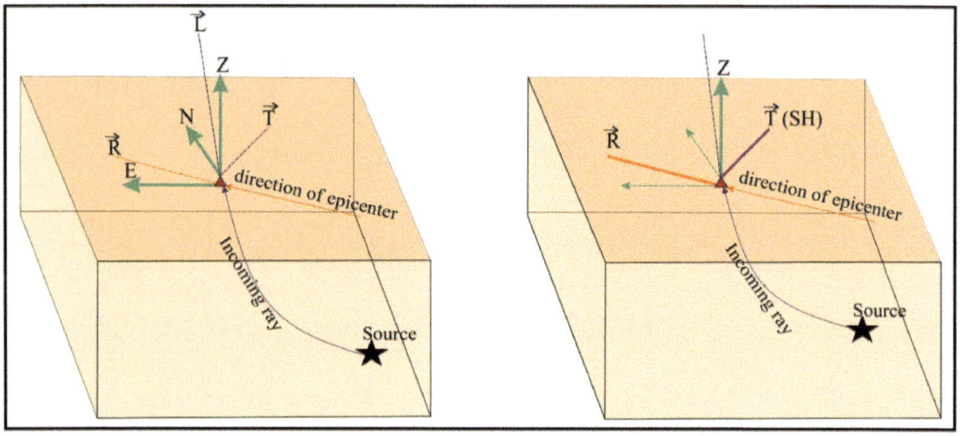

Figure 7.2 Rotation of ZNE coordinate system into ZRT coordinate system for incoming events.

horizontal components of the seismogram into its radial (R) and transverse (T) components. The R and T components are aligned along and perpendicular to the great circle connecting the epicenter and seismograph. The rotation is done about the vertical axis by an angle equal to the back azimuth which converts E-W and N-S components to ZRT coordinate system (Figure 7.2). After rotation, the P wave dominates the vertical component (Z), whereas the Ps converted phases and their multiples are recorded on the radial (SV) component. Rotation of the ZNE system into the ZRT system is based on the theoretical back azimuth of the recorded event (Figure 7.2). Thus, rotation is one of the necessary processing steps to do horizontal components into their right orientation or coordinate system.

Suppose u is the ground vibration motion due to an earthquake. Its components along NS and EW are u_N and u_E respectively and the components along the radial and transverse direction are u_R and u_T.

Rotation of a coordinate system is the multiplication of a rotation matrix with its respective components. It is given by

$$\begin{bmatrix} u_R \\ u_T \end{bmatrix} = \begin{bmatrix} \cos\theta & \sin\theta \\ -\sin\theta & \cos\theta \end{bmatrix} \begin{bmatrix} u_E \\ u_N \end{bmatrix} \quad (7.1)$$

where $\Theta = 3\pi/2-\xi$ is the rotation angle and ξ is the back azimuth at the station.

Equation (7.1) is used to rotate the horizontal component seismogram into ZRT coordinate system.

7.2.2 Filtering

The teleseismic waveforms used for RF analysis are filtered with a low-pass Gaussian filter to remove the high-frequency components of the wave. The Gaussian filters, $G(\omega)$ is expressed as:

$$G(\omega) = \exp\left(-\omega^2 / \left(4a^2\right)\right) \tag{7.2}$$

The Gaussian filter width parameter a controls the frequency content and defines the bandwidth by excluding the high-frequency signals. The filter gains at $\omega = 0$ are unity (a unit-area pulse).

The approximate values of pulse width with frequency are listed in Table 7.1 (http://eqseis.geosc.psu.edu/) by taking $G(\omega) = 0.1$. Low-frequency components of the recorded waveform are examined by a lower value of a, whereas the high-frequency components are examined by a higher value of a (Cassidy, 1992).

Table 7.1 Table shows the values of Gaussian pulse width with frequency at Gaussian width parameter (http://eqseis.geosc.psu.edu/)

Gaussian width parameter a	*Frequency (Hz) at which G(w) = 0.1*	*Approximate Pulse width (s)*
10	4.8	0.50
5	2.4	0.75
2.5	1.2	1.00
2.0	1.0	1.25
1.25	0.6	1.50
1.0	0.5	1.67
0.625	0.3	2.10
0.5	0.24	2.36
0.4	0.2	2.04
0.2	0.1	3.73

Usually, low-frequency waveforms are sufficient to explain the main features of first-order discontinuities like Moho and mid-crustal layers. The vertical resolution of RF depends on the Gaussian width factor (a) used during the computation of RFs.

For an average shear-wave velocity (Vs) 3.6 km s^{-1}, the wavelength (λ) is = 3.71 km corresponds to frequency 0.97 Hz (following the relation $V = n\lambda$). As the minimum vertical resolvable length is $\lambda/4$ (Sheriff and Geldart, 1995), therefore, the layer thickness larger than ~1 km is resolvable.

7.2.3 Restitution

Restitution is defined as the deconvolving of the instrumentation effect from the recorded seismological data. Different instruments have been used for the detection of seismic signals (local, regional and teleseismic signals). Based on natural frequency, seismo-meters are classified as short-period seismometer (0.1–1 s), long-period seismo-meter (0.01–0.1 s) and broad-band seismometer (0.02–1000 s). For receiver function studies, teleseismic wave signals have a frequency in the range of 0.1 Hz–1 Hz. The waveforms recorded by different instruments will be quite dissimilar because of the different instrument responses. Therefore, to isolate the receiver effects of the seismogram recorded at different instruments, the instrument response has to be deconvolved from the receiver functions. The frequency response for the teleseismic signals recorded at broadband seismometers is flat in the useful range of RF analysis. In the short period seismometers, the frequency responses become flatten after applying the restitution process.

7.2.4 Source Equalization/Deconvolution

7.2.4.1 Water-level deconvolution

The water-level deconvolution is the simplest deconvolution method used for the computation of RF (Figures 7.3a, b). It is a frequency domain deconvolution of the radial component with the horizontal component. This is done to avoid numerical instabilities in the deconvolution process. In water-level deconvolution, replacing the small values in the denominator with a fraction of the maximum value of the denominator amplitude helps in avoiding the division by small numbers. This fraction value is termed as the 'water-level parameter', which is chosen by the hit-and-trial method.

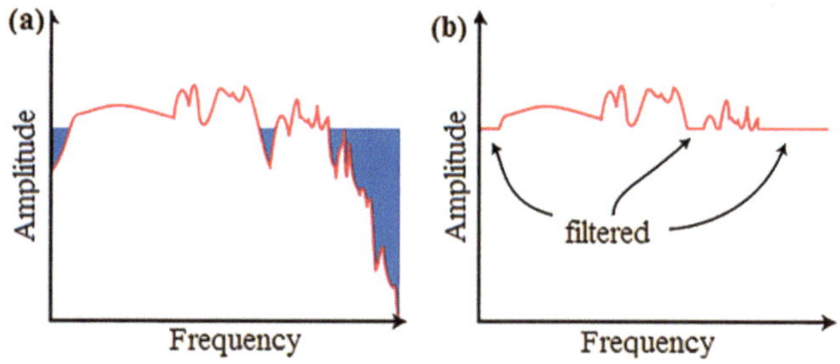

Figure 7.3 Illustration of water-level deconvolution showing (a) amplitude spectrum with water level highlighted by blue colour, (b) amplitude spectrum after water-level deconvolution. (After http://eqseis.geosc.psu.edu)

In practice, a range of *c* values within 1.0–0.0001 is used during deconvolution. The appropriate value of *c* is controlled by the signal-to-noise ratio (SNR) and type of the vertical component seismogram used and is selected by scrutinizing the results of numerous trial runs of water-level fractions.

The acceptable noise levels in the RFs are produced by the lowest water level. Since water level filters can cause distortions of the RF, it is generally attempted to keep the values smaller to obtain better results.

7.2.4.2 Iterative time domain deconvolution

Iterative deconvolution is a time domain deconvolution technique that was initially used for estimation of earthquakes source time function (Kikuchi and Kanamori, 1982) and later adapted for RF computation (Ammon et al., 1990). The detailed mathematical basis of this method is illustrated in Kikuchi and Kanamori (1982). The assumption behind this is that the radial seismogram is a convolution of the vertical component with the Earth's structure. When convolved with the vertical component, the Earth's structure can be obtained by pursuing a timeseries that would approximate its corresponding horizontal component.

RF is a time series that primarily defines the impulse response of the Earth and can be interpreted mathematically as linear combinations of Dirac delta function $\delta(t)$ shifted in time as shown next.

$$\mathrm{RF(t)} = \sum_{i=1}^{n} \mathrm{xi}\,\delta(t - ti) \tag{7.3}$$

where x_i is the amplitude of the Dirac delta function and t_i defines the time with *i*th iteration.

The iterative deconvolution approach is based on the least-squares minimization of the difference between a predicted signal generated by the convolution of an iteratively updated spike train with the vertical component seismogram and the actual horizontal seismogram observed. This method uses a cross-correlation function for the estimation of the time amplitude and lag of the spikes that create the final RF.

7.3 RECEIVER FUNCTION STACKING

7.3.1 Stacking of receiver function

The RFs computed at a station from different events are highly correlated; particularly RFs for earthquakes arriving from a narrow bin of back azimuth (BAZ) and epicentral distance (Δ) are similar. However, some differences in arrival and amplitudes persist among the individual RFs. The stacking of RFs is important to average out these differences. The stacking of RF for events with similar BAZ and Δ enhances SNR. The SNR should go up by a factor of $\sqrt{N}$ if N traces are stacked (Owens et al., 1984). A standard deviation (SD) of ±1 is calculated from the variance of the stacked data and these bounds are used to check the fit of RFs during modelling. Owens et al.(1984) suggested stacking bounds of 20° BAZ and 10° (for the Δ ranges from 45° to 60°) or 15° (over the Δ ranges from 80° to 100°) in the epicentral distance. The processed RFs are stacked with Δ and BAZ bins less than 20° for modeling. At some seismological stations, where BAZ variation is insignificant, the stacking bin is widened for the improvement of the SNR. Stacked RRF calculated from the stacking of individual RRFs is shown in Figure 7.4.

7.3.2 Slant stacking/*H-k* stacking method

RF is an effective technique for extracting information of crust and upper mantle structure. It is primarily used to estimate the thickness of different subsurface layers based on the seismic shear-wave velocities estimated at different depths (Burdick and Langston, 1977; Vinnik, 1977;

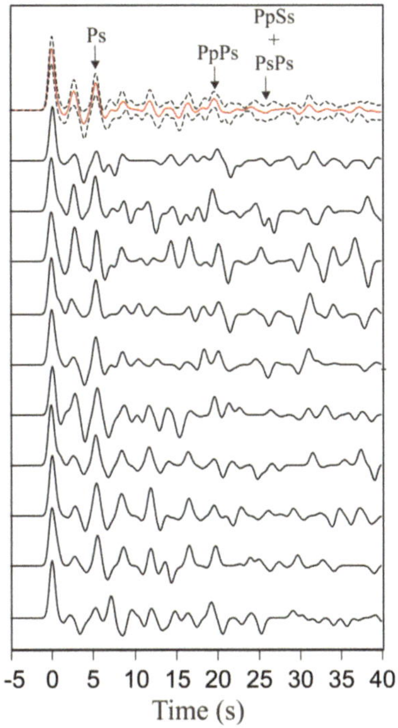

Figure 7.4 Stacking of individual RRFs of teleseismic earthquakes. The individual RFs used for stacking are shown by black waveforms and the stacked RF is shown by red waveform at the top. The waveforms shown by dashes represent ±1 SD bounds.

Langston, 1979; Ammon, 1991). The basis of the RF method is that the incoming near-vertical P wave gets converted into S wave when encounters a velocity discontinuity like Moho discontinuity. A part of this P-to-S (or Ps) converted phase gets reflected that causes reverberated phases or crustal multiples (e.g. PpPs, PsSs + PsPs) (Figures 7.1a, b). Since the S wave travels at a slower velocity than P wave, the converted Ps phases arrive within the P-wave coda following the direct P onset (Pp phase). The difference in velocities between Ps and Pp phases provides a rough estimate of the thickness of the converting layer.

The RF method uses the teleseismic wave for computation and incoming teleseismic wave bears the information of the earthquake source, the path travelled by the seismic wave, instrument response, and response of local Earth's structure beneath the receiver site. The RF is obtained by removing the effects of source, the travel path, and instrument response by deconvolving vertical component seismogram with the radial and tangential components resulting in radial receiver function (RRF) and tangential receiver function (TRF), respectively.

The travel times of converted phases (Ps_1 and Ps_2) and their multiples (P_PP_S, P_PSs, PsP_S) arrive in the coda of P wave can be utilized to measure the thickness (H) of the converting layers and V_P/V_S ratio. The relationship between V_P/V_S and Poisson's ratio, σ, can be defined as:

$$\sigma = 0.5\left[1-\left(k^2-1\right)^{-1}\right]\text{where } k = V_P / V_S \tag{7.4}$$

Let us consider t_1, the differential time between P_S and direct P phase, t_2, the differential time between P_PP_S and P; t_3, the differential time between $P_PS_S + P_SP_S$ and P; then following Zhu and Kanamori (2000), the arrival times t_1, t_2, and t_3 can be expressed as:

$$t_1 = H\left(\sqrt{Vs^{-2} - p^{-2}} - \sqrt{Vp^{-2} - p^{-2}}\right) \tag{7.5}$$

$$t_2 = H\left(\sqrt{Vs^{-2} - p^{-2}} + \sqrt{Vp^{-2} - p^{-2}}\right) \tag{7.6}$$

$$t_3 = 2H\left(\sqrt{Vs^{-2} - p^{-2}}\right) \tag{7.7}$$

where H is the thickness of the converting layer and p is the ray parameter. The radial RFs from different back azimuth distance bins are then stacked to enhance the amplitude of converted and its multiple phases and pick the phases with more clarity. According to equations (7.5), (7.6), and (7.7), it is clear that there is a trade-off between V_P and H. To minimize uncertainty due to this trade-off, the radial RFs of each crustal multiples are then stacked for each station in a gridded H-k domain. The stack function S (H, k) can be defined as:

$$S(H,k) = W_1S(t_1) + W_2S(t_2) - W_3S(t_3) \tag{7.8}$$

where, $S(t_1)$, $S(t_2)$, and $S(t_3)$ are the stack radial RFs at t_1, t_2, and t_3, respectively and the Wj (j = 1, 2, 3) are weighting factors. The values of weighting parameters are based on signal-to-noise ratio of individual phases. To obtain the accurate value of crustal thickness and average V_P/V_S, it is critical to differentiate different crustal phases and their multiples. Usually, the Ps, PpPs, and PpSs + PsPs phases are weighted as 0.7, 0.2, and 0.1, respectively. Based on the energy of Ps, PpPs, and PpSs + PsPs phases, the weighting parameters (W_1, W_2, and W_3) are set as 0.7, 0.2, and 0.1 for converted phases and corresponding two multiple phases, respectively, such that

$$W_1 + W_2 + W_3 = 1.$$

However, these values can be different for different waveforms based on the clarity and signal-to-noise ratio of the individual phases. The optimum value of H and k is considered as those values for which the stacking amplitude is maximum. The uncertainties of crustal thickness and V_P/V_S ratio are calculated by the following relations:

$$\sigma^2{}_H = 2\sigma_s / \left[\partial^2 s / \partial H^2\right] \tag{7.9}$$

$$\sigma^2{}_k = 2\sigma_s / \left[\partial^2 s / \partial \kappa^2\right] \tag{7.10}$$

where σ_s is the estimated variance of $S(H, k)$ from stacking (Zhu and Kanamori, 2000).

7.4 MODELING APPROACH TO DECIPHER CRUST AND UPPER MANTLE STRUCTURE

There are several modeling techniques, for example, time domain-linearized inversion method (Ammon et al., 1990) and the global optimization technique (Shibutani et al., 1996) that are introduced for inversion of RF data to obtain shear-wave velocity structure underneath the recording stations. Many methods are used to solve nonlinear inversion problems like simulated annealing (SA) technique (Holland, 1975, Kirkpatrick et al., 1983) and genetic algorithm (GA) (Stoffa and Sen, 1991). The SA inversion method makes use of control parameters and searches for the best model parameter space, while the GA inversion method uses the initial model space by doing the maximum number of iterations and selects the best-fitted model space, which is less dependent on defined constraints of the model. These search methods are mainly used for the optimization of the misfit function in the finite-dimensional parameter space.

7.4.1 Time-linearized inversion scheme

RF inversion is one of the complex nonlinear geophysical inverse problems. To formulate any geophysical inverse problem, whether it is linear or nonlinear, it is essential to have some possible set of finite-dimensional model parameters and a defined reference model or model space. Each point in a reference model space describes several parameters, for example, physical characteristics of the Earth. To solve the non-uniqueness of RF inversion, Ammon et al. (1990) established a linearized time domain waveform inversion technique that is strongly based on the initial model. The assumption of the time domain inversion technique is to consider the initial velocity model, M_o, which is very close to the true Earth-layered velocity model. The initial velocity model is represented by different model parameters, namely, velocities, thickness, and Poisson's ratio. A grid search approach is followed to search for the best-suitable values of model parameters to have the best-selected model for the crustal structure. The searching approach tends to minimize the difference between the observed and predicted velocity models. The possible solutions of models are obtained by running the inversion for a particular range of preliminary models based on perturbations to the given initial model. Further, the ideal solution is computed by averaging the solutions that fit the data with less misfit error. An example of the time-linearized inversion technique of RFs at a station is shown in Figure 7.5. The inversion technique results provide the estimation of the mid-crustal layer (H_1) and the depth of Moho discontinuity (H_2).

7.4.2 Neighbourhood algorithm

Neighbourhood algorithm is a nonlinear inversion method that represents a derivative-free self-adaptive technique that searches the best-accepted models with suitable data fit in multi-dimensional parameter space (Sambridge, 1999a). It is a random search technique that marks the search of possible sampled model space with a good data misfit of

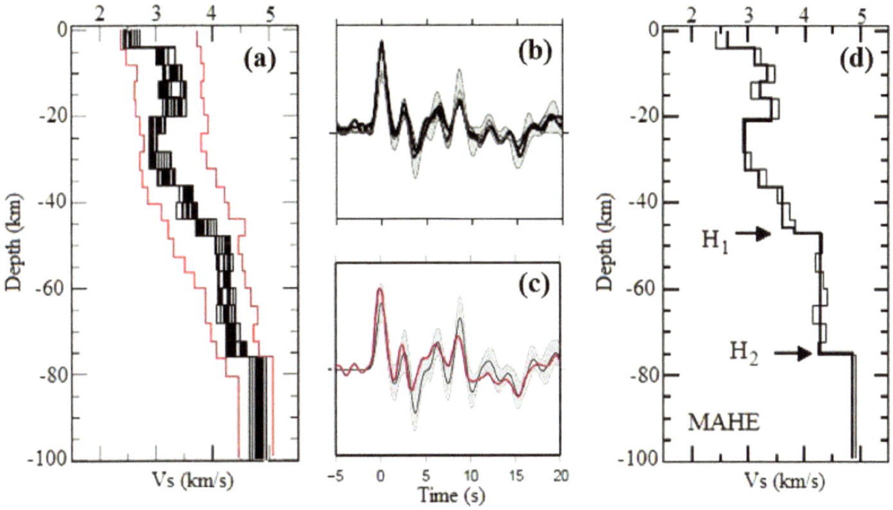

Figure 7.5 Example of inversion results of RFs recorded at a station using time-linearized inversion technique. The inverted velocity models are shown in the left panel (a) as black lines and red lines show the range of starting models. The synthetic RFs corresponding to the range of inverted models in (a) are shown in (b) with their ±SD bound of observed RF. In the lower panel, (c), the synthetic RF (red) and stacked RF (black) are shown with ± SD bounds. In the right panel, (d), the best-fitting inverted (thin black line), as well as final velocity model (thick black line) is shown with the depth of H_I (mid-crustal layer)) and H_2 (Moho discontinuity). (Modified after Hazarika et al., 2014.)

parameter space. This method aims to sample the region of high dimensional model parametric space that has models with suitable data fit. It focuses on the different regions of model space and decreases the misfit function through the neighbourhood sampling defined for the inverse problem. Sambridge (1999a, b) has developed another nonlinear inversion technique called neighbourhood algorithm that uses stochastic sampling to search in the multi-dimensional parameter space. An example of inversion results of RFs recorded at a station using neighbourhood algorithm inversion technique is shown in Figure 7.6. The inversion results infer the depth of Main Himalayan Thrust (H_1) and Moho discontinuity (H_2), respectively.

7.4.3 Voronoi cell

A Voronoi cell is described as a point *V* nearest neighbour region in a parametric space or as a distance norm in model space. The point *V* is

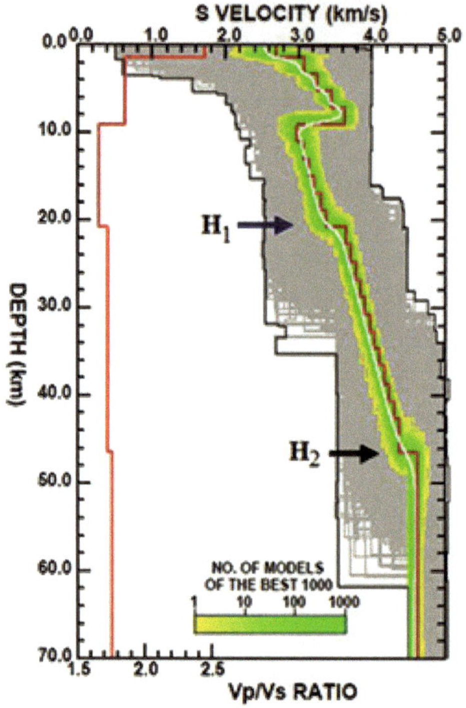

Figure 7.6 Example of inversion results of RFs recorded at a station using neighbourhood algorithm inversion technique. The grey area specifies the range of velocity models (40,020), the green area shows the best 100 models with a minimum error observed and computed RFs. The red line indicates the best fitting and the white line represents the average model. The H_l and H_2 show the depth of Main Himalayan Thrust (MHT) and Moho discontinuity, respectively. (Modified after Hazarika et al., 2017.)

a set of a number of points in a particular model space which can be expressed as:

$$V = \{n_1, n_2 \ldots\ldots\ldots\ldots n_{mp}\}, \text{where } 2 \le m_p \ge \infty$$

and the Voronoi cell can be represented by

$$V(n_i) = \{x \mid \|x - a_j\| \le \|x - b_k\| \text{for } k \ne j \text{ and} (j, k = 1, 2, 3\ldots\ldots, m_p).$$

The randomly distributed points within the finite-dimensional parameter model space are searched with the use of geometrical constructed Voronoi cells (Sambridge et al., 1995). Voronoi cells are found to be developed uniquely from the previous samples and inversely proportional to the distribution of sampling points. The distribution of a different number of sample points with their Voronoi cells is shown in Figure 7.5 in a 2-D space.

Initially, 10 random points (represented as black dots) are distributed in parameterized model space with their drawn Voronoi cells (Figure 7.7a). Further, it generates 10 new random points by doing 10 iterations and the total distributed points ($10 \times 10 = 100$) in a plane is 100 (Figure 7.7b) and the generated randomly distributed points alter the Voronoi

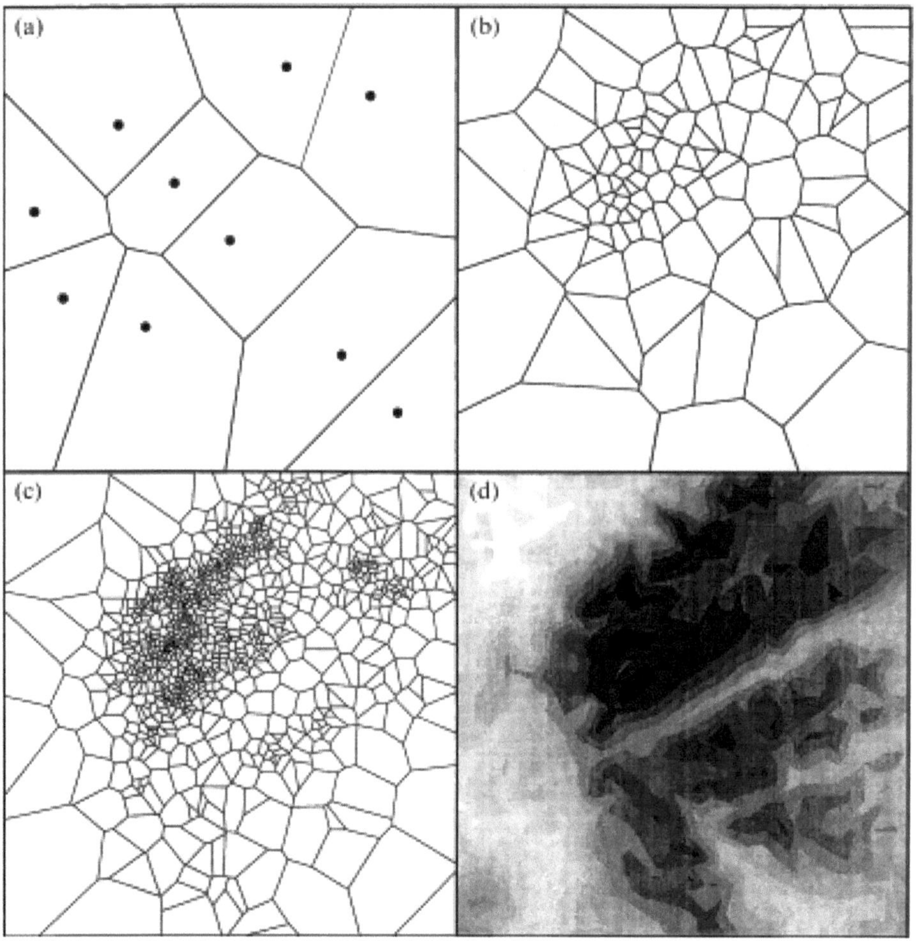

Figure 7.7 Voronoi cells with (a) 10 random sample points, (b) 100 sample points, (c) 1000 sample points, and (d) contours of the test objective function. (Sambridge, 1999a.)

cells locally with their clusters in the model space with the best fit models. In the final step, the 1000 best models are generated by doing 100 iterations 10 times (100 × 10 = 1000) represented in Figure 7.7c, and Voronoi cells are changed locally with their true misfit values and this increases the resolution of the nearest neighbourhood of multi-parameterized model space.

The foundation of the neighbourhood algorithm inversion technique (Sambridge, 1999a) is very simple and unique and based on two simple principles that are stated next:

1. It requires two tuning parameters: one is the numbers of models generated in every iteration (n_s) and the selected number of low misfit models (n_r).
2. It requires the best-fitted models to be evaluated for their relative fit to the data as it forms the use of the only rank of misfit function rather than its numerical value.

7.4.4 Neighbourhood sampling

The nearest neighbourhood inversion method plays a very significant role in finding the solution of nonlinearity and non-uniqueness of RF inversion. The RF inversion is the best data fit procedure that focusses on the estimation of the physical properties of Earth from the analysed RF data by providing the predicted initial model space of the crustal properties. This new nonlinear inversion method is a very simple search behaviour algorithm that requires the following steps:

1. Generation of an initial set of n_s models in defined parameter space (Figure 7.8).
2. Calculation of misfit function for the recently obtained models n_s and determination of the models n_r with the lowest misfit.
3. Generation of new models n_s by searching in the nearest neighbourhood regions called Voronoi cells.
4. Repeating Step 2 until the misfit between the given initial model space and obtained data becomes satisfactory small.

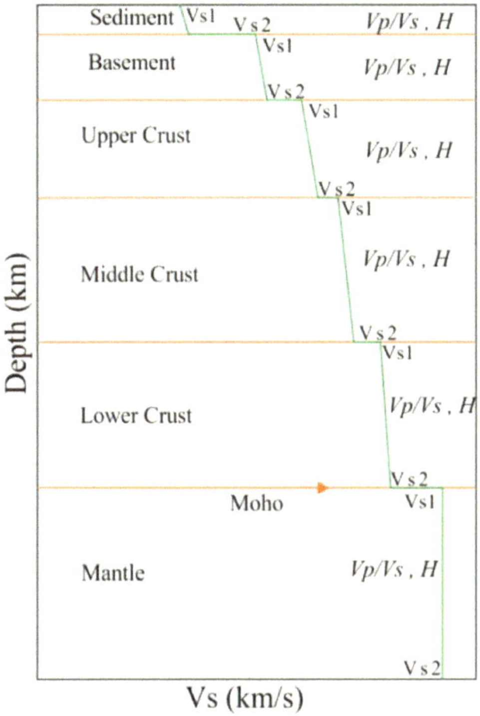

Figure 7.8 Velocity model parameterization for RF inversion indicating six different layers having a thickness of layer (*H*), shear-wave velocity at the top and bottom of the layer (*Vs1* and *Vs2*), and *Vp/Vs* for each layer. The arrow marks the Moho discontinuity.

The neighbourhood algorithm method has successfully formulated the inversion of RF for identifying the crustal shear-wave velocity structure. The Earth's structure is divided into six horizontal layers, viz., 'sediment', 'basement', 'upper crust', 'middle crust' and 'lower crust' and 'mantle' (Figure 7.8).

This finite-dimensional model space is composed of four components on each layer, viz., thickness of layer (*H*), *Vs* (km/s) of the top layer, *Vs* of the bottom layer (km/s), and *Vp/Vs* or Poisson's ratio of the layer. The parameterization uses the large range of Earth models to be represented by multi-dimensional parameter space. The gradient of each layer is defined by its maximum and minimum values that are used as input to a layer. On every parameter of each layer, very loose constraints are applied to obtain the best-fit model for the crustal structure of the Earth.

7.5 CASE STUDY: CRUSTAL THICKNESS AND POISSON'S RATIO IN THE SATLUJ VALLEY, NORTHWEST HIMALAYA

The collision and convergence created large-scale crustal shortening of ~1500 km accommodated mainly by tectonic stacking of upper- and middle-crustal slabs on the northern margin of the Indian plate along several thrust sheets originating from the basal decollement of the Main Himalayan Thrust (MHT) (Yin, 2006; Avouac, 2007). Moreover, collision followed by underthrusting of the Indian plate beneath Asia resulted in large lateral variations in crustal thickness and its composition. In addition, to tracing the depths of the subsurface features, it is essential to investigate the composition of the crust which is an important factor in the understanding of crustal growth in the Himalayas (Gansser, 1964; McKenzie and Sclater, 1971; Yin, 2006).

To interpret the crustal thickness and composition of crust with elastic parameter, Poisson's ratio in the Satluj valley of Northwest Himalaya, a passive source study has been carried out using ten broadband seismological stations (BBS) of Wadia Institute of Himalayan Geology (WIHG), Dehradun, that covers the sub-Himalaya in the south to Tethyan Himalaya in the north (Figure 7.9). Figure 7.9 illustrates the distribution of BBS stations shown by red colour (this study) triangles and blue colour triangles were part of the previous study (Hazarika et al., 2013) in different tectonic units of Satluj valley. Each station is equipped with Taurus Digitizers of dynamic range >138 dB (Make M/S Nanometrics, Canada) and Trillium 240 seismometers of flat velocity response with a frequency range of 0.004–35 Hz. Receiver functions (RFs) are computed using ~3000 teleseismic earthquake waveform data at ten BBS stations by applying the iterative time domain deconvolution technique (Kikuchi and Kanamori, 1982). The *H-k* stacking method (Zhu and Kanamori, 2000) has been applied on RFs to estimate the crustal thickness and Poisson's ratio in Satluj valley.

7.5.1 Crustal thickness along with the seismological profile

The *H-k* stack method of Zhu and Kanamori (2000) has been used to derive the average crustal thickness and *Vp*/*Vs* ratio for each seismological station. Reasonable values for the average P-wave velocity (6.0–6.3 km/s)

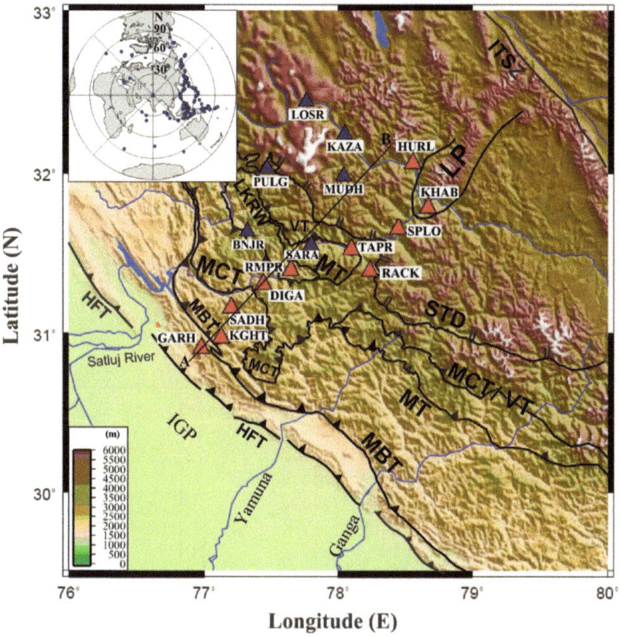

Figure 7.9 A simplified tectonic map of the NW Himalaya. Red triangles are the locations of seismological stations used in the present study whereas blue triangles stations were used for a similar study previously by Hazarika et al. (2013). The background SRTM image shows the topography of the NW Himalaya and surrounding regions with major tectonic features. The inset shows distributions of teleseismic earthquakes (blue dot) used in this study. (After Wadhawan et al., 2017.)

have been obtained from previous studies (Rai et al., 2006; Borah et al., 2015) during *H-k* stacking. The teleseismic earthquake waveforms recorded at ten seismological stations in the studied profile (Figure 7.9) are analysed using the *H-k* stacking method. An example of RFs computed at a station is shown in Figure 7.10a to portray the geometry of Moho discontinuity along with the profile. The remarkable feature in the RFs is two strong positive arrivals after the direct P-wave arrival. Based on a clear record of Ps_1 (at ~3.0 s) and Ps_2 (at ~5.5 s) phases and its crustal multiples (PpP_S and $PpS_S + PsP_S$ or PpS_S), data are selected from a different epicentral distance and back azimuth as shown in Figure 7.10a. The stacked RF is shown as red waveform at the top of Figure 7.10a. The waveform shown by the dashed line indicates upper and lower standard deviation (SD) bound during stacking. The individual and the stacked RFs at the KGHT station show two clear positive phases at ~3 and 5 s and corresponding pairs of multiples at (~9 s and 13 s) and (19 s and 26 s), respectively.

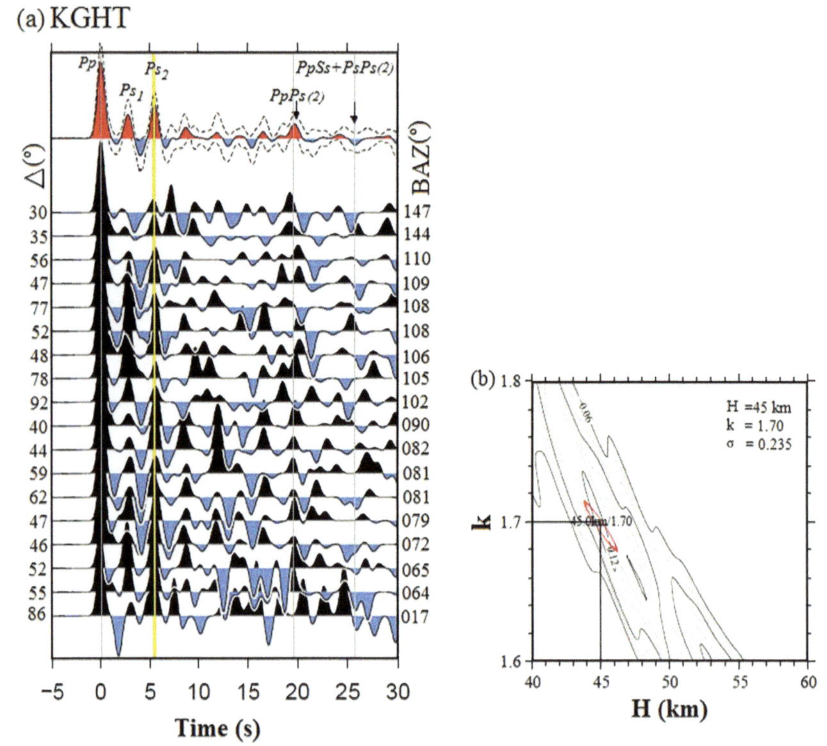

Figure 7.10 Example of *H-k* stacking analysis at KGHT stations for selected RF data. The RFs used for *H-k* stacking analysis are shown in (a) with their stack RF at the top along with +1 SD-bound computed during stacking. The Moho converted Ps_2 phases are marked by a yellow line and its multiples are marked by grey lines at ~19–26 s. The grey line at 0s shows direct *P* arrival (Pp phase). The results of the *H-k* stacking analysis using Ps_2 phase and its multiples are shown in (b). The best-estimated *H* and *k* values are indicated by the centre of the red error ellipse calculated using equations (7.6) and (7.7). The average crustal thickness (*H*) and Poisson's ratio (σ) are obtained as $H = 45$ km, and $\sigma = 0.235$, respectively. (After Wadhawan et al., 2017.)

Similarly, for the rest of the stations, the first positive arrival after the direct P wave or Pp phase indicates a shallow intra-crustal feature which is consistently observed at ~3–3.5 s beneath the stations of sub-, lesser-, and higher Himalaya. On the other hand, this phase is observed at ~5 s or higher at KHAB and HURL stations of the Tethyan Himalaya. The next positive arrival is marked as Moho converted Ps_2 phase. It is observed that there is also a gradual increase in arrival time of Ps_2 phase from ~5.5 s at KGHT station to ~7.5 s at HURL station indicating the dipping nature of the Moho from south to north of the profile in Figure 7.11.

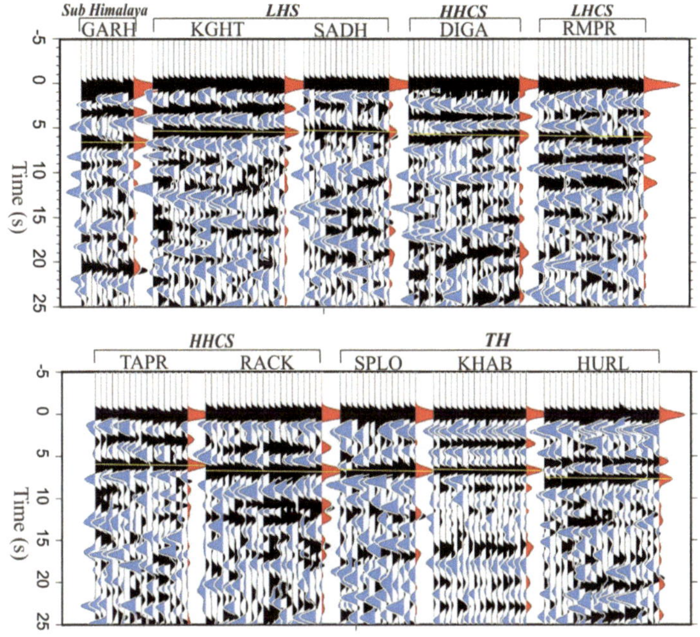

Figure 7.11 Examples of time section plots of radial RFs computed at each station. The plots show well-defined Moho interface (marked by the yellow line). The stack receiver functions at each station are shown in red colour. (After Wadhawan et al., 2017.)

The crustal thickness obtained from *H-k* stacking shows gentle northward dipping of Moho from ~45 km beneath sub-Himalaya (GARH station) to ~48 km beneath the higher Himalaya (TAPR station) as shown in Figure 7.11. The depth of Moho further increases toward the north and reaches ~62 km beneath the Tethyan Himalaya (HURL station) (Hazarika et al., 2017). The results from this profile along with previous studies by Hazarika et al. (2013; 2014) provide complete information of crustal thickness in a long profile starting from sub-Himalaya to Ladakh-Karakoram zone. The gradual increase in crustal thickness from south to north in the NW Himalaya is also reported earlier (Rai et al., 2006; Oreshin et al., 2008; 2011; Chamoli et al., 2011). The variation of crustal thickness beneath the studied profile is comparable with the GPS study (Banerjee and Bürgmann, 2002) that reported the convergence rate of the Indian plate in a northward direction.

By using the Zhu and Kanamori method, the estimated crustal thickness beneath the KGHT station is ~45 km with average V_P/V_S: 1.70 and σ:~0.24. The error ellipse (red) is drawn on the contour plot following equations (7.6) and (7.7), which indicates error estimation of H and V_P/V_S at ~0.9 km and 0.01 respectively (Figure 7.10b). A similar analysis method was applied to the rest of the stations for the estimation of average crustal thickness and Poisson's ratio. The *H-k* stack result at each station is represented in Figure 7.12 and the values of *H*, *k*, and σ are listed in Table 7.2. The range of Poisson's ratios is also marked in the figure.

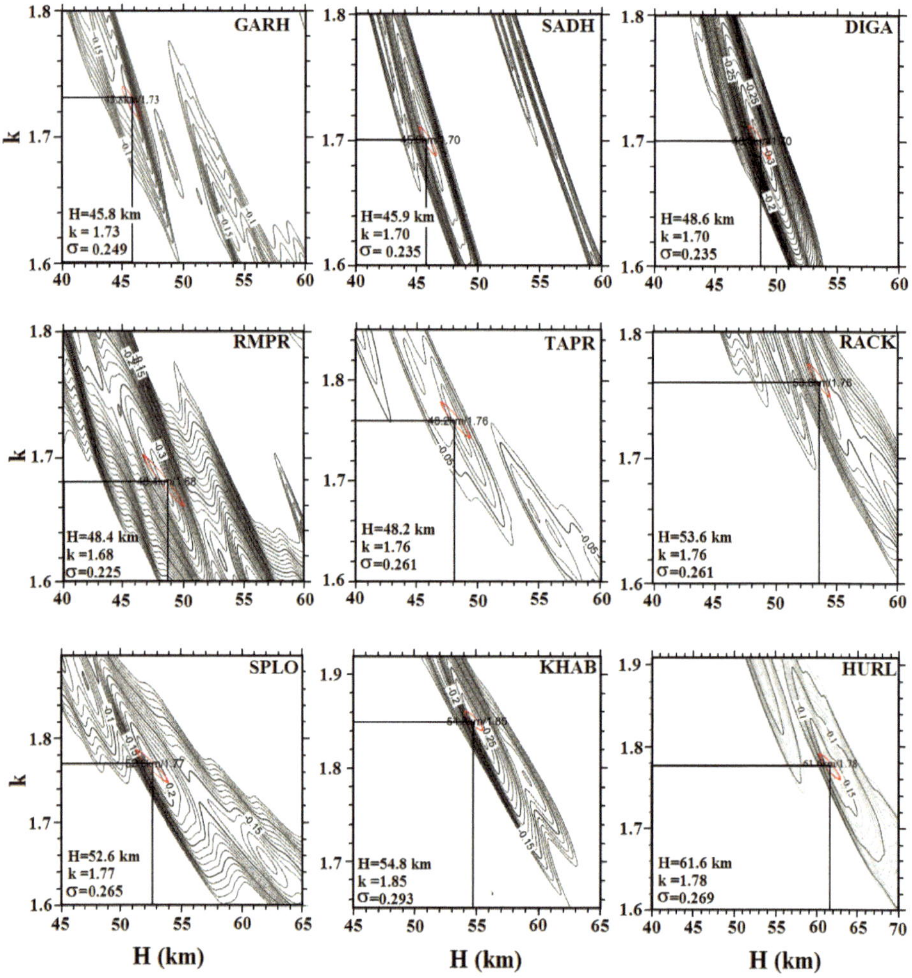

Figure 7.12 Examples of *H-k* stack results at nine stations showing crustal thickness (*H*) and *Vp/Vs* (or *k*) ratio and its corresponding value of Poisson's ratio (σ) at each station from the sub-Himalaya to the Tethyan Himalaya. The best-estimated *H* and *k* values are indicated by the centre of the red error ellipse. (After Wadhawan et al., 2017.)

Table 7.2 Showing the estimated values of crustal thickness *H*, *Vp/Vs* ratio and Poisson's ratio (σ) at each station

Station No	*Station Name with code*	*Latitude (°N)*	*Longitude (°E)*	*Elevation (m)*	*Average crustal thickness (H) with estimated error (km)*	*Average crustal Vp/Vs with estimated error*	*Poisson's ratio (σ)*
1	Garhkal (GARH)	76.97	30.90	1494	46±0.95	1.73±0.011	0.249
2	Kandaghat (KGHT)	77.12	30.97	1229	45±0.9	1.70±0.012	0.235
3	Sadhora (SADH)	77.20	31.15	2111	46±0.6	1.70±0.01	0.235
4	Digadhar (DIGA)	77.43	31.30	1979	49±1.0	1.70±0.009	0.235
5	Rampur (RMPR)	77.64	31.39	920	48±1.2	1.68±0.006	0.225
6	Tapri (TAPR)	78.09	31.52	1679	48±0.95	1.76±0.09	0.261
7	Rackcham (RACK)	78.356	31.393	3129	54±1.0	1.76±0.01	0.261
8	Spillo (SPLO)	78.441	31.650	2390	53±1.2	1.77±0.01	0.265
9	Khab (KHAB)	78.644	31.469	2715	55±1.0	1.85±0.01	0.293
10	Hurling (HURL)	78.551	32.062	3190	62±1.1	1.78±0.01	0.269

7.5.2 Variation of V_P/V_S ratio

The average Poisson's ratio of the crust provides an idea of underlying rock types and mineral composition (Christensen and Fountain, 1975; Zandt and Ammon, 1995); it is tried to correlate with local geology and rock type underneath the seismological stations along with laboratory estimates for common rock types (Christensen, 1996). Poisson's ratio values are sometimes affected by the presence of fluid content and partial melt in the crust (O'Connell and Budiansky, 1974; Owens and Zandt, 1997; Wang et al., 2010). Laboratory estimates suggest that it is low for felsic rocks with high silica content and it is high for mafic rocks (Christensen, 1996).

In this case study, the estimated value of V_P/V_S (Poisson's ratio) shows significant variation from south to north along with the profile. The GARH station located over sub-Himalaya shows close to normal value (V_P/V_S: 1.73; σ: 0.249), whereas these values are low in lesser Himalaya (V_P/V_S: 1.70; σ: 0.225), intermediate beneath the higher Himalaya (V_P/V_S: 1.76; σ ~0.261) and high in the Tethyan Himalaya (V_P/V_S: 1.77–1.85; σ: 0.265–0.293). The results show anomalously low values at the RMPR station (V_P/V_S: 1.68; σ: 0.225) and high values at the KHAB station (V_P/V_S: 1.85; σ: 0.293). The results at the GARH station show a slightly low value of σ compared to the average value (V_P/V_S ~1.75) for the Indian crust (Singh et al. 2015). The main lithology near this seismological station is sandstone and shale (Dubey et al., 1991). The laboratory estimates suggest a wide range of V_P/V_S values (~1.5–1.9) for sandstone, however, this value is minimum for unsaturated sandstone (V_P/V_S ~1.5) (McCormack et al., 1984). On the other hand, shale shows higher values (V_P/V_S ~2.43–2.70) (McCormack et al., 1984; Johnston and Christensen, 1992) also observed higher Vp/Vs values (> 1.90) for shale samples. Considering the previously reported values for sandstone and shale, it can be inferred that sandstone with high quartz content possibly influences the Vp/Vs ratio of the crust beneath the GARH station to a considerable extent.

The low values observed at KGHT and SADH (V_P/V_S ~1.70; σ: 0.235) and local geology suggests a felsic composition of the crust beneath these stations. The main rock types surrounding these stations are sandstone, shale, quartzite, limestone with minor metavolcanics (Dubey and

Bhat, 1991). The low values are possibly due to the presence of quartzite. The Poisson's ratio of quartzite is extremely low (σ: 0.102) (Christensen, 1996). In contrast, V_P/V_S values for limestone ranges between 1.71 and 1.75. A similar low σ value (0.235) is also observed in the Higher Himalayan Crystalline Sequence (HHCS) (DIGA station), which is comprised of high-grade metamorphic gneisses with the thick sequence of amphibolites facies to migmatites paragneisses with minor metabasic rocks (Vannay and Grasemann, 1998; Thakur and Patel, 2012).

The RMPR station located over Larji-KuluRampurWindow (LKRW) of Lesser Himalayan Crystalline Sequence (LHCS) shows the extremely low value (V_P/V_S: 1.68; σ: 0.225). The main lithology in and around this station is basalt undergoing low-grade greenschist metamorphism. Basalt occurs in the association of quartzite (Webb et al., 2007). The V_P/V_S ratio of basalt is normally higher than our observed value at this station (Christensen, 1996). It is interpreted that the occurrence of quartzite in association with basalt is a possible reason for an extremely low value of σ. Moreover, green-schist and amphibolites rocks show a low value of σ (Christensen, 1996).

The TAPR and RACK stations show intermediate values. The observation at the TAPR station is in the conformity with local geology. The common lithology in and around this station is massive granitic Augen gneiss with minor concordant metabasites of LHCS commonly known as 'Wangtu gneiss' (Vannay and Grasemann, 1998). The RACK station is located over HHCS and the lithology near the station mainly consists of pelitic gneisses and granitic gneisses. Xenoliths of pelitic and mafic composition are periodically distributed in these gneissic rocks (Miller et al., 2001; Thakur and Patel, 2012). The results obtained at these stations are similar to laboratory estimation for granite gneisses (Christensen, 1996).

The high values of σ at the SPLO station (V_P/V_S: 1.77; σ: 0.265) located over the Tethyan Himalaya can be explained based on rock composition. Low-grade pelitic rock with mineral assemblage: muscovite-chlorite-biotite-plagioclase-quartz is the main lithology around this station. The estimated values are in good agreement with the published value of low-grade pelitic rock (σ: 0.266; V_P/V_S: 1.77) (Christensen, 1996). On the other hand, the cause of the extremely high value of σ observed at KHAB station (σ: 0.293) and HURL (σ: 0.27) may be due to the presence of metamorphic fluid/partial melt beneath the Tethyan Himalaya. The occurrence of the leucogranite in the vicinity of the station supports the presence of metamorphic fluid beneath the KHAB station. The magnetotelluric (MT) survey (Arora et al., 2007) and seismological studies (Caldwell et al., 2013; Hazarika et al., 2013, 2014; Hajra et al., 2019) also support the presence of fluid/partial melt at mid-crustal depth in the NW Himalaya. The variation of Poisson's ratio along the studied profile suggests that it largely depends on the thickness and sedimentary, metamorphic, and igneous components of the Himalayan belt such as and subducting Indian crust beneath the Eurasian plate. The influence of the Indian crust is maximum at the southern end and it gradually decreases towards the north as the thickness of the crust of the Himalayan belt increases.

7.6 SUMMARY

The results which are obtained from the study of crustal thickness and Poisson's ratio are summarized next:

1. The average crustal thickness gradually increases from ~45 km beneath sub-Himalaya in the south to ~62 km beneath the Tethyan Himalaya suggesting thickening of the crust from south to north.
2. The Poisson's ratio is low in sub- and lesser Himalaya, intermediate in higher Himalaya, and high in the Tethyan Himalaya.
3. The low and intermediate Poisson's ratio values to the south of the South Tibetan detachment is possibly due to local geology and compositions and extremely high Poisson's ratio to the north of South Tibetan detachment is interpreted as due to the presence of fluid/partial melt in the crust.

ACKNOWLEDGEMENT

The authors sincerely thank the director, Wadia Institute of Himalayan Geology (WIHG), Dehradun, India, for providing the necessary support to carry out this work. The technical staff members Mr. Chandan Bora, Mr. Ravinder Singh, and Mr. H.S. Pandey are acknowledged for their support in field work. The author, MW thank the NORSAR, Norway, for providing fellowship under the Indo-Norwegian collaboration program. The Generic Mapping Tool (GMT) is used for the plotting of figures.

REFERENCES

Ammon, C. J., Randall, G. E., & Zandt, G. (1990). On the nonuniqueness of receiver function inversions. *J. Geophys. Res: Solid Earth*, *95*(B10), 15303–15318.

Ammon, C. J. (1991). The isolation of receiver effects from teleseismic P waveforms. *Bull. Seismol. Soc. Am.*, *81*(6), 2504–2510.

Angus, D. A., Wilson, D. C., Sandvol, E., & Ni, J. F. (2006). Lithospheric structure of the Arabian and Eurasian collision zone in eastern Turkey from S-wave receiver functions. *Geophys. J. Int.*, *166*(3), 1335–1346.

Arora, B. R., Unsworth, M. J., Rawat, G. (2007). Deep resistivity structure of the northwest Indian Himalaya and its tectonic implications. *Geophys. Res. Lett.*, *34*(4), L0430. doi:10.1029/2006GL029165.

Avouac, J. P. (2007). Mountain building: From earthquakes to geological deformation. Dynamic processes in extensional and compressional settings. *Treatise on Geophysics*, vol. 6, (pp. 377–439), Amsterdam: Elsevier.

Banerjee, P., & Bürgmann, R. (2002). Convergence across the northwest Himalaya from GPS measurements. *Geophys. Res. Lett.*, *29*(13), 30–1.

Borah, K., Kanna, N., Rai, S. S., & Prakasam, K. S. (2015). Sediment thickness beneath the Indo-Gangetic Plain and Siwalik Himalaya inferred from receiver function modelling. *J. Asian Earth Sci.*, *99*, 41–56.

Bostock, M. G. (1998). Mantle stratigraphy and evolution of the Slave province. *J. Geophys. Res: Solid Earth*, *103*(B9), 21183–21200.

Bostock, M. G., & Rondenay, S. (1999). Migration of scattered teleseismic body waves. *Geophys. J. Int.*, *137*(3), 732–746.

Burdick, L. J., & Langston, C. A. (1977). Modeling crustal structure through the use of converted phases in teleseismic body-wave forms. *Bull. Seismol. Soc. Am.*, *67*(3), 677–691.

Caldwell, W. B., Klemperer, S. L., Lawrence, J. F., & Rai, S. S. (2013). Characterizing the Main Himalayan Thrust in the Garhwal Himalaya, India with receiver function CCP stacking. *Earth Planet. Sci. Lett.*, *367*, 15–27.

Cassidy, J. F. (1992). Numerical experiments in broadband receiver function analysis. *Bull. Seismol. Soc. Am.*, *82*(3), 1453–1474.

Chamoli, A., Pandey, A. K., Dimri, V. P., & Banerjee, P. (2011). Crustal configuration of the northwest Himalaya based on modeling of gravity data. *Pure Appl. Geophys.*, *168*(5), 827–844.

Chevrot, S., Vinnik, L., & Montagner, J. P. (1999). Global-scale analysis of the mantle Pds phases. *J. Geophys. Res: Solid Earth*, *104*(B9), 20203–20219.

Christensen, N. I., & Fountain, D. M. (1975). Constitution of the lower continental crust based on experimental studies of seismic velocities in granulite. *Geol. Soc. Am. Bull.*, *86*(2), 227–236.

Christensen, N. I., & Mooney, W. D. (1995). Seismic velocity structure and composition of the continental crust: A global view. *J. Geophys. Res: Solid Earth*, *100*(B6), 9761–9788.

Christensen, N. I. (1996). Poisson's ratio and crustal seismology. *J. Geophys. Res: Solid Earth*, *101*(B2), 3139–3156.

Clayton, R. W., & Wiggins, R. A. (1976). Source shape estimation and deconvolution of teleseismic body waves. *Geophys. J. Int.*, *47*(1), 151–177.

Dubey, A. K., & Bhat, M. I. (1991). Structural evolution of the Simla area, NW Himalayas: Implications for crustal thickening. *J. Asian Earth Sci.*, *6*(1), 41–53.

Dueker, K. G., & Sheehan, A. F. (1997). Mantle discontinuity structure from midpoint stacks of converted P to S waves across the Yellowstone hotspot track. *J. Geophys. Res: Solid Earth*, *102*(B4), 8313–8327.

Dueker, K. G., & Sheehan, A. F. (1998). Mantle discontinuity structure beneath the Colorado Rocky Mountains and high plains. *J. Geophys. Res: Solid Earth*, *103*(B4), 7153–7169.

Gansser, A. (1964). *Geology of the Himalayas*, vol. 289. New York: Interscience.

Gashawbeza, E. M., Klemperer, S. L., Wilson, C. K., & Miller, E. L. (2008). Nature of the crust beneath northwest Basin and Range province from teleseismic receiver function data. *J. Geophys. Res: Solid Earth*, *113*(B10).

Gilbert, H. J., & Sheehan, A. F. (2004). Images of crustal variations in the intermountain west. *J. Geophys. Res: Solid Earth*, *109*, B03306. doi:10.1029/2003JB002730.

Goldstein, P., Dodge, D., Firpo, M., & Minner, L. (2003). SAC2000: Signal processing and analysis tools for seismologists and engineers. In *The IASPEI International Handbook of Earthquake and Engineering Seismology*, vol. 81, (pp. 1613–1620), London: Academic Press

Gurrola, H., & Minster, J. B. (1998). Thickness estimates of the upper-mantle transition zone from bootstrapped velocity spectrum stacks of receiver functions. *Geophys. J. Int.*, *133*(1), 31–43.

Hajra, S., Hazarika, D., Bankhwal, M., Kundu, A., & Kumar, N. (2019). Average crustal thickness and Poisson's ratio beneath the Kali River Valley, Kumaon Himalaya. *J. Asian Earth Sci.*, 173, 176–188.

Hammersley, J. M., & Handscomb, D. C. (1964). Percolation processes. In *Monte Carlo Methods* (pp. 134–141). Dordrecht, Netherlands: Springer.

Hazarika, D., Kumar, N., & Yadav, D. K. (2013). Crustal thickness and Poisson's ratio variations across the northwest Himalaya and eastern Ladakh. *Acta Geophys.*, *61*(4), 905–922.

Hazarika, D., Sen, K., & Kumar, N. (2014). Characterizing the intracrustal low velocity zone beneath northwest India–Asia collision zone. *Geophys. J. Int.*, *199*(3), 1338–1353.

Hazarika, D., Wadhawan, M., Paul, A., Kumar, N., & Borah, K. (2017). Geometry of the Main Himalayan Thrust and Moho beneath Satluj valley,

northwest Himalaya: Constraints from receiver function analysis. *J. Geophys. Res: Solid Earth*, *122*(4), 2929–2945.
Helffrich, G. (2006). Extended-time multitaper frequency domain cross-correlation receiver-function estimation. *Bull. Seismol. Soc. Am.*, *96*(1), 344–347.
Holland, J. H. (1975). *Adaptation in natural and artificial systems. An introductory analysis with application to biology, control, and artificial intelligence.* Ann Arbor, MI: University of Michigan Press.
Johnston, J. E., & Christensen, N. I. (1992). Shear wave reflectivity, anisotropies, Poisson's ratios, and densities of a southern Appalachian Paleozoic sedimentary sequence. *Tectonophysics*, *210*(1–2), 1–20.
Kennett, B. L. N., & Engdahl, E. R. (1991). Traveltimes for global earthquake location and phase identification. *Geophys. J. Int.*, *105*(2), 429–465.
Kern, H. (1982). Elastic-wave velocity in crustal and mantle rocks at high pressure and temperature: the role of the high-low quartz transition and of dehydration reactions. *Phys. Earth Planet*, *29*(1), 12–23.
Kikuchi, M., & Kanamori, H. (1982). Inversion of complex body waves. *Bull. Seismol. Soc. Am.*, *72*(2), 491–506.
Kind, R., & Vinnik, L. P. (1988). The upper mantle discontinuities underneath the GRF array from P-to-S converted phases. *J. Geophys.*, *62*(13), 147.
Kirkpatrick, S., Gelatt, C. D., & Vecchi, M. P. (1983). Optimization by simulated annealing. *Science*, *220*(4598), 671–680.
Kosarev, G., Kind, R., Sobolev, S. V., Yuan, X., Hanka, W., & Oreshin, S. (1999). Seismic evidence for a detached Indian lithospheric mantle beneath Tibet. *Science*, *283*(5406), 1306–1309.
Langston, C. A. (1979). Structure under Mount Rainier, Washington, inferred from teleseismic body waves. *J. Geophys. Res: Solid Earth*, *84*(B9), 4749–4762.
Levin, V., & Park, J. (1997). P-SH conversions in a flat-layered medium with anisotropy of arbitrary orientation. *Geophys. J. Int.*, *131*(2), 253–266.
Levin, V., & Park, J. (2000). Shear zones in the Proterozoic lithosphere of the Arabian Shield and the nature of the Hales discontinuity. *Tectonophysics*, *323*(3), 131–148.
Ligorría, J. P., & Ammon, C. J. (1999). Iterative deconvolution and receiver-function estimation. *Bull. Seismol. Soc. Am.*, *89*(5), 1395–1400.
Lodge, A., & Helffrich, G. (2006). Depleted swell root beneath the Cape Verde Islands. *Geology*, *34*(6), 449–452.
McCormack, M. D., Dunbar, J. A., & Sharp, W. W. (1984). A case study of stratigraphic interpretation using shear and compressional seismic data. *Geophysics*, 49(5), 509–520.
McKenzie, D., & Sclater, J. G. (1971). The evolution of the Indian Ocean since the Late Cretaceous. *Geophys. J. Int.*, *24*(5), 437–528.
Miller, C., Thöni, M., Frank, W., Grasemann, B., Klötzli, U., Guntli, P., & Draganits, E. (2001). The early Palaeozoic magmatic event in the Northwest

Himalaya, India: Source, tectonic setting and age of emplacement. *Geol. Mag.*, *138*(3), 237–251.
Nábělek, J., Hetényi, G., Vergne, J., Sapkota, S., Kafle, B., Jiang, M., & Huang, B. S. (2009). Underplating in the Himalaya-Tibet collision zone revealed by the Hi-CLIMB experiment, *Science*, *325*(5946), 1371–1374.
Nyblade, A. A., Owens, T. J., Gurrola, H., Ritsema, J., & Langston, C. A. (2000). Seismic evidence for a deep upper mantle thermal anomaly beneath east Africa. *Geology*, *28*(7), 599–602.
O'Connell, R. J., & Budiansky, B. (1974). Seismic velocities in dry and saturated cracked solids. *J. Geophys. Res.*, *79*(35), 5412–5426.
Oreshin, S., Kiselev, S., Vinnik, L., Prakasam, K. S., Rai, S. S., Makeyeva, L., & Savvin, Y. (2008). Crust and mantle beneath western Himalaya, Ladakh and western Tibet from integrated seismic data. *Earth Planet. Sci. Lett.*, *271*(1–4), 75–87.
Oreshin, S. I., Vinnik, L. P., Kiselev, S. G., Rai, S. S., Prakasam, K. S., & Treussov, A. V. (2011). Deep seismic structure of the Indian shield, western Himalaya, Ladakh and Tibet. *Earth Planet. Sci. Lett.*, *307*(3), 415–429.
Owens, T. J., Zandt, G., & Taylor, S. R. (1984). Seismic evidence for an ancient rift beneath the Cumberland Plateau, Tennessee: A detailed analysis of broadband teleseismic P waveforms. *J. Geophys. Res: Solid Earth*, *89*(B9), 7783–7795.
Owens, T. J., & Zandt, G. (1997). Implications of crustal property variations for models of Tibetan plateau evolution. *Nature*, *387*(6628), 37–43.
Park, J., & Levin, V. (2000). Receiver functions from multiple-taper spectral correlation estimates. *Bull. Seismol. Soc. Am.*, *90*(6), 1507–1520.
Park, J., Yuan, H., & Levin, V. (2004). Subduction zone anisotropy beneath Corvallis, Oregon: A serpentinite skid mark of trench-parallel terrane migration? *J. Geophys. Res: Solid Earth*, *109*(B10). doi:10.1029/2003JB002718.
Phinney, R. A. (1964). Structure of the Earth's crust from spectral behavior of long-period body waves. *J. Geophys. Res: Solid Earth*, *69*(14), 2997–3017.
Rai, S. S., Priestley, K., Gaur, V. K., Mitra, S., Singh, M. P., & Searle, M. (2006). Configuration of the Indian Moho beneath the NW Himalaya and Ladakh. *Geophys. Res. Lett.*, *33*(15), L15308. doi:10.1029/2006GL02607.
Saikia, S., Chopra S., Baruah, S., Baidya, P. R., & Singh, U. K., 2016. Crustal imaging of the Northwest Himalaya and its foredeep region from teleseismic events. *Geomatics, Nat. Hazards Risk.*, *7*(4), 1265–1286.
Sambridge, M., Braun, J., & McQueen, H. (1995). Geophysical parametrization and interpolation of irregular data using natural neighbours. *Geophys. J. Int.*, *122*(3), 837–857.
Sambridge, M., & Kennett, B. (1996). Genetic algorithm inversion for receiver functions with application to crust and uppermost mantle structure. *Geophys. Res. Lett.*, *23*(14), 1829–1832.
Sambridge, M. (1999a). Geophysical inversion with a neighbourhood algorithm searching a parameter space—I. Searching a parameter space. *Geophys. J. Int.*, *138*, 479–494.

Sambridge, M. (1999b). Geophysical inversion with a neighbourhood algorithm—II. Appraising the ensemble. *Geophys. J. Int.*, *138*(3), 727–746.

Savage, M. K. (1998). Lower crustal anisotropy or dipping boundaries? Effects on receiver functions and a case study in New Zealand. *J. Geophys. Res: Solid Earth*, *103*(B7), 15069–15087.

Schulte-Pelkum, V., Monsalve, G., Sheehan, A., Pandey, M. R., Sapkota, S., Bilham, R., & Wu, F. (2005). Imaging the Indian subcontinent beneath the Himalaya. *Nature*, *435*(7046), 1222–1225.

Sheehan, A. F., Abets, G. A., Jones, C. H., & Lerner-Lam, A. L. (1995). Crustal thickness variations across the Colorado Rocky. *J. Geophys. Res: Solid Earth*, *100*(B10), 20391–20404.

Sheehan, A. F., Shearer, P. M., Gilbert, H. J., & Dueker, K. G. (2000). Seismic migration processing of P-SV converted phases for mantle discontinuity structure beneath the Snake River plain, western United States. *J. Geophys. Res: Solid Earth*, *105*(B8), 19055–19065.

Sheriff, R. E., & Geldart, L. P. (1995). *Exploration Seismology*. Cambridge, UK: Cambridge University Press.

Shibutani, T., Sambridge, M., & Kennett, B. (1996). Genetic algorithm inversion for receiver functions with application to crust and uppermost mantle structure beneath eastern Australia. *Geophys. Res. Lett.*, *23*(14), 1829–1832.

Stoffa, P. L., & Sen, M. K. (1991). Nonlinear multiparameter optimization using genetic algorithms: Inversion of plane-wave seismograms. *Geophysics*, *56*(11), 1794–1810.

Thakur, S. S., & Patel, S. C. (2012). Mafic and pelitic xenoliths in the Kinnaur Kailash Granite, Baspa River valley, NW Himalaya: Evidence of pre-Himalayan granulite metamorphism followed by cooling event. *J. Asian Earth Sci.*, *56*, 105–117.

Vannay, J. C., & Grasemann, B. (1998). Himalayan inverted metamorphism in the High Himalaya of Kinnaur (NW India): Petrography versus thermobarometry. *SchweizerischeMineralogische und PetrographischeMitteilungen*, *78*, 107–132.

Vernon, F. L., Fletcher, J., Carroll, L., Chave, A., & Sembera, E. (1991). Coherence of seismic body waves from local events as measured by a small-aperture array. *J. Geophys. Res: Solid Earth*, *96*(B7), 11981–11996.

Vinnik, L. P. (1977). Detection of waves converted from P to SV in the mantle. *Phys. Earth Planet*, *15*(1), 39–45.

Vinnik, L. P., & Kosarev, G. L. (1981). Determination of crustal parameters from observations of teleseismic body waves. *Proc. Acad. Sci. USSR*, *261*, 1091–1095.

Wadhawan, M., Hazarika, D., Paul, A., Kumar, N., Thakur, S. S., & Gupta, V. (2017). Crustal thickness and Poisson's ratio variation in the Satluj Valley, Northwest Himalaya. *Himalayan Geology*, *38*(1), 38–48.

Wang, C. Y., Zhu, L., Lou, H., Huang, B. S., Yao, Z., & Luo, X. (2010). Crustal thicknesses and Poisson's ratios in the eastern Tibetan Plateau and their tectonic implications. *J. Geophys. Res: Solid Earth, 115*, B11301. doi:10.1029/2010JB007527.

Webb, A. A. G., Yin, A., Harrison, T. M., Célérier, J., & Burgess, W. P. (2007). The leading edge of the Greater Himalayan Crystalline complex revealed in the NW Indian Himalaya: Implications for the evolution of the Himalayan orogen. *Geology, 35*(10), 955–958.

Yin, A. (2006). Cenozoic tectonic evolution of the Himalayan orogen as constrained by along-strike variation of structural geometry, exhumation history, and foreland sedimentation. *Earth Sci. Rev. 76*(1–2), 1–131.

Zandt, G., & Ammon, C. J. (1995). Continental crust composition constrained by measurements of crustal Poisson's ratio. *Nature, 374*(6518), 152–154.

Zhang, J., & Langston, C. A. (1995). Dipping structure under Dourbes, Belgium, determined by receiver function modeling and inversion. *Bull. Seismol. Soc. Am., 85*(1), 254–268.

Zhu, L., & Kanamori, H. (2000). Moho depth variation in southern California from teleseismic receiver functions. *J. Geophys. Res: Solid Earth, 105*(B2), 2969–2980.

Chapter 8

Lg wave attenuation in special context to Indo-Burman mobile belt of North-East India

Bijit K. Choudhury

8.1 INTRODUCTION

The North-East India (NEI) is categorized as one of the most seismically hazardous regions in India as well as in the global scenario. It stems from the fact that India-Asia collisional tectonics is going on in the northern part of the region with subduction tectonics in the eastern flank along India-Myanmar (Burma), popularly known as Indo-Burman mobile belt (IBMB) (Satyabala, 2003), apart from the intraplate tectonics of the Shillong-Mikir Hills Plateau (Kayal, 2008). These are also the three major seismic sources and tectonic domains that contribute significantly to the high level of seismicity in this region. Although many seismological studies have been carried out in the past with the aim of seismic hazard assessment for NEI, a lacuna remains as the characteristics of the regional seismic wavefield and its implication for the seismic hazard assessment of NEI are not well understood and offer ample scope of study. Lateral variation in crustal heterogeneities has a great bearing on the regional seismic wavefield (Furumura et al., 2014). In various tectonically active regions of the world, it has been observed that the character of regional seismic waves may be grossly modified by the variation in heterogeneous structure from the source area to regional distances, which may cause abnormal amplification of ground motion amplitudes resulting in large scale damage. This may be the consequence of anomalous source, path, and/or site effects (Furumura and Singh, 2002).The impact of the source mechanism and its depth of occurrence on the seismic waves and their attenuation characteristics in low- and high-frequency wavefields is critical in comprehending the strong ground motion in the wake of a destructive earthquake, and it has its own implications in estimating the damage to be triggered by impending earthquakes (Furumura and

DOI: 10.1201/9781003177692-8

Singh, 2002). Lg is one of the strongest regional seismic phases which are affected by the rapid changes in lithospheric heterogeneities (Zhang and Lay, 1995; Kennett and Furumura, 2001). This seismic phase propagates in the continental crust with robust amplitudes that are hardly affected by the source mechanism (Hong, 2010), and it relates very well to the regional crustal structure (Miguel et al., 1992). Therefore, Lg is routinely engaged in studying the crustal attenuation defined by Q_{Lg} that is directly used to gauge the attenuation by the crust (Sargeant and Ottemöller, 2009).

This work is the first of its kind in terms of Lg to comprehend the degree of attenuation in IBMB and its surrounding areas in NEI. In view of the unique characteristics of the Lg waves and their wide applicability with regards to crustal structure, Lg has been employed in this study to work out the crustal quality factor of IBMB such that an insight may be achieved into the seismic hazard posed by IBMB to the NEI due to the effects of the regional wavefield generated by the highly energetic Lg waves. This study will also provide a priori information to investigate further the distribution of lateral crustal heterogeneities in the study region in order to differentiate the paths of high Lg attenuation from the low ones through seismic imaging of Lg attenuation structures. From this study, it is understood that Lg attenuation in the IBMB region is grossly impacted by the frequency of the waves. It is further deduced that the energy of Lg waves is scattered by the heterogeneities of the lithospheric structure of IBMB. The high degree of heterogeneity in the lithosphere underneath IBMB is the consequence of the ongoing subduction tectonics in the study region (Figure 8.1).

8.2 AN OVERVIEW ON LG WAVE AND ITS CHARACTERISTICS

Tectonic stress in the lithosphere has its own bearing on the physical and rheological aspects of its constituent geologic matters, and, thus, has a great bearing on the propagation of seismic waves. This is clearly witnessed in the path attenuation of seismic waves, measured by the crustal quality factor Q. Q may be ascribed as a parameter impacted by the heterogeneities encountered by the seismic waves in their paths of propagation and plays a cardinal role in deciphering the crustal structure. In general, Q is an index of the geothermal gradient and the presence of fluids and/

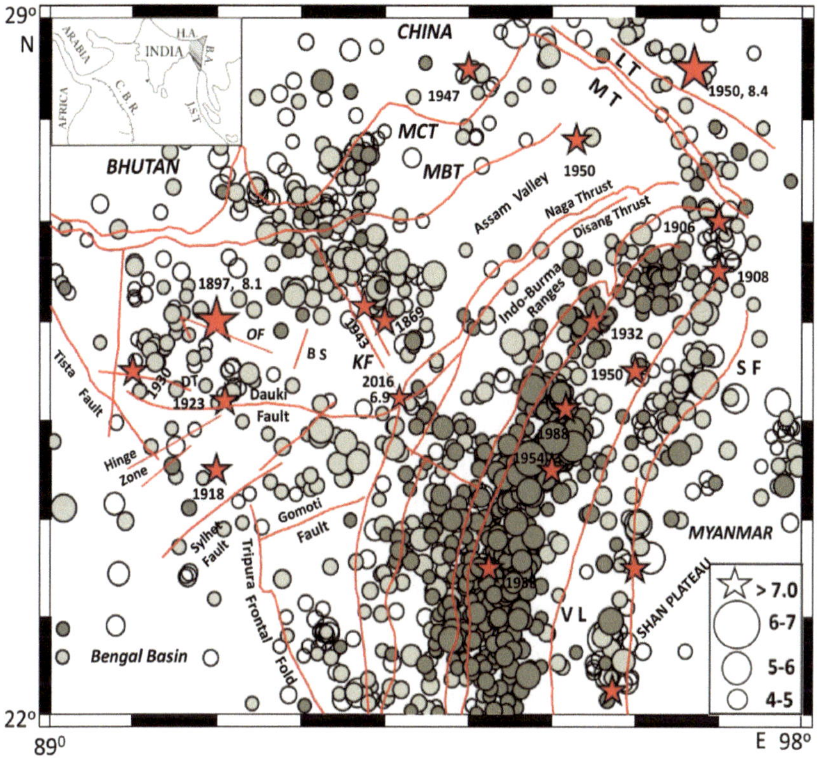

Figure 8.1 Seismicity map of NER of India (Sharma et al., 2017) showing IBR as one of the tectonic entities. The map shows recent seismicity (1990–2016) and major tectonic features of NER of India region. Seismic data are taken from the catalogue of broadband network of CSIR-NEIST-(Jorhat) and CSIR-NGRI-Hyderabad, and the tectonic features are from the GSI (2000). The 2 great (Mw ≥ 8.0) and some 17 large (Mw ≥ 7.0) earthquakes are shown with year of occurrence by stars respectively. Depth ranges are indicated by open circle: 0–30 km; light grey: 30–60 km; dark grey: > 60 km. Abbreviations used for different tectonic features are MBT: Main Boundary Thrust; MCT: Main Central Thrust; DT: Dapsi Thrust; BS: Barapani Shear Zone; LT: Lohit Thrust; MT: Mishmi Thrust; SF: Sagaing Fault; OF: Oldham Fault; KF: Kopili Fault; VL: Volcanic Line. Scale: 1° = 111 Km.

or liquid phases (e.g. partial melt) in the rocks (Sato and Sacks, 1990; Haberland and Rietbrock, 2001; Ojeda and Ottemöller, 2002). Lg is one such seismic phase that has been thoroughly studied by researchers to understand the attenuation characteristics in terms of regional crustal features (e.g. Kennett, 1986; Mitchell and Hwang, 1987; Bowman and Kennett, 1991; Campillo and Plantet, 1991; Mitchell et al., 1998; Baumgardt, 2001; Sargeant and Ottemöller, 2009). Lg waves are shear waves which travels through the crustal channels or guides through multiple reflection and comprises of superposed phases which have undergone post-critical reflections between the crust and the Mohorovicic discontinuity (Moho) (Furumura et al., 2014). The amplitudes of Lg travelling in continental paths are under the influence of the structure of the crust and the physical properties of the geologic material (Mitchell, 1995; Hosseini et al., 2015). Compared to the stable tectonic blocks, Lg is strongly attenuated in tectonically unstable regions (Aki, 1980a, b; Zhou et al., 2011). Lg has a conspicuous arrival from 0.5 to 5.0 Hz frequency and travels with around 2.8–3.7 km/s group velocity. In particular, the attenuation of Lg waves is influenced by fractures, velocity discontinuities, geologic material, crustal heat, and thickness (Ojeda and Ottemöller, 2002). Lg has been numerically modelled and studied as surface waves with high modes (Herrmann and Kijko, 1983a, b). Lg propagation has been profoundly studied by adopting different techniques, including numerical simulation, by many researchers in different regions worldwide (Ruzaikin et al., 1977; Campillo and Plantet, 1991; Campillo et al., 1993; Chazalon et al., 1993; McNamara et al., 1996; Shi et al., 1996; Benj et al., 1997; Rial and Ritzwoller, 1997; Rodgers et al., 1997; Mellors et al., 1999; Calvert et al., 2000; Ottemöller et al., 2002; Ottemöller, 2002; Ojeda and Ottemöller, 2002). The Lg wave attenuation plays a pivotal role in deciphering the age and thickness of the crust, crust-mantle transition, sediment thickness, and so on (Sargeant and Ottemöller, 2009). The magnitudes of regional events can be properly estimated by employing Lg (Hong, 2010). Lg has been exploited well to obtain the source parameters and site response (Atkinson and Mereu, 1992). Lg is significantly attenuated when it encounters an oceanic crust, as well as when it travels in thin continental paths or negotiates distorted Moho geometry (Kennett, 1986; Campillo, 1987; Campillo et al., 1993). Lg gets trapped in basins with thick sediments, causing high dissipation of its energy, which results in abnormally high ground motion amplification (Knopoff et al., 1979; Zhang and Lay, 1995; Shapiro et al., 1996).

It may be noted that Q_{Lg} is governed by the relation given by Fedotov and Boldyrev (1969) and Aki (1980a, b) that establishes its relation to frequency and may be expressed as follows:

$$Q_{Lg}(f) = Q_o (f / f_o)^{\eta}$$

Conventionally, f_o equals 1 Hz, and Q_o is taken as Q at 1 Hz (Shi et al., 1996). Q_0 is known as the coefficient of attenuation, and η defines the degree of dependence on frequency. These parameters differ depending upon the nature of the heterogeneities of the paths of propagation and the degree of present tectonism (Aki, 1980a; Aki, 1980b; Shi et al., 1996). Q_0 is inversely related to attenuation (Q_0^{-1}). η characterizes the dependence of attenuation on the frequency of the seismic wave. This attenuation relationship has a great bearing on having an idea about the present level of tectonism and the type of mechanism by which attenuation of seismic waves occurs. The relationship has a direct connection to the physical characteristics of the geological materials encountered by the seismic waves. It has been established through experimental observation that for tectonically active regions like subduction regions or regions marked by plate collisions, η attains a higher value (1 or more) (Aki, 1980a; Aki, 1980b) relative to the tectonically stable regions of the world (Gupta and Kumar, 2002). Q_0 is generally observed to be low in tectonically unstable regions and basins, which means a higher level of attenuation. In the tectonically active regions, the lithosphere is highly heterogeneous due to the fracturing of the crust, which scatters the wave energy, and hence results in high attenuation. Whereas, liquid phases entrapped in the upper crustal rocks and thick sedimentary basins plummet the Q_0 in basins, causing high attenuation. Physiographic features such as mountainous terrain and uplifted regions, which are composed of old rigid rocks and wherein the basement rocks appear on the surface, show high values of Q_0. Therefore, attenuation of seismic waves is low since the higher elastic terrain traversed by the waves results in less dissipation of the seismic energy.

Aki in 1980 (a, b) hypothesized that in the tectonically unstable regions of the world, seismic waves attain a high value of attenuation around frequencies of 0.5–1 Hz. This difference in attenuation between unstable and stable areas fades away around 25 Hz. From this experiment, it is validated that attenuation (Q^{-1}) proportionately increases with frequency (f) in the order of f^n. Also, it is observed that Q^{-1} varies from place to place around the period of 1 second depending upon the intensity of the current tectonic activity but declines to a lower value at higher frequencies, independent of the current degree of tectonic activity. Similarly, the power of frequency dependence η approaches 1 or may be more for places with a high intensity of tectonic activities. Mukhopadhyay and Tyagi (2007) also advocated that Q_o and η are indicative of the heterogeneity and tectonism of a region. Many authors (e.g., Aki, 1980; Pulli and Aki, 1981; Roecker et al., 1982; Van Eck, 1988; Akinci et al., 1994; Hellweg et al., 1995; Gupta et al., 1998; Biescas et al., 2007) have opined an intricate relationship between η and the varying intensity of tectonism in different regions of the world. A higher value of η has been associated with tectonically unstable regions of the world, whereas a lower value is seen in the relatively less active regions of the world.

The following important observations have been reported worldwide and summarized as follows:

1. In most regions of the world, either η is higher when Q_o or vice versa (Xie and Mitchell, 1990; Shi et al., 1996; Zhu, 2014);
2. Low Q_o and high η are witnessed in seismically active regions; examples may be cited from the southern Sierra Nevada, California (Paul et al., 1996); southern California (Singh and Herrmann, 1983; Frankel et al., 1990; Benz et al., 1997); the subduction zone in Mexico (Domínguez et al., 1997); the Basin and Range Province (Chávez and Priestley, 1986; Benz et al., 1997; Erickson et al., 2004); the East African Rift (Ferdinand, 1998); northern California, the Pacific Northwest, and the mountain states (Erickson et al., 2004); the Tibetan Plateau (McNamara et al., 1996; Xie, 2002; Fan and Lay, 2003b); the southeast basin of the North China region (Zhu and Chen, 2012); and northern Baja California (Domínguez and Rebollar, 1997);
3. High Q_o and low η have been observed in the seismically inactive regions of the world, such as Norway (Sereno et al., 1988); eastern Canada (Chun et al., 1987); central Appalachian, New England Appalachian, Adirondack Mountains, Erie-Ontario Lowlands, coastal basins and highlands, and the Adirondack Mountains in the United States (Shi et al., 1996); central United States and northeast United States (Erickson et al., 2004), and
4. Intermediate values of Q_o and η are observed in some regions of the world where the level of tectonic activity and seismicity are intermediate as compared to the seismically and tectonically active and inactive regions of the world. Some examples are central France

(Campillo et al., 1985); the Great Basin (Chávez and Priestley, 1986); eastern Kazakhstan (Sereno, 1990); the southern part of northeast China (Chun et al., 2009); North Korea (Chun and Henderson, 2009), (data produced from underground nuclear test).

5. Lower Q_0 values do point to the presence of lateral heterogeneity in the crustal structure (Hazarika et al., 2009). The reason being, low Q_0 is a characteristic of tectonically active regions that introduce substantial heterogeneities into the lithosphere, causing scattering of seismic wave energy.

The aforementioned facts demonstrate that the variation of Q is critically dependent upon the tectonic make-up of a region, and draw home the point that high Q_0 values and weak frequency dependence of η are characteristic of tectonically stable regions, whereas tectonically active blocks of the world show low Q_0 and high η values.

8.3 INDO-BURMAN MOBILE BELT (IBMB) OF NORTH EAST INDIA (NEI) CASE STUDY

8.3.1 Tectonics and seismicity of the IBMB

The Indo-Burman arc, or the mobile belt, came into being around early to mid-Eocene times. This westward convex arc is 1,100 km long along which a pronounced oblique subduction of the Indian plate beneath the Burmese platelet occurs (Satyabala, 2003). The mobile belt is described by a folded belt known as the Indo-Burman Ranges (IBR) on the western margin of this belt, and on its east lie the Burmese low lands and the Chindwin basin of Myanmar (Burma). On the eastern flank of the arc, the Sagaing fault runs parallel to IBR. The eastern Himalayas to the north and the IBMB are bridged by the Mishmi massif in the northern periphery of the arc. The Andaman Spreading Ridge (ASR) occurs at the southern extremity of this arc (Kayal, 2008). Starting from the north of the arc, the Naga Hills, followed by the Chin Hills, and then Arakan Yoma, are chains of hills which comprise the arc (Kayal, 2008). Lithologically, flysch and sub-flysch, which dates from Eocene to Oligocene, occurs chiefly combined with Palaeozoic to Mesozoic sediments followed by the sporadic presence of metamorphics with ophiolites in patches (Nandy, 2001). Seismically, this arc has been reckoned among the intercontinental regions of the world, famous for intermediate depth earthquakes and with large concentration of the events defined by a value Bouguer gravity anomaly which is negative. This arc registered ten large earthquakes with magnitudes greater than 7 since 1897, among which the notable one is the 6 August 1988 event of Mw 7.5 (Kayal, 2008).

8.3.2 Data

Broadband seismic data from the CSIR-NEIST seismic network for the period 2008–2012, recorded by 12 seismic stations, was employed in this study. Sensors of Trillium-40 make and 24-bit Reftek 130 recorders have been installed in these seismic stations, which record data in the continuous form at the rate of 100 samples per second. The recording sites are equipped with GPS in order to have a precise time of record of the seismic signals. Lg phases have been picked in the group velocity range of 3.0–3.7 km s^{-1}. The frequency chosen to pick the Lg phase varied from 0.5 to 5 Hz, such as to avoid *Sn* contamination by Lg phases (Ojeda and Ottemöller, 2002). The Jeffrey-Bullen travel timetable was referred for cross-checking the proper identification of Lg. In order to ensure clear picking, an epicentral range of 200–600 km was chosen with the magnitude of the events ranging from 3.1 to 4.6 so as to avoid interference due to rupture complexity caused by larger events (Ottemöller, 2002). Since the thickness of the crust is on average around 35–45 km (Kayal, 2008; Nayak et al., 2008; Borah et al., 2016), only those crustal events with a depth of ≤ 35 km were chosen. The final dataset had events with a depth ranging from 6.79 to 35.2 km, which helped to identify 66 Lg paths from 19 regional events (Figure 8.2), which were used to compute the spectral decay of Lg displacement amplitudes. The location details of these 19 earthquakes have been tabulated in Table 8.1.

8.3.3 Method for determination of average Q_{Lg}

The spectral decay of displacement amplitudes of Lg is computed over a range of frequencies for different distances to compute the Q_{Lg}. This method was propounded by Ottemöller et al. (2002). The following

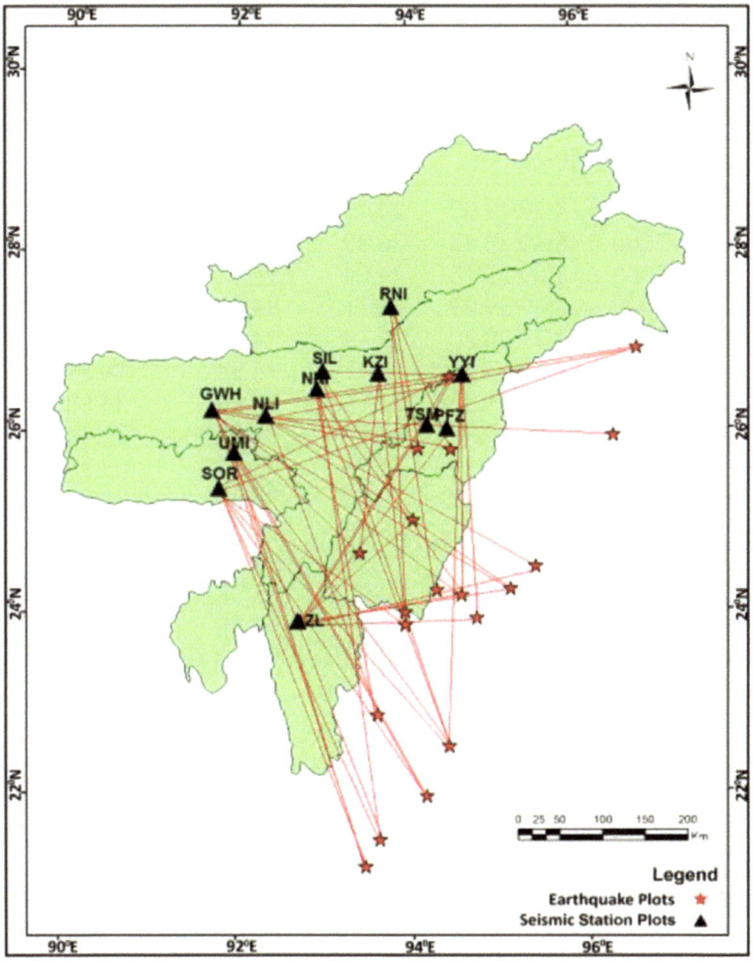

Figure 8.2 A map of North Eastern region (NER) of India showing the 19 IBR earthquakes by a star symbol and the location of 12 seismic stations represented by filled triangles along with 66 source-receiver paths shown by solid lines. The latitudes are shown along the vertical axis (°N) on the left and the longitudes (°E) along the horizontal axis at the bottom.

expression gives the Lg displacement spectrum amplitude after clearing off the instrumental response. In this relation, $A_{kl}(f)$ stands for the event k, l refers to the recording site.

$$A_{kl}(f) = S_k(f) L_l(f) G(R) \exp\left(-\pi f R Q_{Lg}^{-1}/v\right) \tag{8.1}$$

The other terms involved may be denoted as $S_k(f)$ for the source term, $L_l(f)$ refers to local site term, R stands for hypocentral distance, v is

Table 8.1 Hypocentre parameters of 19 events used for Lg attenuation studies in IBR

SL No.	*Event Date*	*Origin Time*	*Latitude (in ° N)*	*Longitude (in ° E)*	*Depth of the hypocenter (in km)*	*Magnitude (M_d)*
1	12-01-2008	14:39:49.00	24.12	95.27	11.50	4.20
2	12-01-2008	16:40:30.45	26.89	96.80	17.30	3.20
3	14-01-2008	13:00:22.18	24.04	94.67	25.40	3.70
4	17-01-2008	13:49:39.03	22.31	94.53	16.80	3.10
5	10-02-2008	08:39:08.20	26.54	94.53	26.26	3.47
6	14-02-2008	01:35:17.28	25.71	94.54	14.99	3.56
7	25-02-2008	09:46:47.32	23.85	93.98	34.68	3.48
8	08-03-2008	16:51:29.54	24.10	94.37	21.44	3.98
9	26-03-2008	04:38:54.23	24.52	93.43	08.30	3.80
10	28-03-2008	06:01:17.49	23.78	94.86	21.20	3.60
11	03-04-2008	22:12:50.17	25.88	96.51	35.20	4.20
12	05-04-2008	18:42:42.00	24.38	95.58	24.30	4.10
13	23-02-2011	21:38:28.27	23.70	93.98	27.30	3.70
14	10-07-2011	00:40:20.12	21.23	93.68	15.80	4.40
15	10-07-2011	23:15:50.00	20.92	93.50	19.50	4.60
16	08-08-2011	18:54:39.49	22.66	93.64	08.80	3.20
17	09-08-2011	17:43:34.28	21.73	94.25	31.80	4.40
18	11-01-2012	22:23:25.02	25.72	94.14	06.79	4.03
19	03-01-2012	06:35:24.08	24.90	94.08	23.04	3.69

Lg group velocity (3.5 km s^{-1}), Q_{Lg} symbolizes Lg quality factor, and $G(R)$ stands for geometrical spreading (Herrmann and Kijko, 1983), which is expressed as:

$$G(R)=\begin{cases} R^{-1} & R \le R_x \\ (R_x R)^{-1/2} & R > R_x \end{cases} \tag{8.2}$$

From the aforementioned relation (8.2), it may be understood that body waves dominate for $R \le Rx$ and for surface waves, $R > Rx$ holds true. Herrmann and Kijko (1983a) and Herrmann and Kijko (1983b) demonstrated that Rx is twice the crustal thickness of the region under study. In this study, Rx has been taken to be 100 km. Shin and Herrmann (1987) gave the observed geometrical spreading for the Lg wave in the time domain $R^{-0.83}$, which equals $R^{-0.5}$ in the frequency domain.

After taking the log of relation (8.1) for distances greater than Rx, the following relation is obtained:

$$\log A_{kl}(f) + 0.5\log(R_x R) = \log S_k(f) + \log L_l(f) - (\pi f R \log(e)/v)/Q_{Lg}^{-1} \quad (8.3)$$

To introduce a constraint for the site term it was required

$$\sum_l \log L_l(f) = 0 \quad (8.4)$$

Relation (8.3) may be expressed in the form of a matrix as

$$d = G\hat{m} \quad (8.5)$$

Here, d refers to the data vector, G stands for kernel matrix, and $\hat{m}$ denotes the estimated model vector. The matrix represented in (8.5) is a function of the source terms, the site terms, and the average inverse quality factor. The dimension of the Kernel matrix G is $M \times N$, where M is the number of all model parameters, and N is the number of travel paths. Following Menke (1989) and Press et al. (1986), the linear overdetermined inverse problem may be worked out using singular value decomposition.

8.4 RESULTS AND DISCUSSIONS

This study clearly points to the fact that Lg attenuation in IBMB and its vicinity region of NEI is governed by a power law (Choudhury, 2020). To compute Q_{Lg} values for frequencies 0.5–5 Hz, the waveforms of 19 regional events from the study area were employed. For each earthquake, log A_{kl} (f) + 0.5 log (R_xR) versus R is plotted from equation (8.3), which represents the equation of a straight line. Therefore, linear regression for each of the plots corresponding to each central frequency (0.5–5 Hz) was performed, and Q_{Lg} was estimated from the slope of the line, which equals (Πf log (e)/v) $Q^{-1}{}_{Lg}$ from equation 8.3. Q for Lg for 0.5–5 Hz frequency was computed as shown in Figure 8.3. A power fit of these Q values gives an average Q_{Lg} value for the study region. The standard errors were calculated using the STDEV algorithm.

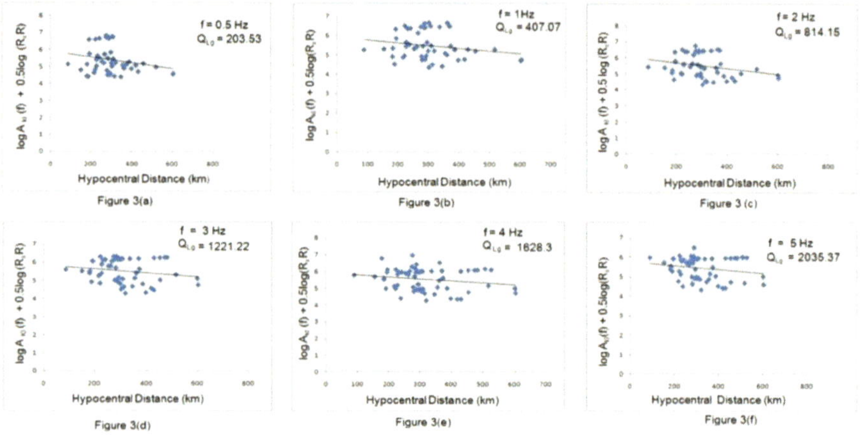

Figure 8.3 Q_{Lg} values for IBR region shown for central frequencies 0.5–5 Hz.

The Q_{Lg} model obtained is shown here by relation (8.6) and in Figure 8.4:

$$Q_{Lg} = (407 \pm 0.00082)\, f^{(1 \pm 0.1864)} \tag{8.6}$$

The corresponding attenuation model ($Q_{Lg}{}^{-1}$) is shown in relation (8.7) and in Figure 8.5:

$$Q_{Lg}^{-1} = 0.002\, f^{-1} \tag{8.7}$$

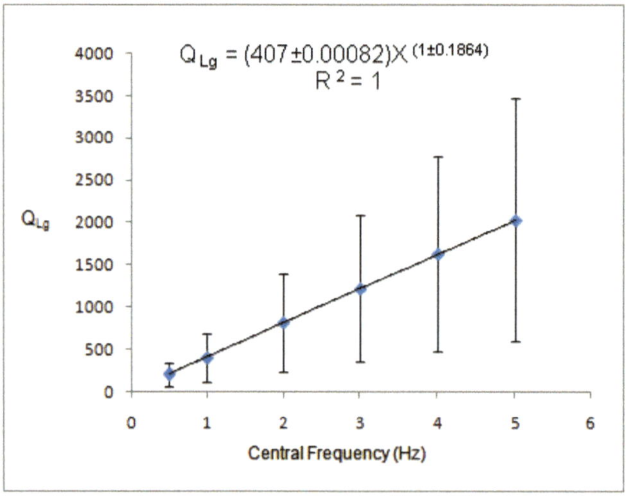

Figure 8.4 Variation of Q_{Lg} for IBR from central frequencies 0.5–5 Hz. (Modified after Choudhury and Duarah, 2017.)

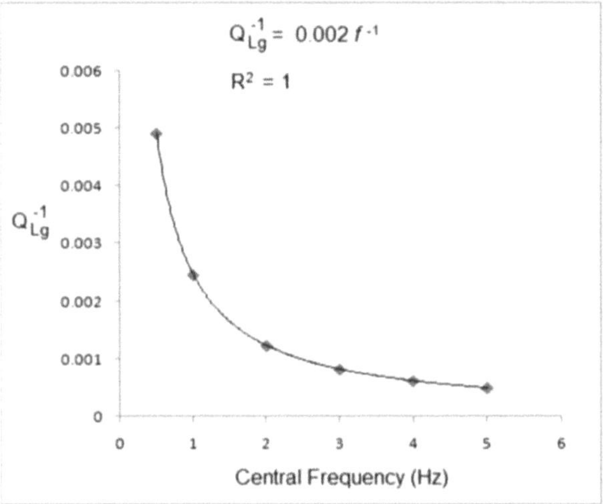

Figure 8.5 Variation of Lg attenution (Q_{Lg}^{-1}) for IBR frequencies 0.5–5 Hz.

As mentioned previously, from Aki's (1980a, b) experiment, we have the peak value of Q_s^{-1} arriving at 0.5 ~1 Hz in tectonically unstable regions. From Figure 8.5, it is conspicuous that peak attenuation is seen at around 0.5 Hz for the study region, which declines gradually towards the high frequencies of the curve. It is a significant observation, which is in concurrence with Aki's observation and points to the fact that IBMB, along with its adjoining areas in NEI, is currently undergoing active tectonic activity. Moreover, the regional average Q_0 value of 407 computed for the study area is parallel to some of the most tectonically unstable regions in the world, for example, Lake Baikal ($Q_o = 407$; $\eta = 0.27$) (Shih et al., 1994) and the Tibetan Plateau ($Q_o = 366$; $\eta = 0.45$) (McNamara et al., 1996). Unlike these regions, η is very high for IBMB ($\eta = 1$), which means a profound influence of frequency content on the attenuation of Lg. Since the scattering mechanism of attenuation is grossly frequency-dependent and not the intrinsic attenuation mechanism (Dainty, 1981; Mayeda et al., 1992), it may be deduced that Lg attenuates in IBMB, dominantly by the scattering mechanism. The reason for the scattering mechanism of attenuation may be attributed to the active subduction mechanism of IBMB, which causes the lithosphere to fracture and introduces lots of heterogeneity to the crust (Choudhury and Duarah, 2017), which causes the scattering of the Lg energy. Thus, this means that it is another proof of the ongoing subduction process in the IBMB region. Compared to some of the major tectonically unstable regions of the world that are also highly attenuating, such as southern Mexico ($Q_{Lg} = 204 f^{0.85}$) (Ottemöller et al., 2002), Central America ($Q_{Lg} = 182 f^{0.84}$) (Ottemöller, 2002), and Colombia ($Q_{Lg} = 230f^{0.6}$) (Ojeda and Ottemöller, 2002), the IBMB with $Q_o = 407$ seems to be moderately attenuating. Given the ongoing subduction tectonics of IBMB, the moderate value of attenuation is a deviation from the anticipated high attenuation in IBMB. The reason for this deviation of attenuation value in IBMB could be the distribution of some seismic stations into the adjoining Assam valley, which has a different degree of lithology and structure. Most likely, there may be some interference in the results due to the paths traversed by Lg in the Assam valley. If IBMB had been considered in isolation with more densification of the seismic network, especially in the southern regions of IBMB, the Q_o would have been much lower or attenuation would have shown a higher value. It may be summarized that the attainment of high frequency-dependent ($\eta = 1$) Lg attenuation model and the peak attenuation ($Q_{Lg}{}^{-1}$) attained at around 0.5 Hz substantiate the fact that IBMB is unequivocally active tectonically. This is also supported by the fact that $Q_o = 407$ is comparable to some of the tectonically unstable regions of the world like Lake Baikal ($Q_0 = 407$), Central Asia ($Q_0 = 398$), the Tibetan Plateau ($Q_0 = 448$), South Asia ($Q_0 = 430$), and the Himalayas ($Q_0 = 484$). The future scope to decipher the actual lateral distribution of heterogeneities in IBMB and its bearing on the seismic hazard posed by this tectonic domain to NEI in terms of attenuation of regional seismic phases like Lg calls for detailed Lg attenuation tomographic imaging of the lithosphere and regional wavefield studies in correlation and corroboration with other geological and geophysical aspects as well.

8.5 CONCLUSION

In this study, the Lg attenuation model that has been developed shows a high degree of frequency dependence ($\eta = 1$), which indicates the fact that Lg attenuation is highly frequency-dependent in the IBMB region. It is well established that the scattering mechanism is more frequency-dependent as compared to the intrinsic mechanism of attenuation. Therefore, it may also be concluded that the Lg attenuation in the study region occurs by mostly scattering attenuation mechanism. There is a valid reason for attenuation to occur by a scattering mechanism, given the fact that IBMB is a subduction zone. The ongoing subduction process causes fracturing of the lithosphere, which scatters the Lg energy.

But the coefficient of attenuation (Q_o = 407) rather points towards a moderately attenuating lithosphere compared to other tectonically active regions of the world. This may be attributed to the fact that some constraints are imposed on the results by the Lg paths which traverse the terrane outside the IBMB region. The structure of the lithosphere and the lithology of the Lg paths, which are outside the purview of IBMB, must have interfered with the values of Q_o. Otherwise, if only the Lg paths traversing within IBMB region were taken into account; with a densified seismic network within IBMB, then a higher value of Lg attenuation could have been expected. The model of Lg attenuation developed through this study will be useful in estimating the ground motion to be triggered by an impending earthquake in the IBMB region. This will contribute to the adoption of safe civil engineering practices for a more rational seismic disaster management and mitigation exercise.

ACKNOWLEDGEMENT

The original work on the case study was carried out by the author using the seismological data from the CSIR-NEIST, Jorhat seismic network. Thus, the author is highly obliged to the honourable director of CSIR-NEIST, Jorhat, for giving approval to carry out the work and for providing the necessary seismological data to accomplish the work successfully. The author is also thankful to Dr. Saurabh Baruah, Chief Scientist, CSIR-NEIST, Jorhat and Prof. (Dr.) Bhagawat Pran Duarah, Dept. of Geological Sciences, Gauhati University, Guwahati, for their invaluable suggestions given while pursuing the original work on attenuation of Lg in IBMB of NEI. Special thanks are due to Mr. Goutam Kashyap Boruah for the help rendered in the preparation of Figure 8.2.

REFERENCES

Aki, K.T. (1980a). Attenuation of shear-waves in the lithosphere from frequencies from 0.05 to 25 Hz. *Phys. Earth Planet. Inter.*, 21, 50–60.

Aki, K.T. (1980b). Scattering and attenuation of shear-waves in the lithosphere. *J.Geophys. Res.*, 85, 6496–6504.

Akinci, A., Taktak, A.G. and Ergintav, S. (1994). Attenuation of coda waves in Western Anatolia. *Phys. Earth Planet. Inter.*, 87, 155–165.

Atkinson, G.M. and Mereu, R.F. (1992). The shape of ground motion attenuation curves in southeastern Canada. *Bull. Seismol. Soc. Am.*, 82, 2014–2031.
Baumgardt, D.R. (2001). Sedimentary basins and the blockage of Lg wave propagation in the continents. *Pure Appl. Geophys.*, 158, 1207–1250.
Benz, H.M., Frankel, A. and Boore, D.M. (1997). Regional Lg attenuation for the continental United States. *Bull. Seismol. Soc. Am.*, 87, 606–619.
Biescas, B., Rivera J., Zapata, A.J. (2007). Seismic attenuation of coda waves in the eastern region of Cuba. *Tectonophysics*, 429, 99–109.
Borah, K., Bora, D.K., Goyal, A. and Kumar, R. (2016). Crustal structure beneath northeast India inferred from receiver function modeling. *Phys. Earth Planet. Inter.*, 258, 15–27.
Bowman, J.R. and Kennett, B.L.N. (1991). Propagation of Lg waves in the North Australian craton: Influence of crustal velocity gradients. *Bull. Seismol. Soc. Am.*, 81, 592–610.
Calvert, A., Sandvol, E., Seber, D., Barazangi, M., Vidal, F., Alguacil, G. and Jabour, N. (2000). Propagation of regional seismic phases (Lg and Sn) and Pn velocity structure along the Africa Ibeia plate boundary zone: Tectonic implications. *Geophys. J. Int.*, 142, 384–408.
Campillo, M. (1987). Lg wave propagation in a laterally varying crust and the distribution of the apparent quality factor in Central France. *J. Geophys. Res.*, 92, 12604–12614.
Campillo, M. and Plantet, J.L. (1991). Frequency dependence and spatial distribution of seismic attenuation in France: Experimental results and possible interpretations. *Phys. Earth Planet. Inter.*, 67, 48–64.
Campillo, M., Feignier, B., Bouchon, M. and Béthoux, N. (1993). Attenuation of crustal waves across the Alpine Range. *J. Geophys. Res.*, 98, 1987–1996.
Campillo, M., Plantet, L. and Bouchon, M. (1985). Frequency-dependent attenuation in the crust beneath central France from Lg waves: Data analysis and numerical modelling. *Bull. Seismol. Soc. Am.*, 75, 1395–1411.
Chávez, D.E. and Priestley, K.F. (1986). Measurement of frequency dependent Lg attenuation in the Great Britain. *Geophys. Res. Lett.*, 16, 551–554.
Chazalon, A., Campillo, M., Gibson, R. and Carreno, E. (1993). Crustal wave propagation anomaly across the Pyrenean Range: Comparison between observations and numerical simulations. *Geophys. J. Int.*, 115, 829–838.
Choudhury, B.K and Duarah, B.P. (2017). A comparative study of Lg wave attenuation in the Arunachal Himalaya and Mishmi Massif and the Indo-Burman Ranges of northeast India. *IOSR-JAGG*, 5(5), 14–20.
Choudhury, B.K. (2020). Attenuation of regional seismic Lg waves and its spectral characteristics in and around the Indo-Burman Ranges of North Eastern region of India. *Acta Geod Geophys.*, 55(1), 133–150.
Chun, K.-Y. and Henderson, G.A. (2009). Lg attenuation near the North Korean border with China, Part II: Model development from the 2006 nuclear explosion in North Korea. *Bull. Seismol. Soc. Am.*, 99, 3030–3038.

Chun, K.-Y., Wu, Y. and Henderson, G.A. (2009). Lg attenuation near the north Korean border with China, Part 1: Model development from regional earthquake sources, *Bull. Seismol. Soc. Am.*, 99, 3021–3029.

Chun, K-Y., West, G.F., Kokoski, R.J. and Samson, C. (1987). A novel technique for measuring Lg attenuation. Results from Eastern Canada between 1 to 10 Hz. *Bull. Seismol. Soc. Am.*, 77, 398–419.

Dainty, A. (1981). A scattering model to explain seismic Q observations in the lithosphere between 1 and 30 Hz. *Geophys. Res. Lett.*, 8, 1126–1128.

Domínguez, T. and Rebollar, C.J. (1997). Regional variations of seismic attenuation from coda and Lg waves in northern Baja California. *J. Geophys. Res.*, 102, 15259–15268.

Domínguez, T., Rebollar, C.J. and Castro, R.R. (1997). Regional variations of seismic attenuation of Lg waves in southern Mexico. *J. Geophys. Res.*, 102, 27501–27509.

Erickson, D., McNamara, D.E. and Benz, H. (2004). Frequency-dependent Lg Q within the continental United States. *Bull. Seismol. Soc. Am.*, 94,1630–1643.

Fan, G.-W. and Lay, T. (2003b). Strong Lg attenuation in the Tibetan Plateau. *Bull. Seismol. Soc. Am.*, 93, 2264–2272.

Fedotov, S.A. and Boldyrev, S.A. (1969). Frequency dependence of the body-wave absorption in the crust and the upper mantle of the Kuril-island chain. *Izv. Acad. Sci. USSR Solid Earth*, 1, 553–562.

Ferdinand, R.W. (1998). Average attenuation of 0.7-5.0 Hz Lg waves and magnitude scale determination for the region bounding the western-branch of the East African Rift. *Geophys. J. Int.*, 134, 818–830.

Frankel, A., McGarr, A., Bicknell, J., Mori, J., Seeber, L. and Cranswick, E. (1990). Attenuation of high-frequency shear waves in the crust: Measurement from New York State, South Africa and Southern California. *J. Geophys. Res.*, 95, 17441–17457.

Furumura T., Hong, T.K. and Kennett, B.L.N. (2014). Lg wave propagation in the area around Japan: Observations and simulations. *Prog. in Earth and Planet. Sci.*, 1, 10.

Furumura, T. and Singh, S.K. (2002). Regional wave propagation from Mexican subduction zone earthquakes: The attenuation functions for interplate and inslab events. *Bull. Seismol. Soc. Am.*, 92(6), 2110–2125.

Gupta, S.C. and Kumar, A. (2002). Seismic wave attenuation characteristics of three Indian regions: A comparative study. *Curr. Sci.*, 82(4), 407–413.

Gupta, S.C., Teotia, S.S., Rai, S.S. and Gautom, N. (1998). Coda Q estimates in the Koyna region, India. *Pure Appl. Geophys.*, 153, 713–731.

Haberland, C. and Rietbrock, A. (2001). Attenuation tomography in the western central Andes: A detailed insight into the structure of a magmatic arc. *J. Geophys. Res.*, 106, 11151–11167.

Hazarika, D., Baruah, S. and Gogoi, N.K. (2009). Attenuation of coda waves in the northeastern region of India. *J. Seismol.*, 13, 141–160.

Hellweg, M., Spudich, P., Fletcher, J.B. and Baker, L.M. (1995). Stability of coda Q in the region of Parkfield, California: View from the U.S. Geological Survey Parkfield Dense Seismograph Array. *J. Geophys. Res.*, 100, 2089–2102.

Herrmann, R.B. and Kijko, A. (1983a). Modeling some empirical vertical component Lg relations. *Bull. Seismol. Soc. Am.*, 73, 157–171.

Herrmann, R.B. and Kijko, A. (1983b). Short-period Lg magnitudes: Instrument, attenuation and source effects. *Bull. Seismol. Soc. Am.*, 73, 1835–1850.

Hong, T-K. (2010). Lg Attenuation in a region with both continental and oceanic environments. *Bull. Seismol. Soc. Am.*, 100(2), 851–858.

Hosseini, M., Pezeshk, S., Soltani, A.H. and Chapman, M. (2015). Investigation of attenuation of the Lg-wave amplitude in the Caribbean region. *Bull. Seismol. Soc. Am.*, 105(2a), 734–744. doi:10.1785/0120140006.

Kayal, J.R. (2008). *Microearthquake Seismology and Seismotectonics of South Asia*. New Delhi: Capital Publishing Company.

Kennett, B.L.N. (1986). Lg waves and structural boundaries. *Bull. Seismol. Soc. Am.*, 76, 1133–1141.

Kennett, B.L.N. and Furumura, T. (2001). Regional phases in continental and oceanic environments. *Geophys. J. Int.*, 146(2), 562–568.

Knopoff, L., Mitchel, R.G., Kausel, E.G. and Schwav, F. (1979). A search for the oceanic Lg phase. *Geophys. J. Int.*, 56, 211–218.

Mayeda, K., Koyanagi, S., Hoshiba, M., Aki, K. and Zeng, Y. (1992). A comparative study of scattering, intrinsic, and coda Q^{-1} for Hawaii, Long Valley and Central California between 1.5 and 15.0 Hz. *J. Geophys. Res.*, 97, 6643–6659.

McNamara, D.E., Owens, T.J. and Walter, W.R. (1996). Propagation characteristics of Lg across the Tibetan Plateau. *Bull. Seismol. Soc. Am.*, 86, 457–469.

Mellors, R.J., Camp, V.E., Vernon, F.L., Al-Amri, A.M.S. and Chalib, A. (1999). Regional waveform propagation in the Arabian Peninsula. *J. Geophys. Res.*, 104, 20221–20235.

Menke, W. (1989). *Geophysical Data Analysis: Discrete Inverse Theory*. California: Academic Press.

Miguel, F. De., Ibáñez, J.M., Alguacil, G., Canas, J.A., Vidal, F., Morales, J., Peña, J.A., Posadas, A.M. and Luzón, F. (1992). 1-18 Hz Lg attenuation in the Granada Basin (southern Spain). *Geophys. J. Int.*, 111, 270–280.

Mitchell, B.J., Baqer, S., Akinci, A. and Cong, L. (1998). Lg coda Q in Australia and its relation to crustal structure and evolution. *Pure Appl. Geophys.*, 153, 639–653.

Mitchell, B.J. and Hwang, H.J. (1987). Effect of low Q sediments and crustal Q on Lg attenuation in the United States. *Bull. Seismol. Soc. Am.*, 77, 1197–1210.

Mitchell, B.J. (1995). Anelastic structure and evolution of the continental crust and upper mantle from seismic surface wave attenuation. *Rev. Geophys.*, 33, 441–462.

Mukhopadhyay, S. and Tyagi, C. (2007). Lapse time and frequency-dependent attenuation characteristics of coda waves in Northwestern Himalayas. *J. Seismol.*, 11(2), 149–158.

Nandy, D.R. (2001). *Geodynamics of Northeastern India and the Adjoining Region*. Calcutta: ABC Publications.

Nayak, G.K., Rao, V.K., Rambabu, H.V. and Kayal, J.R. (2008). Pop-up tectonics of the Shillong Plateau in the great 1897 earthquake (Ms 8.7): Insights from the gravity in conjunction with the recent seismological results. *Tectonics*, 27, 1–8.

Ojeda, A. and Ottemöller, L. (2002). QLg tomography in Colombia. *Phys. Earth Planet. Inter.*, 130, 253–270.

Ottemöller, L. (2002). Lg wave Q tomography in Central America. *Geophys. J. Int.*, 150, 295–302.

Ottemöller, L., Shapiro, N.M., Singh, S.K. and Pacheco, J.F. (2002). Lateral variation of Lg wave propagation in southern Mexico. *J. Geophys. Res.*, 107(B1).

Paul, A., Jongmans, D., Campillo, M., Malin, P. and Baumont, D. (1996). Amplitude of regional seismic phases in relation to crustal structure of the Sierra Nevada, California. *J. Geophys. Res.*, 101, 25243–25254.

Press, W.H, Flannery B.P., Teukolsky, S.A. and Vetterling, W.T. (1986). *Numerical Recipes*. New York: Cambridge University Press.

Pulli, J.J. and Aki, K. (1981). Attenuation of seismic waves in the lithosphere: Comparison of active and stable areas. In: J.E. Beavers (Ed.), *Earthquakes and Earthquake Engineering: The Eastern United States*. Ann Arbor, pp. 129–141.

Rial, J.A. and Ritzwoller, M.H. (1997). Characteristics of Lg propagation across South America. *Geophys. J. Int.*, 131, 401–408.

Rodgers, A.J., Ni, J.F. and Hearn, T.M. (1997). Propagation characteristics of short-period Sn and Lg in the Middle East. *Bull. Seismol. Soc. Am.*, 87, 396–413.

Roecker, S.W., Tucker, B., King, J. and Hatzfeld, D. (1982). Estimates of Q in Central Asia as a function of frequency and depth using the coda of locally recorded earthquakes. *Bull. Seismol. Soc. Am.*, 72, 129–149.

Ruzaikin, A., Nersesov, I.L. and Khalturin, V.I. (1977). Propagation of Lg and lateral variations in the crustal structure of Asia. *J. Geophys. Res.*, 82, 307–316.

Sargeant, S. and Ottemöller, L. (2009). Lg wave attenuation in Britain. *Geophys. J. Int.*, 179, 1593–1606.

Sato, H. and Sacks, I.S. (1990). Magma generation in the upper mantle inferred from seismic measurements in peridotite at high pressure and temperature. In: M.P. Ryan (Ed.), *Magma Transport and Storage*. New York: Wiley, pp. 277–292.

Satyabala, S.P. (2003). Oblique plate convergence in the Indo-Burma (Myanmar) subduction region. *Pure Appl. Geophys.*, 160, 1611–1650.

Sereno, T.J. (1990). Frequency-dependent attenuation in eastern Kazakhstan and implications for seismic detection thresholds in the Soviet Union. *Bull. Seismol. Soc. Am.*, 80, 2089–2105.

Sereno, T.J., Bratt, S.R. and Bache, T.C. (1988). Simultaneous inversion of regional wave spectra for attenuation and seismic moment in Scandinavia. *J. Geophys. Res.*, 93, 2019–2035.

Shapiro, N., Béthoux, N., Campillo, M. and Paul, A. (1996). Regional seismic phases across the Ligurian Sea: Lg blockage and oceanic propagation. *Phys. Earth Planet. Inter.*, 93, 257–268.

Sharma, A., Baruah, S., Piccinni, D., Saikia, S., Phukan, M.K., Chetia, M. and Kayal, J.R. (2017). Crustal seismic anisotropy beneath Shillong plateau – Assam valley in North East India: Shear-wave splitting analysis using local earthquakes. *Tectonophysics*, 717, 425–432.

Shi, J., Kim, W.-Y. and Richards, P.G. (1996). Variability of crustal attenuation in the northeastern United States from Lg waves. *J. Geophys. Res.*, 101, 25231–25242.

Shih, X.R., Chun, K.Y. and Zhu, T. (1994). Attenuation of 1-6 s Lg waves in Eurasia. *J. Geophys. Res.*, 99, 23863–23875.

Shin, T.-C. and Herrmann, R.B. (1987). Lg attenuation and source studies using 1982. Miramichi data. *Bull. Seismol. Soc. Am.*, 77, 384–397.

Singh, S. and Herrmann, R. (1983). Regionalization of crustal coda Q in the continental United States. *J. Geophys. Res.*, 88, 527–538.

Van Eck, T. (1988). Attenuation of coda wave in the Dead Sea region. *Bull. Seismol. Soc. Am.*, 78, 770–779.

Xie, J. K. and Mitchell, B.J. (1990). Attenuation of multiphase surface waves in the Basin and Range province, part 1: Lg and Lg coda. *Geophys. J. Int.*, 102, 121–137.

Xie, J.K. (2002). Lg Q in the Eastern Tibetan Plateau (short notes). *Bull. Seismol. Soc. Am.*, 92, 871–876.

Zhang, T.R. and Lay, T. (1995). Why the Lg phase does not traverse oceanic crust. *Bull. Seismol. Soc. Am.*, 85(6), 1665–1678.

Zhou, L., Zhao, C., Chen, Z. and Zheng, S. (2011). Amplitude tomography of Lg waves in Xinjiang and its adjacent regions. *Bull. Seismol. Soc. Am.*, 101(3), 130–131.

Zhu, X.Y. and Chen, Y.T. (2012). An inversion of Lg-wave attenuation and site response in the North China region. *Bull. Seismol. Soc. Am.*, 102, 2600–2610. doi: 10.1785/0120110294.

Zhu, X-Y. (2014). An inversion of Lg-wave attenuation and site response from seismic spectral ratios in the Eastern China region. *Bull. Seismol. Soc. Am.*, 104(3), 1389–1399.

Chapter 9

Analysis of fluctuation in radon concentration and its correlation with geophysical phenomena of a seismically active region

T. Thuamthansanga, B.K. Sahoo, and R.C. Tiwari

9.1 INTRODUCTION

According to the seismic hazard zonation map of India (BIS, 2002), northeast India lies at zone V (highest seismic level). The region was also reported to be one of the six most seismically active regions of the world along with California, Taiwan, Japan, Mexico and China (Tiwari, 2002). The study area, Mizoram in particular, consists of several fault lines of which Mat fault, Tuipui fault and Chite fault were amongst the prominent faults (Jaishi et al., 2013, 2014, 2014a, 2014b, 2015; Singh et al., 2014, 2016, 2017). Few literatures based on passive measurement of radon for earthquakes precursory study were available from Mat fault and Chite fault (Jaishi et al., 2013, 2014, 2014a, 2014b, 2015; Singh et al., 2014, 2016, 2017). The available literatures lack the real-time nature and no continuous radon data has been generated until 2017 when the author starts monitoring continuous and *in-situ* data at Mizoram University and Mat fault (Thuamthansanga et al., 2019, 2020, 2020a, 2020b, 2021) (Figure 9.1). In the present study, the authors generate continuous and *in-situ* online data at Mizoram University and Chite fault, in a hope to reveal the real-time response of radon anomaly to geophysical phenomena of the region. Faults lines are regions, where the soil is loosely packed and form an easy pathway for terrestrial gases, such as radon, CH_4, CO_2 and so on, while escaping to the atmosphere. The exhalation process of some of these gases was altered and causes anomaly in their concentration during geophysical phenomena. Radon is a radioactive nuclear gas and

DOI: 10.1201/9781003177692-9

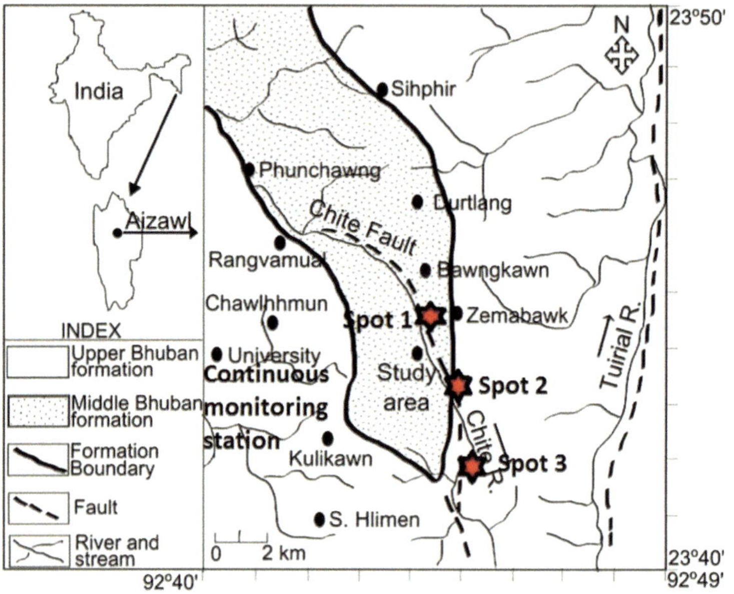

Figure 9.1 Geological set-up of the study area showing Chite fault and the measuring spots. (Modified after Singh et al., 2014.)

has three isotopes radon (^{222}Rn, $T_{1/2}$ = 3.825 days), thoron (^{220}Rn, $T_{1/2}$ = 55.6 s) and actinon (^{219}Rn, $T_{1/2}$ = 3.6 s). The latter two often get neglected due to their short half-life and as well in the present study. The first official record of radon anomaly was found in 1966 in a Russian well before the Tashkent earthquake (Ulomov and Mavashev, 1967). This observation marks radon as a potential premonitory gas for earthquake and instantly ignites its monitoring all over the country (Dubinchuk, 1993; Ghosh et al., 2009; Igarashi and Wakita, 1990; Iwata et al., 2018; Kawada et al., 2007; King, 1978; Omori et al., 2007; Shapiro et al., 1981; Singh et al., 1991; Singh et al., 2017; Thuamthansanga et al., 2021; Virk and Singh, 1993; Wakita, 1996; Walia et al., 2009; Yasuoka and Shinogi, 1997; Zmazek et al., 2002). Though it has been continuously studied from the past decades, monitoring of radon as a premonitory gas for earthquakes still has lots of uncertainties. One such being uncertainties due to meteorological influences on the radon data which is also presented in the present study. Besides radon anomaly studies, radon and thoron concentrations, their respective fluxes and estimated ^{238}U and ^{232}Th content of the region are also presented in detail.

9.2 MATERIALS AND METHODS

A continuous monitoring station for radon was set up at Mizoram University, Aizawl, Mizoram (India). It was a shed with 4 m^3 size design to minimize meteorological influence on radon data. The accumulator chamber (3.1×10^{-5} m^3) was connected with the radon monitor in a closed loop manner using a rubber tube and placed at the centre of the monitoring station (Figure 9.2a). The monitor was operated with 15 min cycle

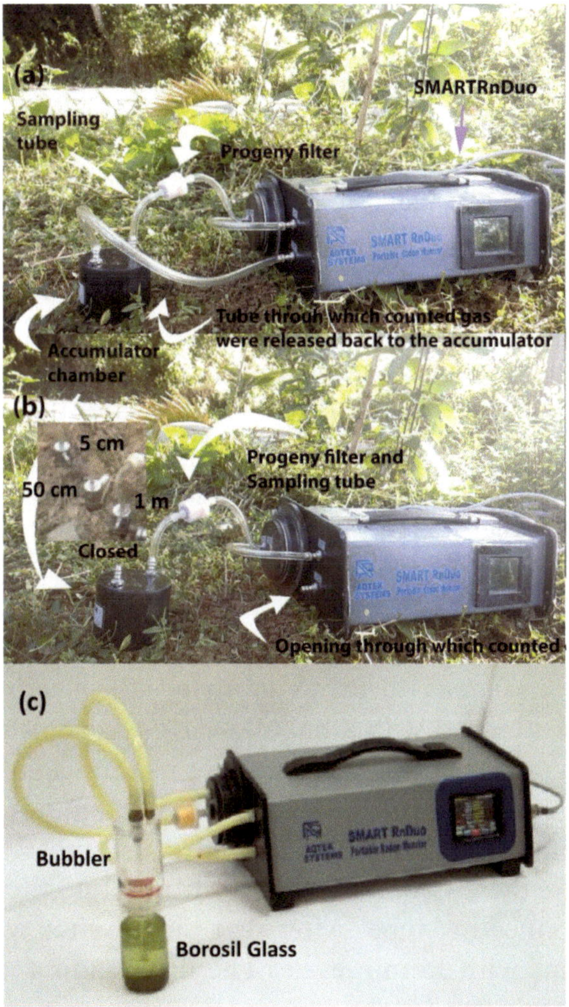

Figure 9.2 Arrangement of the SMARTRnDuo for (a) measurement at continuous monitoring station, Mizoram University, (b) measurement at the soil-air interface, 5 cm, 50 cm and 1 m and (c) measurement in water.

circling sample gas within the chamber continuously for 24 hours. In this manner, anomaly in the radon concentration was easily detected. The monitor was a ZnS(Ag) alpha-based scintillator counter developed and calibrated by Bhabha Atomic Research Centre, Mumbai, India. It was equipped with an in-built pump capable of drawing in the sample gas at the rate of 0.5–0.7 L/min. Detail features and functions of the instrument are available at Thuamthansanga et al. (2019, 2020, 2020a, 2020b, 2021).

On the other hand, at Chite fault (Figure 9.1), three measuring spots were selected along the fault line from where radon data was measured once a month between February 2018 and July 2018. An accumulator chamber (3.1×10^{-5} m^3) and soil probe of length 5 cm, 50 cm and 1 m were deployed to measure radon at the soil-air interface, 5 cm, 50 cm and 1 m depth along with the monitor (Figure 9.2b). The operating technique for the monitor was the same as that of the monitoring station except that the tubing connection was an open system. Such that the counted gas was released into the atmosphere in order to retrieved the *in-situ* online radon data. To measure radon data at the soil-air interface, the accumulator chamber was placed at the ground surface of the first spot and the monitor was operated in thoron monde for 15 min. After that the accumulator chamber was replaced by a soil probe of length 5 cm, 50 cm and 1 m, to measure the respective depths. In this manner, radon data of four different depths were generated at the first spot. After that it was proceeded to the second spot and so on.

For obtaining radon and thoron data of water, water sample near the vicinity of the previous three spots were collected in a glass bottle (2.2×10^{-4} m^3) attached with a bubbler. The glass bottle was connected to the monitor in place of the accumulator as shown in Figure 9.2c. The monitor pump was then turned ON for 3 min to agitate the sample water through the bubbler. This agitation will cause most of the dissolved radon rise above the water and gradually diffuse into the scintillation cell through the connecting tube. After that the monitor was run in radon mode for 1 hour with 15 min cycle. The first reading was neglected to avoid corruption in the data by external factors. Average of the last three reading was taken as radon concentration of the sample water. The same procedure was followed for retrieving thoron concentration in water except that the monitor was operated in thoron mode.

9.3 RESULT AND DISCUSSION

It has been reported that radon exhalation was not only influenced by geophysical phenomena but also by other parameters like meteorological factors (Iwata et al., 2018; Jaishi et al., 2014, 2014a, 2014b, 2015; Singh et al., 2014, 2016, 2017; Thuamthansanga et al., 2019, 2020, 2020a, 2021). Hence, before correlating with geophysical phenomena, it was first correlated with meteorological data to observe to what extent it was effected by it. Details of the correlation are given in Table 9.1. From Table 9.1, a significant influence by pressure and humidity were observed. The influence due to wind speed may be regarded as masking effect due to the other parameters which has been continuously observed in our previous study (Thuamthansanga et al., 2019, 2021). Hence care must be taken while correlating the radon data with geophysical phenomena as the result can be deviated by meteorological factors. At the same time, it was already shown that section of the continuous data at Mizoram University, refer in the present study, has negligible correlation with all meteorological data due to the provided shading (Thuamthansanga et al., 2019, 2021). It was also used to categorize radon data at Chite fault into anomaly and non-anomaly period data. Figure 9.3 shows normalized value of the continuous data between January 15, 2018, and August 15, 2018, with mean at 1. The normalization was done using a software named "radon data processor" developed by Bhabha Atomic Research Centre, Mumbai (India). Since radon data of the continuous monitoring station was negligibly influenced by meteorological factors, anomaly in its concentration was confidently regarded as due to geophysical phenomena. In the present study, radon anomaly was observed at $\pm 2\sigma$ from the mean (Figure 9.3). Such that any radon data generated at Chite fault by the time radon data at continuous monitoring station lies beyond the $\pm 2\sigma$ line were taken as anomaly period data (Figure 9.3). But if the radon data at Chite fault were generated by the time radon data at the continuous monitoring station lies within the $\pm 2\sigma$, they were regarded as non-anomaly period data.

Table 9.1 Details of correlation between radon and meteorological data at Chite fault.

	Temperature (°C)	*Pressure (Pa)*	*Rainfall (mm)*	*Humidity (%)*	*Wind speed (Kmh^{-1})*
^{222}Rn	−0.1	−0.7	0.2	0.7	−0.5

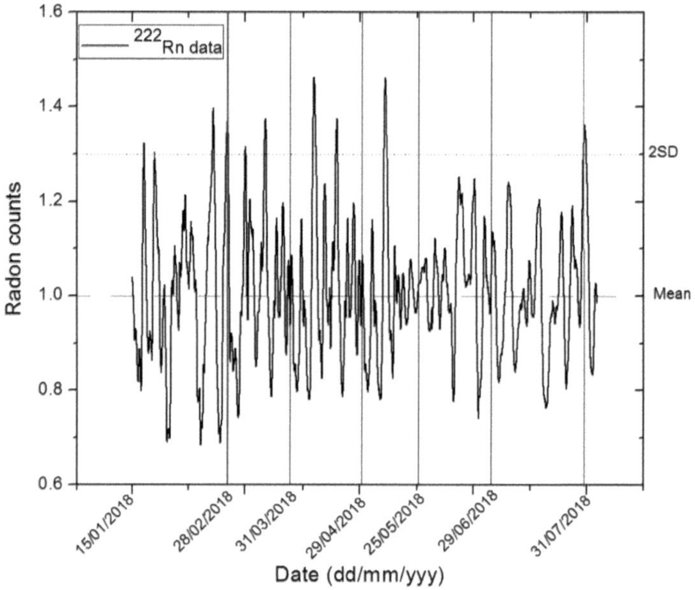

Figure 9.3 Normalized 15 min cycle radon data of the continuous monitoring station at Mizoram University categorizing anomaly (represented by vertical red line) and non-anomaly period data (represented by vertical black line). The date of *in-situ* measurement at Chite fault were indicated by vertical red-and-black line on its date of measurement.

At Chite fault, radon data were generated on February 28, 2018; March 31, 2018; April 29, 2018; May 25, 2018; June 29, 2018; and July 31, 2018. As per the aforementioned criterion, radon data of February 28, 2018, and July 31, 2018, were considered anomaly period data. As they were generated at the fault by the time the monitoring station data fluctuated beyond $\pm 2\sigma$. While that of March 31, 2018; April 29, 2018; May 25, 2018, and June 29, 2018, were considered non-anomaly period data since they were measured by the time the monitoring station data stays within the $\pm 2\sigma$. Now anomaly and non-anomaly period data of each measuring depth and water of all the measuring spots were compared. From Figure 9.4a, radon data at the soil-air interface shows higher radon counts during anomaly period than that of non-anomaly period at all the measuring spots. The same was observed for radon data at 5 cm, 50 cm and 1 m depth as well as in water (Figures 9.4b–e). From Figure 9.4 besides the higher radon counts, spot 2 has the highest deviation during anomaly period compared to the other spots. The observation reveals that the region, in particular Chite fault, was seismically active. Above all, it approves radon

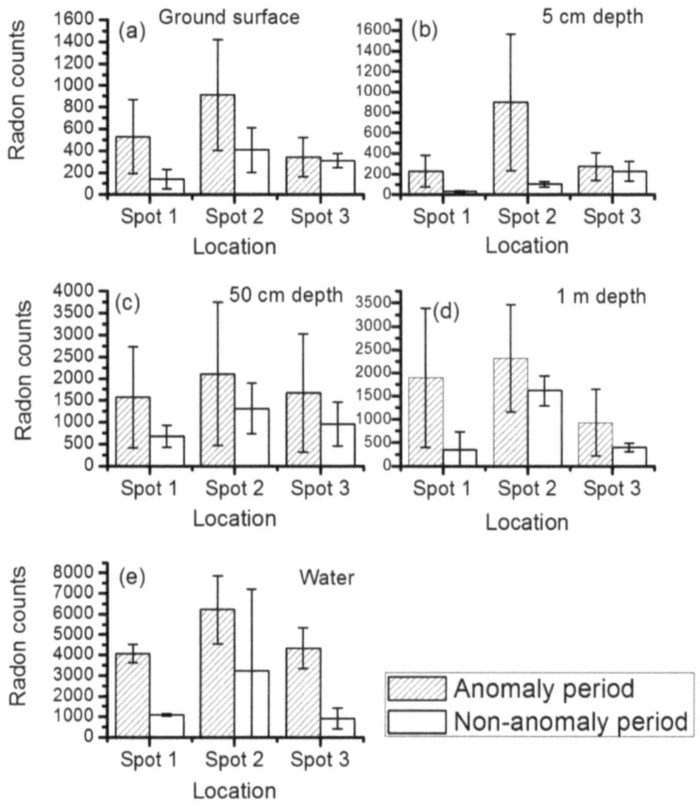

Figure 9.4 Plot of anomaly and non-anomaly period radon data at (a) the soil-air interface (b) 5 cm depth, (c) 50 cm depth, (d) 1 m depth and (e) in water.

monitoring as a potential premonitory gas for earthquake forecasting in the region. Hence by accumulating enough radon data, the study suggested that geophysical phenomena of the region may be predicted with high accuracy.

To estimate uranium and thorium content, respectively, from radon and thoron, first the radon and thoron concentrations of the region were obtained by equation (9.1).

$$C_{Rn} = \frac{C}{3EVe^{-\lambda t}} \tag{9.1}$$

where C is the net count rate (count per second, s^{-1}) of ^{222}Rn or ^{220}Rn, E is the efficiency of counting, V is volume of the sampler (m^3), λ is the decay constant of ^{222}Rn or ^{220}Rn, *t* is the time delay post-sampling (*s*) and 3 represent the three alphas in the respective decay chain of ^{222}Rn and ^{220}Rn.

The obtained radon and thoron concentrations were used to obtain their respective fluxes at the soil-air interface. The radon and thoron fluxes were obtained by substituting the radon concentration gradient $C(t)$ and equivalent thoron concentration C_{eq} in equation (9.2) and (9.3), respectively.

$$C(t) = C_0 + k\frac{A}{V}ft \tag{9.2}$$

where C_0 is the initial concentration (Bqm^{-3}), k is the factor by which the initial flux drops while the gas inside the accumulator passes through the state of uniform mixing prior to deployment to the state of diffusive mixing post deployment, A is the surface area of the opening of the accumulator (m^2), V is the effective volume of the sampling device (m^3), f is the flux of radon ($Bqm^{-2}s^{-1}$) and t is the measurement time (s).

$$f = \frac{C_{eq}V\lambda}{A} \tag{9.3}$$

where C_{eq} is the equilibrium ^{220}Rn concentration (Bqm^{-3}), V is the effective volume, λ the thoron decay constant and A is the surface opening area of the accumulator (m^2).

The radon concentration gradient was obtained by generating the radon data with 15 min cycle for 3 hrs in each of the measuring spot in radon mode. After that the obtained radon data were least fitted and averages of three spots were taken to obtain $C(t)$. The equilibrium thoron concentration on the other hand was obtained by measuring thoron data with 15 min cycle for 1 hrs in each of the spot in thoron mode. The first readings of each spot were removed as they could be affected by external factors. Now averages of the last three reading from the three spots were taken to obtain C_{eq}.

Now the radon and thoron fluxes obtained from equations (9.2) and (9.3) were substituted in equation (9.4) along with conventional parameters given in IAEA (2013).

$$f = \lambda LR\rho E \tag{9.4}$$

Table 9.2 Details of radon and thoron concentration and fluxes along with estiumated ^{238}U and ^{232}Th content of Chite fault

	^{222}Rn	^{220}Rn	Ratio ($^{220}Rn/^{222}Rn$)
Bqm^{-3}	5164.6	8968.1	1.7
$Bqm^{-2}s^{-1}$	0.02	2.4	120
	^{238}U	^{232}Th	Ratio ($^{232}Th/^{238}U$)
$Bqkg^{-1}$	32.9	44.9	1.4

where f is the radon or thoron flux at the soil-air interface, λ is the radon or thoron decay constant, L is the diffusion length of radon or thoron in the soil, R is the ^{226}Ra or ^{224}Ra content in the soil based on whether parameters were substituted for radon or thoron, ρ is the density of the soil matrix and E is the emanation coefficient of radon and thoron.

The ^{226}Ra and ^{224}Ra content obtained from equation (9.4) were in equilibrium concentrations with their parent nuclei ^{238}U and ^{232}Th, respectively. Hence their data might be used for depicting concentrations of their respective parent nuclei. After applying the aforementioned equations on measured radon and thoron counts, the radon and thoron concentrations of the region were observed to be 5164.6 Bqm^{-3} and 8968.1 Bqm^{-3}, respectively. The radon and thoron fluxes were also found to be 0.02 $Bqm^{-2}s^{-1}$ and 2.4 $Bqm^{-2}s^{-1}$, respectively. Finally, the ^{238}U and ^{232}Th content of the region were estimated to be 32.9 $Bqkg^{-1}$ and 44.9 $Bqkg^{-1}$, respectively. The observed results are given in Table 9.2 and compared with those reported global averages. The worldwide averages reported by UNSCEAR (1982, 2000) and IAEA (2004, 2013) were as follows: (1) radon and thoron averages = 10^3–10^5 Bqm^{-3} in soil (IAEA, 2013), (2) radon flux = 15–20 $mBqm^{-2}s^{-1}$, thoron flux = 1–1.9 $Bqm^{-2}s^{-1}$ (UNSCEAR, 1982) and (3) ^{238}U content of soil = 35 $Bqkg^{-1}$, ^{232}Th content of soil = 30 $Bqkg^{-1}$ (UNSCEAR, 2000). The given worldwide averages and the data obtained at Chite fault were in agreement with high accuracy. Besides the observed uranium and thorium content were below the critical value (1000 $Bqkg^{-1}$) reported by IAEA (2004), imposing no radiological risk for the region.

9.4 CONCLUSION

The study shows that the region was geophysically active. The radon data at the soil-air interface, 5 cm, 50 cm and 1 m depth simultaneously exhibits radon anomaly during geophysical phenomena with the highest anomaly at spot 2. The same was observed for radon data of water where anomaly period data were higher than the non-anomaly period data. The observation clearly reveals that the region was seismically active and radon data might be confidently sampled from soil or water to forecast it. Despite the observed meteorological influence through correlation analysis, the radon data significantly manage to show anomaly inflicted by geophysical phenomena. The radon and thoron concentration of Chite fault and their respective fluxes were in agreement and fall within the reported worldwide average (IAEA, 2013). Also the estimated concentration of their parent nuclei (^{238}U and ^{232}Th) were also within the worldwide averages given by UNSCEAR (2000). No radiological risk was observed from ^{238}U and ^{232}Th as per critical value assigned by IAEA (2004). The studied fault was located at the heart of Aizawl, the capital of Mizoram state (India), with highest human population in the state. Hence, it might also be concluded that no radiological risk was there for the people living in the state capital and probably its surroundings. On the other hand, as Chite fault, located at its heart was observed to be active, initiating preventive measurement is a must to avoid possible future disaster due to seismic activity. At the same time, the authors were continuously monitoring radon data, upon accumulating enough data we hope to predict seismic activity of the region with high accuracy. We hope to communicate soon in near future.

ACKNOWLEDGEMENT

This work was supported financially by DAE-BRNS, BARC, Mumbai, India [Sanction Order No. 36(4)/14/66/2014-BRNS/36024 Dt.26.02.2016.]

Compliance with ethical standards

Disclosure of potential conflicts of interest

Conflict of Interest: The authors declare that they have no conflict of interest.

Research involving human participants and/or animal

Ethical approval: This chapter does not contain any studies with human participants or animals performed by any of the authors.

Informed consent

Informed consent: Not applicable as the studies do not contain any human participants or animals performed by any of the authors.

REFERENCES

BIS. 2002. *Indian Standard criteria for earthquake resistant design of structure part 1-general provisions and buildings.* BIS IS 1893 (Part 1) New Delhi.

Dubinchuk, V. T. 1993. *Radon as a precursor of earthquakes; isotopic and geochemical precursors of earthquakes and volcanic eruptions.* IAEA-TECDOC-726.

Ghosh, D., A. Deb, S. Haldar, S. R. Sahoo, and R. Sengupta. 2009. Radon time series and earthquake signals—a study by SSNTD at Matigara (Darjeeling), India. *Earth Science India* 2: 76–82.

IAEA. 2004. *Safety Standards Series No. RS-G-1.7: Application of the concepts of exclusion, exemption and clearance. Safety Guide.* Vienna: International Atomic Energy Agency.

IAEA. 2013. *Measurement and calculation of radon releases from NORM residues.* Vienna: International Atomic Energy Agency. TRS 474: pp. 1–74.

Igarashi, G., and H. Wakita. 1990. Groundwater radon anomalies associated with earthquakes. *Tectonophysics* 180(2–4): 237–254.

Iwata, D., H. Nagahama, J. Muto, and Y. Yasuoka. 2018. Non-parametric detection of atmospheric radon concentration anomalies related to earthquakes. *Scientific Reports* 8(1): 1–9.

Jaishi, H. P., S. Singh, R. P. Tiwari, and R. C. Tiwari. 2013. Radon and thoron anomalies along Mat fault in Mizoram, India. *Journal of Earth System Science* 122(6): 1507–1513.

Jaishi, H. P., S. Singh, R. P. Tiwari, and R. C. Tiwari. 2014. Temporal variation of soil radon and thoron concentrations in Mizoram (India), associated with earthquakes. *Natural Hazards* 72(2): 443–454.

Jaishi, H. P., S. Singh, R. P. Tiwari, and R. C. Tiwari. 2014a. Correlation of radon anomalies with seismic events along Mat fault in Serchhip District, Mizoram, India. *Applied Radiation and Isotopes* 86: 79–84.

Jaishi, H. P., S. Singh, R. P. Tiwari, and R. C. Tiwari. 2014b. Analysis of soil radon data in earthquake precursory studies. *Annals of Geophysics* 57(5): S0544.

Jaishi, H. P., S. Singh, R. P. Tiwari, and R. C. Tiwari. 2015. Soil-gas thoron concentration associated with seismic activity. *Chiang Mai Journal of Science* 42(4): 972–979.

Kawada, Y., H. Nagahama, Y. Omori, Y. Yasuoka, T. Ishikawa, S. Tokonami, and M. Shinogi. 2007. Time-scale invariant changes in atmospheric radon

concentration and crustal strain prior to a large earthquake. *Nonlinear Processes in Geophysics* 14(2): 123–130.
King, C. Y. 1978. Radon emanation on San Andreas fault. *Nature* 271(5645): 516–519.
Omori, Y., Y. Yasuoka, H. Nagahama, Y. Kawada, T. Ishikawa, S. Tokonami, and M. Shinogi. 2007. Anomalous radon emanation linked to preseismic electromagnetic phenomena. *Natural Hazards and Earth System Sciences* 7(5): 629–635.
Shapiro, M. H., J. D. Melvin, T. A. Tombrello, M. H. Mendenhall, P. B. Larson, and J. H. Whitcomb. 1981. Relationship of the 1979 Southern California radon anomaly to a possible regional strain event. *Journal of Geophysical Research: Solid Earth* 86(B3): 1725–1730.
Singh, S., H. P. Jaishi, R. P. Tiwari, and R. C. Tiwari. 2014. Variations of soil radon concentrations along Chite fault in Aizawl district, Mizoram, India. *Radiation Protection Dosimetry* 162(1–2): 73–77.
Singh, S., H. P. Jaishi, R. P. Tiwari, and R. C. Tiwari. 2016. A study of variation in soil gas concentration associated with earthquakes near Indo-Burma subduction zone. *Geoenvironmental Disasters* 3(1): 1–8.
Singh, S., H. P. Jaishi, R. P. Tiwari, and R. C. Tiwari. 2017. Time series analysis of soil radon data using multiple linear regression and artificial neural network in seismic precursory studies. *Pure and Applied Geophysics* 174(7): 2793–2802.
Singh, M., R. C. Ramola, B. Singh, S. Singh, and H. S. Virk. 1991. Subsurface soil gas radon changes associated with earthquakes. *International Journal of Radiation Applications and Instrumentation. Part D. Nuclear Tracks and Radiation Measurements* 19(1–4): 417–420.
Thuamthansanga T., B. K. Sahoo, and R. C. Tiwari. 2020a. Study of the influencing nature of meteorological factors air temperature and relative humidity on the exhalation process of $^{222}Rn/^{220}Rn$ gases at Mat fault. *Journal of Applied and Fundamental Sciences* 6(1): 41.
Thuamthansanga T., R. C. Tiwari, B. K. Sahoo, and D. Datta. 2020. *Analysis of Meteorological Influence on Exhalation of 222Rn and 220Rn Gases at Mat Fault.* Radon: Detection, Exposure and Control. USA: Nova Science Publishers, Inc.
Thuamthansanga T., R. C. Tiwari, R. P. Tiwari, and B. K. Sahoo. 2020b. Correlation study of ^{222}Rn production rate and exhalation rate with geophysical process at Mat fault in Mizoram. *Journal of International Academy of Physical Sciences* 24(1): 83–93.
Thuamthansanga, T., B. K. Sahoo, R. C. Tiwari, and B. K. Sapra. 2019. A study on the anomalous behaviour of radon in different depths of soil at a tectonic fault and its comparison with time-series data at a distant continuous monitoring station. *SN Applied Sciences* 1(7): 1–13.
Thuamthansanga, T., B. K. Sahoo, R. C. Tiwari, and R. P. Tiwari. 2021. Study of meteorological influence on the count of ^{222}Rn and ^{220}Rn gases and its

possibility for a forecasting gas. *Radiation Environment and Medicine* 10(1): 37–47.

Tiwari, R. P. 2002. Status of seismicity in the northeast India and earthquake disaster mitigation. *ENVIS Bulletin* 10(1): 11–21.

Ulomov, V. I., and B. Z. Mavashev. 1967. A precursor of a strong tectonic earthquake. *Doklady Akademii Nauk:Russian Academy of Sciences* 176(2): 319–321.

UNSCEAR. 1982. *Ionizing Radiation: Sources and Biological Effects. Annex D: Exposures to Radon and Thoron and their decay products.* United Nations.

UNSCEAR. 2000. *Sources and effects of ionizing radiation. Annex B: Exposures from natural radiation sources.* United Nations.

Virk, H. S., and B. Singh. 1993. Radon anomalies in soil-gas and groundwater as earthquake precursor phenomena. *Tectonophysics* 227(1–4): 215–224.

Wakita, H. 1996. Geochemical challenge to earthquake prediction. *Proceedings of the National Academy of Sciences* 93(9): 3781–3786.

Walia, V., S. J. Lin, W. L. Hong, C. C. Fu, T. F. Yang, K. L. Wen, and C. H. Chen. 2009. Continuous temporal soil-gas composition variations for earthquake precursory studies along Hsincheng and Hsinhua faults in Taiwan. *Radiation Measurements* 44(9–10): 934–939.

Yasuoka, Y., and M. Shinogi. 1997. Anomaly in atmospheric radon concentration: A possible precursor of the 1995 Kobe, Japan, earthquake. *Health Physics* 72(5): 759–761.

Zmazek, B., M. Živčić, J. Vaupotič, M. Bidovec, M. Poljak, and I. Kobal. 2002. Soil radon monitoring in the Krško Basin, Slovenia. *Applied Radiation and Isotopes* 56(4): 649–657.

Chapter 10

Seismicity pattern during violent roof movements in longwall mines

Debjeet Mondal, P.N.S Roy, and Manoj Kumar

10.1 OVERALL OUTLINE

A very crucial role is played by coal in driving India's economy, which contributes to about 64% of the country's primary energy need. According to BP Statistical Review on World Energy (2017), the total annual coal consumption of India was about 411.9Mt in the financial year 2016–2017, which was almost 15.3Mt greater than the previous year (BP Statistical Review of World Energy, 2017). Coal is abundantly available in India, and it has a vast reserve of about 94769Mt. The abundant availability and high demand have made India one of the leading producers of coal in the world after USA and China. Indian coal is mostly obtained through opencast mining, which contributes to around 80% share of the total annual coal production. Thus, only 20% of coal comes from underground mining. Despite such low contribution, the mine fatalities reported from underground mining is very much higher compared to opencast mining. Previous studies on Indian coal mine accidents have shown that a total of about 612 accidents (almost 50% of the total reported accidents) were reported in various underground mines during 1989–1997, which was much higher than the accidents reported in opencast mines (213 accidents) and other surface-related issues (198 accidents) (Mandal and Sengupta, 2000). Therefore, special attention is required towards the safety of the underground mines.

The accidents occurring in the underground mines are mostly related to roof/side fall, winding, HEMMs, haulage, explosives, electricity, dust/gas, fall of object/person, inundation and other miscellaneous reasons. Out of the aforesaid mentioned reasons, roof fall has been mainly responsible for maximum mine-related causalities. A ten-year study (1989–1999) carried out by the Indian Statistical Institute (Kolkata) over country's

DOI: 10.1201/9781003177692-10

major coal mines showed that out of total 655 reported incidents, around 375 accidents (almost 56.79%) were related to roof falls, hence making it the most serious threat in the coal mine (Mandal and Sengupta, 2000). Therefore, the real-time monitoring of the roof strata movement and temporal forecasting of untimely roof falls is necessary towards ensuring safety in the coal mines. Previously, studies related to the monitoring of mine-induced microseismicity has proved to be effective in studying the behaviour of overlying strata during underground coal excavation (Ghose and Sivakumar, 2018, Mondal et al., 2020). The objectives of the microseismic-monitoring technique were to mitigate several mine-related issues like fracture monitoring, rock burst assessment, gas burst monitoring and so forth. The present study also uses this robust technique in a longwall panel of Central India (overlain by thick hard sandstone bed) for studying the behaviour of distressed zone during coal excavation.

Longwall is very effective and safest coal mining and is widely accepted worldwide. But in India, the scenario is quite different, where very little technical advancement has been seen in the longwall technology. The output of most of the Indian longwall panels is far below the desired level of productivity (Ghose, 2003). Irregular caving, which is often followed by sudden roof collapse, is a common and most serious problem faced in the Indian underground coal mines. During 2006–2010, the highest roof fall-related incidences were reported in Indian coal mines as compared to other major coal-producing countries (Harris et al., 2014). Therefore, an Indian retreating longwall panel overlain by a thick hard sandstone bed has been especially taken up for the present study to investigate the stress condition of roof strata during coal excavation. The microseismic monitoring was earlier used in the Indian longwall mine for getting precursory signatures of the sudden rock falls in terms of Event Release Ratio (ERR) but their spatio-temporal forecasting and proper identification of stressed zones could not be done. The present study includes the estimation of fractal and *b*-value (from passive mine microseismicity), which has been earlier used in several studies related to earthquakes, fracture monitoring, identification of stressed zones, and magnitude study (Ouchi and Uekawa, 1986; Roy and Mondal, 2009). The fractal dimension and *b*-value have been used in the present study for locating the stressed zones in the overlying strata of an active longwall panel and spatio-temporal forecasting of roof fall in real time for enhancing safety in the working area.

The presence of adverse roof conditions in the form of massive hard sandstone beds has been a major issue faced by most of the Indian underground mines, where safe extraction of coal becomes a great challenge. Thus, the present study was conducted over a Central Indian longwall panel having similar geological problems related to the presence of a very stable immediate roof in the form of a thick sandstone bed. The previous details of the panel showed that irregular caving has been a major issue faced during coal extraction, which resulted in time delays of several man-hours, low productivity and also increased the chances of roof fall-related accidents. The microseismic data acquisition has been carried out by two different monitoring instruments, namely NIRM Microseismic Monitoring System (NMMS) and Integrated Seismic System (ISS), which were capable of recording microseismic activity (ERR) and various source parameters such as event location, magnitude, energy index, apparent stress, apparent volume, seismic energy and wave velocities (P and S wave) in real time.

The inputs for the present study were event coordinates (in *x*, *y*, *z*) and their corresponding magnitude (*m*). The raw data was made noise-free by applying suitable filters and the filtered data was then divided into uniform time windows having suitable lengths such that linearity is maintained between scaling factor (*r*) and correlation integral (*c*) without missing the minor events. The sampled data is then used for calculating the correlation integral (*c*), that is, fraction of points available between given scale ranges (*r*). The slope of the curve (with best fit) plotted between correlation integral and scale ranges (on logarithmic scale) provides the value of fractal dimension (*d*). Thus, the fractal dimension has been calculated from the microseismic events acquired from the longwall mine for the entire investigation period (September 2001–February 2002). The variation of fractal dimension is jointly studied with chronological sequence reported from the longwall panel, which consisted of productive working days (*pw*), non-productive working days (*npw*), non-productive non-working days (*npnw*), local falls (*lf*), roof falls (*rf*), and surface blasting (*sb*). The fractal dimension estimated from passive seismicity had a good correlation with the chronological sequence of the aforesaid mining activities/events. The change in the fractal dimension has been shown in terms of fractal gradient, which helped in monitoring the abrupt stress in roof strata during mining. During the investigation period, a total of 98 major rock bursts were reported in the mine, which were also recorded in ISS and named as MS-ISS-RF1 to MS-ISS-RF98. The estimated fractal values were carefully studied for identifying the pattern of graph before major roof falls. It was seen that the fractal dimension values have gone below 0.4 before the impending rock bursts making $d \le 0.4$ as a precursory signature roof fall forecasting. The event windows corresponding to low fractal values were characterized as the stress accumulation events and the microseismic events corresponding to those windows represented the location of stressed zone in the roof strata. It was also seen that the aforesaid critical values of fractal dimension ($d \le 0.4$) were mostly obtained within 24 hours prior to major roof falls, thus making it a very significant result towards monitoring roof displacements and enhancing safety.

The aforesaid study was followed by analysis of microseismic activity and magnitude study based on calculated seismic *b*-value. The Gutenberg-Richter Frequency-Magnitude relation between event magnitude and frequency of occurrence helped in estimation of seismic activity (*a*) and *b*-value. The temporal variation of *b*-value was also calculated using the maximum likelihood method over each event windows. The increase in event magnitude has resulted in lowering of *b*-values before major roof falls and hence showed a similar trend as that of fractal dimension. Apart from roof fall forecasting, the application of aforesaid parameters was then extended towards the characterization of mine events especially the local falls, roof falls and surface blasting. The aforesaid 98 roof falls along with some other minor falls have been further classified into 64 local falls (MS-LF1 to MS-LF64) and 76 roof falls (MS-RF1 to MS-RF76), which were reported in the mine. Surface blasting was carried out to artificially release the excess accumulated stress from the roof strata named as MS-SB1 to MS-SB15. The average values of fractal dimension (*d*) and *b*-value (*d*-avg, *b*-avg) obtained for minor (*lf*) and major falls (*rf*) were (0.462, 0.378) and (0.342, 0.374), respectively. Thus, it can be seen that *d*-avg(*lf*) > *d*-avg(*rf*) and *b*-avg(*lf*) > *b*-avg(*rf*), which is an expected result as more stress is required to initiate a major fall and microseismic events involved are more clustered and have higher magnitude. The average values aforesaid seismic parameters were also estimated for surface blasting, which were around 0.616 (*d*-avg) and 0.459 (*b*-avg), respectively. Since, microseismic events corresponding to surface blasting were more dispersed in space and had higher magnitude, which resulted in higher fractal dimension and lower *b*-value. Thus, the present study also helped in characterization of different seismic activities based on fractal dimension and *b*-value.

Apart from roof fall forecasting and characterization of mining events, the present study also helped in understanding the temporal variation of stress state of longwall roof during mining. The estimated values of critical fractal dimension before roof fall gradually increased with time and gradient of fractal dimension became smoother, which showed that microseismic events became more dispersed in space as the state of roof rock moved away from its virgin/stable state. Since, the fractal dimension values estimated at the time of surface blasting showed a positive correlation with the explosive charge size, and, therefore, it also helped in estimation of explosive size to be required for releasing the stress further making the explosion very economic and safe. The fractal dimension values also helped in monitoring the fractures induced in the destressed zone (fractured and caved zone) due to surface blasting and roof failure. The zonation of stress conditions in the roof strata during various phases of mining was also carried out based on variation of both the seismic parameters (fractal dimension and *b*-value). Lastly, analysis of deviation from expected result was also successfully done, which showed that lack of microseismic data was mainly responsible for the ambiguity.

Thus, the present study helped in mitigating several issues related to the safety of underground mines. Most important, the aforesaid study can be carried out in any form of underground mines (coal as well non-coal), which will be very helpful in enhancing its safety and smooth operation.

10.2 INTRODUCTION

10.2.1 General

Mining refers to the excavation of minerals through surface as well as subsurface operations, which can neither be naturally grown nor synthetically manufactured. From ancient times, mining has been associated with human civilization for extraction of natural resources, and with the beginning of industrial revolution it has become an important part of global scenario. The increasing demand of primary energy in power and manufacture sector has made coal mining one of the principal mining activities today. Coal mining started thousands of year back, where early archaeological evidences was seen in China dating back to 3490 BC (Dodson et al., 2014). Evidences from several ancient civilizations (such as the Aztecs of the Americas) showed the utilization of coal as a source of fuel for various activities. Coal mining increased during the Roman times in the 2nd century and coal mining at small scale especially for domestic fuel was initiated by 13th century in most of the coalfields of England, Scotland, and Wales (Smith, 1997). The gradual increase in demand led to several advances in mining technologies during the 17th century, which included application of boreholes for exploration of reserves and chain pumps water drainage from collieries. The importance of coal drastically increased during the industrial revolution around the 19th century, where it was extensively used for powering up the steam engines for transportations and generating electricity.

Depth and geology of the seams act as prime deciding factors during the selection of suitable mining methods. On the basis of the mode of operation (surface/subsurface), coal mining is broadly categorized into opencast and underground mining. The opencast mining is generally carried out for the coal seams present at shallower depth usually below 50m. Usually underground mining is preferred for coal seams present at the depth of 50–100m, but, in some cases, opencast mining is also opted for the coal seams with low stripping ratio. Coal seams below 100m are mostly extracted using underground mining methods as stripping ratio becomes very high (Christman et al., 1980). The economic feasibility of the coal mines is decided in terms of its stripping ratio (*SR*, Eq. 10.1), which is given in terms of total overburden to be removed for mining of unit volume of coal. There are different types of stripping ratios used in mining, which includes volumetric/geometric stripping ratio (*VSR*), thickness ratio (*TR*), breakeven stripping ratio (*BESR*), limiting stripping ratio (*LSR*) and instantaneous stripping ratio (*ISR*), overall stripping ratio (*OSR*), and average stripping ratio (*ASR*).

$$SR = \frac{Overburden(volume)}{Coal(weight)} \tag{10.1}$$

Opencast mining operations are more economic at lower stripping ratio, as less overburden is removed for per tonne extraction of coal.

10.2.2 Underground coal mining and roof fall hazards

The underground mines are considered to be more effective and economic as compared to opencast operations. In fact, most of the minable coal seams around the world are situated at greater depths hence making opencast mining is not economically feasible. About 60% of the world's total coal production comes from underground mining (World Coal Institute, 2009), which is carried out through two modes of operation: (1) room and pillar mining and (2) longwall mining (World Coal Association, 2014). In the conventional room and pillar mining, the coal seams are drawn into successive networks of pillars and interconnecting galleries also called as the room. Once these galleries are constructed, the depillaring method is carried out to extract the left coal from the pillars. Room and pillar mining is very effective for coal seams (recovery: ~60%) up to the depth of 1000ft, whereas at high depth (greater than 1000ft), the pillar sizes also increase leading to low coal recovery during depillaring. Longwall technique is considered to be more advanced and sophisticated form of underground coal mining, where coal seams are drawn into cuboidal sections known as panel and a shearer is deployed at the face of the panel under the protection of powered supports. The longwall technique has very high recovery rate (~80%) and provides continuous supply of coal.

Though coal mining has been a common practice since ancient times but due to its long history of accidents, it has been considered as a risky activity. The accidents associated with coal mining (especially underground mining) are considered to more hazardous than other hard rock mining due to its adverse geology and gassiness of the presence of trapped methane gas in the coal seams. The accidents in the opencast mines are mostly related to bench/stope failures, HEMM (Heavy Earth Moving Machinery) collision and bench blasting, whereas in underground mines accidents occur mainly due to suffocation, gas contamination, roof collapse, rock outbursts and gas explosions (from mainly methane gas). Since accidents in underground mines occur in a closed environment, therefore, are considered to be more catastrophic as compared to accidents in the opencast mines. The country-wise analysis of coal mine accidents showed that higher number of incidents and related fatalities were occurred in underground mines (MacNeill, 2008).

Among all the coal mining-related hazards, roof falls are considered to be the most serious threats faced by coal industries globally and responsible for highest fatalities in the underground mines. Similar results were seen in the underground mines of Tabas and Kerman region (Iran), where sudden roof falls were responsible for most of the accidents in the mines during 2003–2007 (Oraee and Bakhtavar, 2009). The scenario of roof fall accidents and fatalities, as studied by various agencies in coal mines of some of the leading coal producers of the world (such as USA, China and India) has been mentioned later in the text. As per the report of Mine Safety and Health Administration (United States), a total of 75 fatalities and 5,941 injuries occurred in various underground coal mines of USA between 1999 and 2008. The Indian coal mines also faced severe problems of roof falls. As per the report of Directorate General of Mine Safety (DGMS), nearly 32% of fatalities in the coal mines were reported due to the roof fall during 1998–2010 (DGMS, 2011; Yerpude and Walke, 2014). In China, the State Administration (Coal Mine Safety) showed that about 4,653 accidents related to roof fall were reported in 2001–2008 (~51.5%) causing 6,173 deaths (~26.4%) (Min-Xiao et al., 2011).

10.2.3 Roof falls hazards in India

The higher dependence of Indian economy on coal has resulted in its mass extraction from both opencast as well as underground mines. The Indian coal mostly comes from the opencast mines, which are also considered to be safer as compared to underground mines. The analysis of fatal accidents in Indian coal mines showed that during 1989–1997, around 612 accidents (~60%) occurred in underground mines out of total 1,023 reported accidents (Mandal and Sengupta, 2000). The cause-wise distribution of the Indian coal mine accidents has been classified into several categories namely roof/side fall, winding, HEMM accident, haulage, explosives, electricity, dust/gas, fall of object/person, inundation, and other miscellaneous reasons (Mandal and Sengupta, 2000). The cause-wise statistical analysis of Indian coal mine accidents during 1989–1999 was carried out by Mandal and Sengupta (2000), which showed that out of total 655 reported accidents, around 372 (~56.79%) were related to roof collapse (Mandal and Sengupta, 2000). Moreover, a four-year (2006–2010) comparative study of the coal mine accidents showed that among the top coal-producing countries of the world, highest fatalities due to roof fall was reported in India (fatalities: 114) (Harris et al., 2014). As per the study of Masabattula (2010), roof fall was responsible for 25% of the total accidents and 47% of the total underground accidents in Indian coal mines during 1996–2007 (Masabattula, 2010a). Though the number of reported accidents have decreased with time but still roof fall remained the primary cause of mine disasters in India. Some of the major roof fall accidents reported in Indian coal mines were Dhemomain Mine (year: 1952; fatality: 12), Kessurgarh Mine (year: 1975; fatality: 11), Topa Mine (year: 1982; fatality: 16), and GDK-8A Mine (year: 2003; fatality: 10). Therefore, special attention was required towards monitoring of the sudden rock bursts for ensuring the safety in the underground coal mines.

10.2.4 Longwall scenario in India

As the present study deals with the longwall mining, therefore, it's very essential to understand the past experiences of India in the aforesaid mining technique. Due to high recovery, longwall is considered to be the most effective coal mining method, which is widely accepted and practiced in the major coal-producing countries of the world (Linden, 2000). Though longwall technology started very early (around 1870s) in the Indian coal industry, but it took almost a century to introduce the caved longwall mining in the country. The first fully mechanized powered support longwall face developed in Moonidih colliery (Bharat Coking Coal Limited, India) in 1978 (Ghose, 1988), but the expected performance could not be achieved due to the adverse strata conditions in the form of igneous intrusions (dykes) and geological disturbances (step faults). Similar problems were also seen in the Seetalpur face of Jhanjra mine (Eastern Coalfields Limited, India), which started in 1985. The longwall technology was also introduced in the mines of Dhemomain (Eastern Coalfields Limited, India) and Pathakhera colliery (Western Coalfields Limited, India) of Coal India Limited in 1982, and some other mines of Singareni Collieries Company Limited (India), such as GDK-7, GDK-9, GDK-11A (Masabattula, 2010b). In India, the longwall technology was mostly applied in the shallower coal seams with depth of occurrence less than 200m, where majority of the minable seams were overlain by thick sandstone cover (Ghose, 2003; Mondal et al., 2017a). The sandstone in the immediate roof reduced its caving capacity, which increases the risk of fatal roof fall accidents in the longwall face. Presence of massive sandstone (in clean and mixed form) in destressed zone is a serious threat to underground coal mine, as it was more that 51% of the mine disaster during 1997–2006 (Masabattula, 2010a). The increase in the hanging load over the powered supports due to untimely caving sometimes leads to its total failure supports and the entire longwall face gets sandwiched between the rock mass. The accidents result in fatalities and total loss to the instruments, which cannot be recovered. Similar incidents were reported in Churcha mine (South Eastern Coalfield Limited, India) in 1989 and Kothadi mine (Eastern Coalfield Limited, India) in 1997 (Ghose, 2003; Masabattula, 2010b; Mondal et al., 2017a). The underperformance and frequent accidents of majority of the Indian longwall mines have drawn attention of government and several other agencies to investigate its root cause, and it was seen that inadequate knowledge about the geological and geotechnical parameters was the prime reason (Ghose, 2003). The longwall panel under investigation was considered to be suitable for present study as it also faced similar problems related to adverse geological conditions, where minable coal seam was overlain by massive sandstone bed having average thickness of about 40m. Therefore, fractal and *b*-value study over the passive microseismicity has been carried out for monitoring the overlying strata dynamics during the various phases of the coal extraction.

10.2.5 History of mine microseismic technology

Mining activity disturbs the equilibrium stress distribution in the rock strata. Since earth acts as anisotropic media, therefore, stress gets concentrated leading to non-uniform distribution and then released though the weaker sections of the rock mass. These radiated seismic waves consists of wide range of energy spectrum from micro earthquakes (~10.5J) to tremors (~10^9J) (Cook, 1976). The lower energy earthquakes having local magnitude less than 2.5 are known as microseismic events (Xu et al., 2012). Usually in coal mines, microseismic energy is released prior to rock failure thus making it an important tool for monitoring the in-situ stress variation. The microseismic technology was introduced in the early 1900s, where first monitoring of mine-induced microseismicity was carried out in Bochum mine of Germany during 1908 (Xu et al., 2012). By the 1940s, the aforesaid technique was extensively used by the American scientists for monitoring the seismicity in the underground mines (Jiang et al., 2008), but the earlier instruments were expensive and had relatively lower precision level. However, significant results were achieved when microseismic instruments were installed to monitor the induced seismicity in the East Rand Proprietary Mines in South Africa during the 1960s (Cook, 1964). The microseismic monitoring in underground mines of USA was introduced by Bureau of Mines in late 1900s (1980–1990) for coal mine bump prediction (Rowell, 1989), but the hardware used for the automated instruments had certain limitations related to frequency content and signal/noise ratio (Rowell, 1989). The PC-based monitoring system was introduced by NIOSH in late 1990s, which was used to investigate the strata behavior in deep longwall mine (Swanson, 2001; Ellenberger et al., 2001). The aforesaid instrument was used to study the seismic source parameters in coal mines of Utah were carried out by the University of Utah (Arabasz et al., 1996). The microseismic monitoring was also carried out in Germany by Ruhr University (Bochum) for monitoring seismicity in the vicinity of the coal mines (Swanson et al., 2016), and gathered data was further used for various studies related to magnitude, *b*-value, source mechanism of longwall events and moment tensor (Bischoff et al., 2010). Presently, around 60 microseismic monitoring stations are actively operating in various underground coal mines of Germany for carrying out strata-related studies (Baltz and Hucke, 2008). In China, microseismic-monitoring technique was developed by Institute of Geophysics (Chinese Academy of Science) in 1960s for monitoring the rock bursts in the underground coal mines (Geng, 2008) and after a decade (in 1970s), portable microseismic monitoring instrument was developed by Changsha Institute of Mining Research for measuring the acoustic emissions (Qin et al., 2012). Since then the aforesaid technique has been extensively carried out in several coal mines as well as reservoir for monitoring the induced seismicity along different provinces of China (Li et al., 2007). In Australia, short-term microseismic studies were carried out in several longwall mines for delineating the caving-related fractures (Luo and Hatherly, 1998) and monitoring the windblasts and hydraulic fracture-related risks (Hayes, 2000). In India, microseismic monitoring was also carried out for rock burst prediction in

Rajendra Mine (South Eastern Coalfield Limited) by calculating the ERR in real time (Srinivasan et al., 2005; Sivakumar et al., 2008). Presently, microseismic technique has been widely accepted for carrying out several studies related to safety and smooth running of underground mines, such as (a) outburst monitoring (using acoustic emission and gas concentration) (Jie and Tai, 1988), (b) water inrush monitoring (by measuring the stress distribution and deformation of rock during water inrush) (Tie and Hong-Guang, 2010), (c) rock burst monitoring (with the help of activity rate and energy released during fracturing) (Srinivasan et al., 2005; Sivakumar et al., 2008), (d) monitoring in-situ combustion zones and rock cracking in coal mines (Qin et al., 2012), and (e) monitoring hydraulic fracturing. Thus, it can be inferred that microseismic monitoring is a robust technique and has vast range of applications.

10.3 STUDY AREA, INSTRUMENTATION, METHODOLOGY AND APPLICATION

10.3.1 Geology of the area

The sedimentary basins of India consists of about 60600Mt of proved reserve, which contributes to about 64% towards country's primary energy demand (BP Statistical Review of World Energy, 2016). These vast sedimentary basins have been subdivided into relatively smaller region known as coalfields. The area of investigation for present study is situated in the Sohagpur Coalfield, which is situated in Central India and considered to be the country's one of the major coal-bearing areas. The aforesaid coalfield (refer Figure 10.1) is situated in the states of Chhattisgarh and Madhya Pradesh within coordinates 23°05′–23°05′ (LAT) and 81°13′–82°12′ (LONG). The total areal extent of Sohagpur Coalfield covers an area of about 3000km^2 with 1000m thick sedimentary deposit having an estimated coal reserve of about 4064Mt (Shahdol, 2011; Jasper et al., 2016; Mondal et al., 2021). The entire coalfield has been divided into two major parts (North and South Sohagpur Coalfield) by the Bahmani-Chilpa fault with throw towards north (N→S). The stratigraphic succession of Sohagpur Coalfield (as provided by Geological Survey of India, refer Figure 10.2) showed the presence of several minable coal seams with reserves belonging to Supra Barakar Formation of Gondwana Supergroup (Raja Rao, 1983; Pareek, 1987;

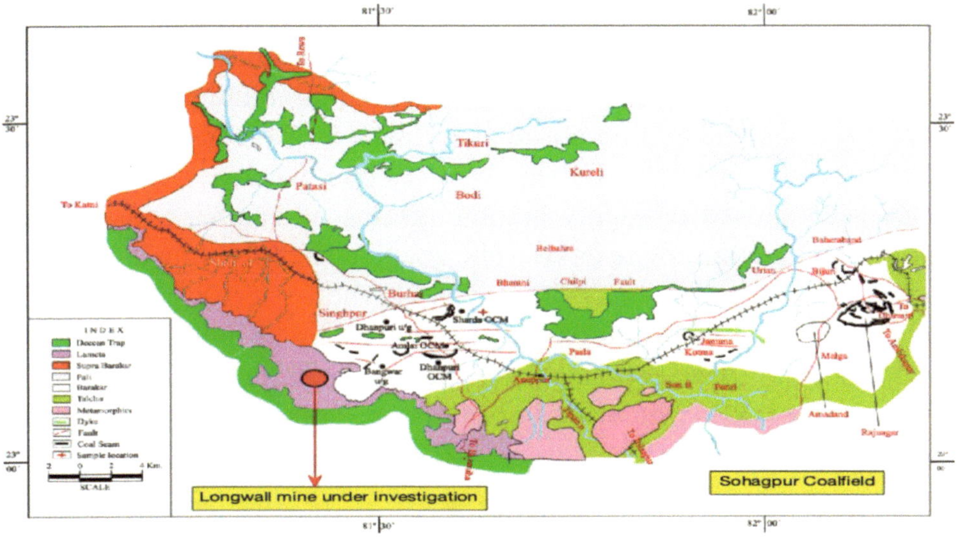

Figure 10.1 Sohagpur coalfield. (Mondal et al., 2020.)

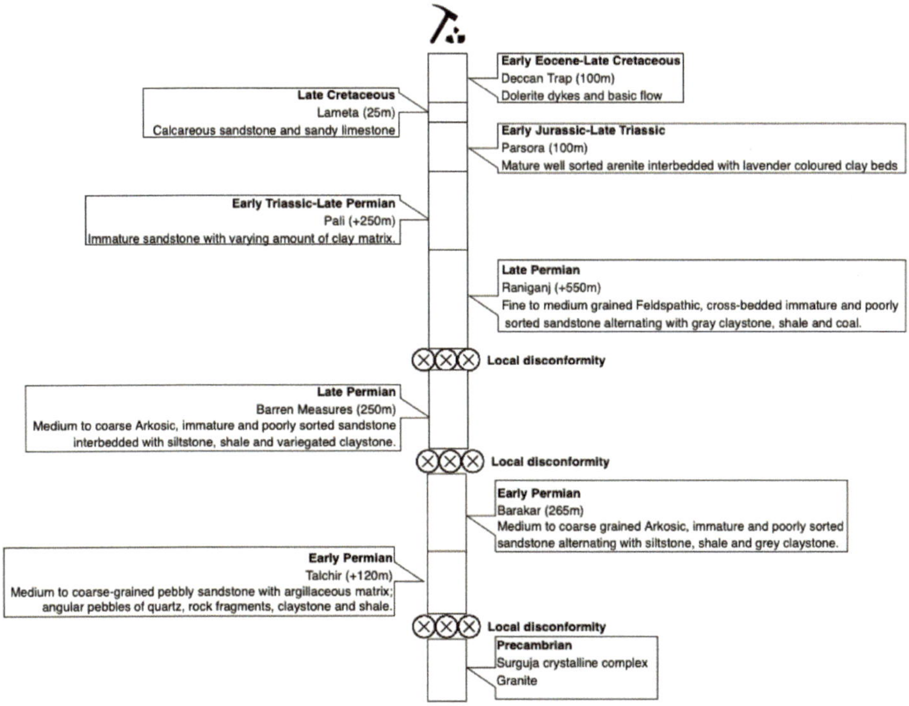

Figure 10.2 Geology of Sohagpur coalfield. (Mondal et al., 2017a; Mondal et al., 2019; Mondal et al., 2020.)

Dhanam et al., 2013; Gautam et al., 2016). The Barakar Formation, which is the major coal-bearing deposit of the area had medium- to coarse-grained sandstone (which are rich in quartz and plagioclase) and dark grey shale. The Lameta of the Upper Cretaceous Age consisted of calcareous sandstone and sandy limestone, whereas the Calcareous Matrix composed of highly angular grains of quartz, feldspar, rock fragments and muscovite flakes. The coal block containing the longwall mine under investigation is situated along the western part of the coal-field. The Talchir Formation had greenish grey to light grey-coloured shale and medium to coarse-grained pebbly sandstone with argillaceous matrix. The basic components of sandstone consisted of quartz, shale, clay clasts and lithic fragments, which are interlaminated with clay and garnet-rich laminae. The Deccan Trap mainly consisted of igneous intrusions in the form of sills and dykes. The stratigraphic succession of coal block having the longwall panel (under investigation) has been shown in Figure 10.3. The detailed lithology along the longwall mine has been studied from core samples obtained from four boreholes (Bh1, Bh2, Bh3 and Bh4), which were drilled at various positions (Bh1: 75m, Bh2: 150m, Bh3: 300m, Bh4: 600m from initial face position) along the panel. The detailed lithology of the aforesaid boreholes (refer Figure 10.4) showed the presence of two minable coal seams: S1 (thickness: 1m, depth: 27m) and S2 (thickness: 3m, depth: 69m) separated by thick sandstone bed (thickness: 30m). The detailed lithology also showed that the coal seams were sandwiched between thick hard sandstone beds. Longwall mining was setup in coal seam S2 having strike direction 30°N and dipping

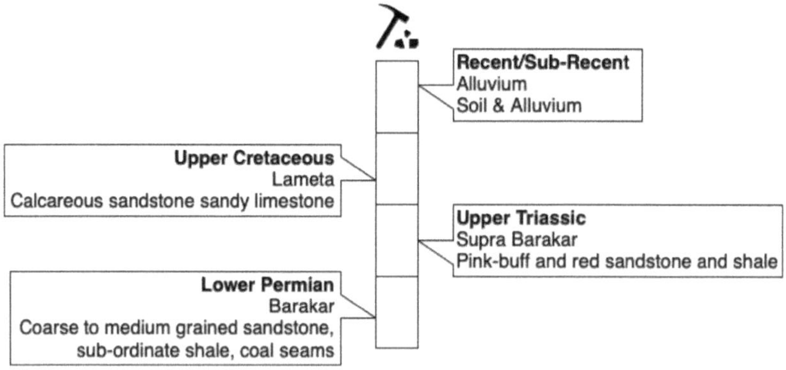

Figure 10.3 Geology of the coal block. (Mondal et al., 2017a; Mondal et al., 2019; Mondal et al., 2020.)

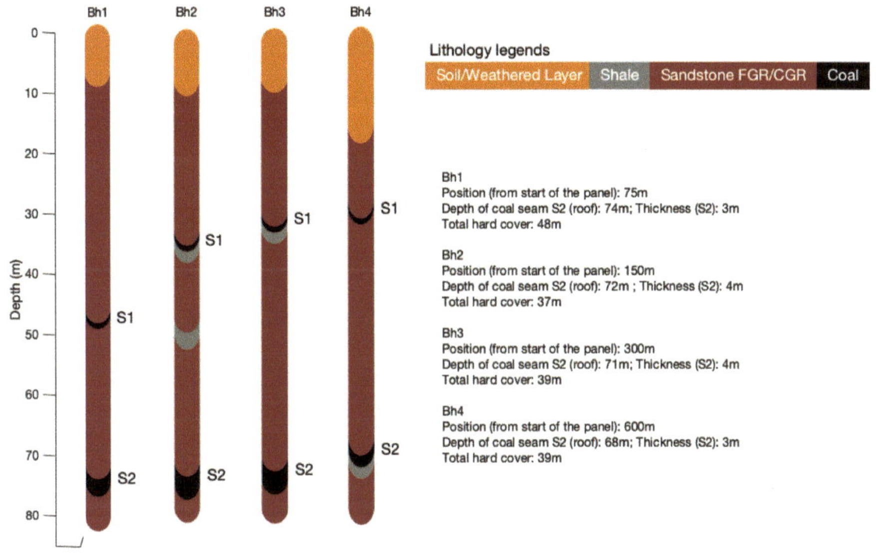

Figure 10.4 Detailed lithology of Bh1, Bh2, Bh3, and Bh4. (Mondal et al., 2019; Mondal et al., 2020.)

towards 120°N (Mondal et al., 2017a, 2017b; Mondal and Roy, 2019). Three rectangular panels have been drawn across seam S2, which are named as P1, P2, and P3 (Figure 10.5). The mining activities along with microseismic monitoring were carried out in panel P2. Mining activities were completed in P1, whereas Panel P3 was under development for further mining. As mining activities were restricted to Panel P2 only,

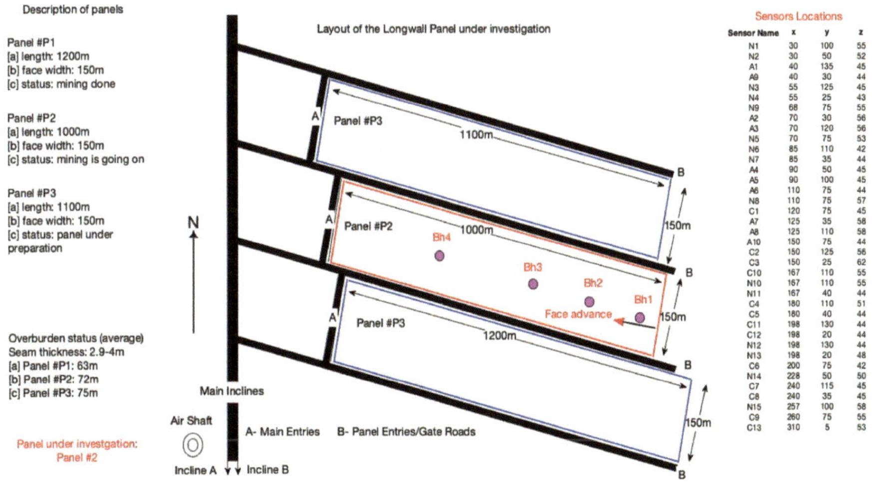

Figure 10.5 Panel layout. (Mondal et al., 2017a; Mondal et al., 2019; Mondal et al., 2020.)

Table 10.1 Panel layout and instrumentation (Mondal et al., 2017a; Mondal et al., 2019)

Description of the Mining Area		*Description of Powered Supports*	
[a] Working height	2–3 m	[a] Capacity	4*503 t
[b] Face length	150 m	[b] Number of legs	4
[c] Length of panel	1000 m	[c] Operating range	1.7–3.0 m
[d] Coal available	0.54 Mt (avg. thickness 3 m)	[d] Total number of supports	104
[e] Face started on	24/08/2001		
shearer specifications			
[a] Minimum height	1800–3920 mm	[i] Voltage	1140V
[b] Minimum depth	630 mm	[j] Haulage force	322 kN–534 kN
[c] Seam tilt angle	≤35°	[k] Haulage speed	Upto 12.8 m/s
[d] Drum diameter	1800 mm	[l] Machine height	1571 mm
[e] Drum speed	36 rpm	[m] Dinting depth	316 mm
[f] Cutting arm length	2600 mm	[n] Total machine weight	60 t
[g] Slewing centre distance of cutting arm	7380 mm	[o] Dust suppression	Inner and Outer Spray
[h] Total power	2*300+2*40+20 kW	[p] Cable collection	Auto Mode
armored face conveyor			
[a] Capacity	800 t/h	[d] Line pan length	1500 mm
[b] Face conveyor length	156 m	[e] Line pan weight	1320 kg
[c] Line pan width	764*290 mm		

therefore, it was only considered as source for generation of the micro-seismic events. The layout of the panel and instrumentation setup for excavation has been summarized in Table 10.1.

10.3.2 A brief about longwall technology

Longwall mining is the most sophisticated and advanced form of underground coal mining method, where seams are drawn in rectangular sections (known as panel) for extracting reserve through single slices (Ghose, 2003). Based on the mining direction, longwall mining is divided into two broad sub-categories: (a) longwall advancing and (b) longwall retreating. A schematic representation of longwall (retreating) mining method is given in Figure 10.6. The aforesaid mining method is suitable for those

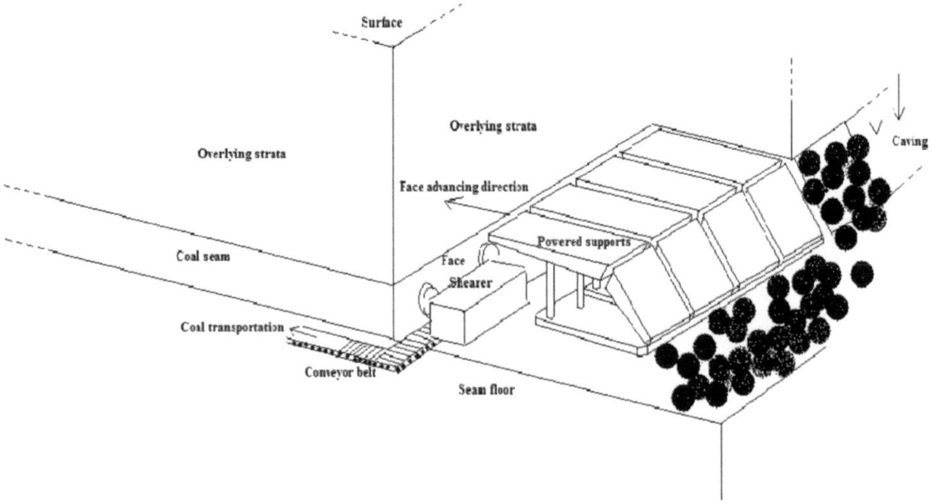

Figure 10.6 Schematic representation of longwall (retreating) mining method.

coal seams, which are not intersected by any structural disturbances (like faults, dykes etc.) and having a nearly uniform thickness. The length and width (face length) of longwall panel can vary from 1000–3500m to 150–300m, respectively, depending upon the geology of the coal seams of the area. The extraction of coal from longwall mines is preceded by formation of roadways by Continuous Miners. These roadways provide passage for supply of workers, instruments, ventilation, electrical supply, compressed air and gas drainage lines, and are commonly known as the main headways. Additional passages (known as development headways) are constructed at both ends of the panel, which are connected with each other. The extraction of coal from the panel is carried out by deploying a shearer at the face, which is then supplied to the surface through a series of conveyor belts. The factors influencing the rate of excavation (viz. face advance) depends on geology of coal seam and capacity of deployed instruments, which usually varies between 50 and 100m. Some portion of coal is left in the pillars (known as the chain pillars) between main headings and development headings in order to protect the roadways. Block diagram of a longwall along with the headings is shown in Figure 10.7. The powered supports are installed in front of the longwall face for preventing the roof strata from falling above the shearer. The working of the shearer under the powered supports across the longwall face is shown in Figure 10.8. The face advance due to massive extraction of coal leads

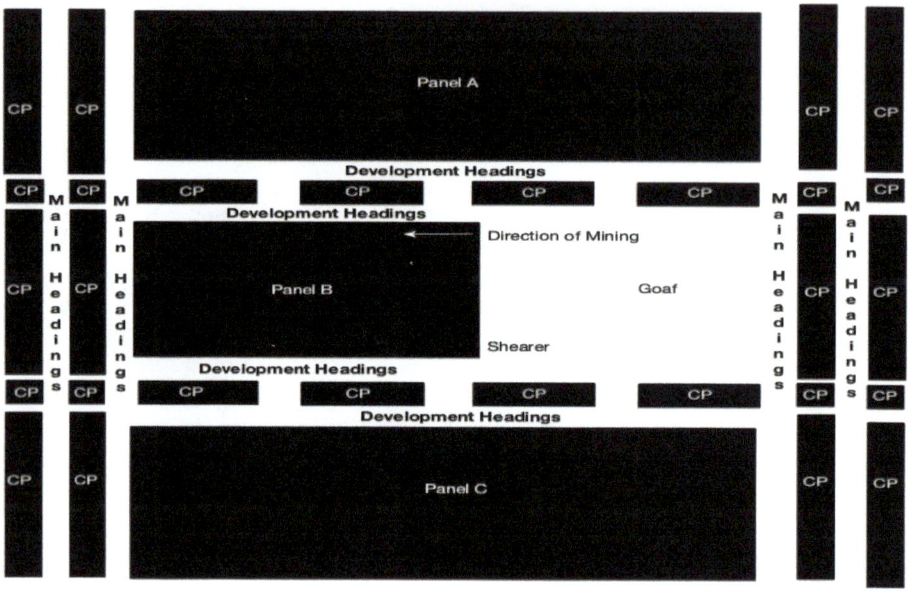

Figure 10.7 Block diagram of longwall panel (chain pillars, main headings and development headings).

to creation of a large void space (known as goaf). The immediate roof behind the powered support is allowed to fall in goaf at regular intervals (called roof caving, refer Figure 10.9) to protect the powered supports from the excess load. Longwall technique is much popular in some of the major coal-producing countries of the world like USA and China.

Figure 10.8 Shearer under-powered supports across the longwall face.

Figure 10.9 Roof caving behind the powered supports (in goaf).

10.3.3 Instrumentation setup for microseismic monitoring

Mining environments (both opencast and underground) are highly influenced by microseismicity created due to various activities related to blasting, instrument movements (HEMMs, conveyors and shearer) and rock fracturing. Massive coal extraction leads to extreme micro-fracturing of rock resulting in microseismic emission, which are basically the elastic energy released due to creation of seismic waves of very smaller magnitude. The presence of mine-induced microseismicity was first noticed and studied in South Stanford coalmine (England) in 1738 and later on was carried out in several countries of the world (Li et al., 2007). Earlier microseismic monitoring was confined to only underground coal mines, but with time it was realized that it is associated with all forms of mining activities, and also considered as a potential hazard (Li et al., 2007). Microseismic monitoring involves tracking the elastic energy released due to micro fracturing of rock mass in order to study the rock mass stability (Yang et al., 2007). Similar instrument for monitoring the microseismic tremors was first installed in Upper Silesia coalmine (Poland) in the late 1920s (Gibowicz and Kijko, 1994), and later on this technique was extensively used for carrying out various studies in coal mines of Poland, South Africa, Canada, China, India and other mining countries (Gibowicz and Kijko, 1994; Sivakumar et al., 1998; Cai et al., 2001; Young and Collins, 2001; Ge, 2005; Li, 2009; Zhao et al., 2015). The instrument for real-time monitoring of mine microseismicity was developed by ISS South Africa for studying the ground pressure activity in Dongguashan copper mine (Yang et al., 2007; Tang and Xia, 2010). The aforesaid instrument was also installed in Hongtoushan copper mine for prediction of mine dynamic disasters (Zhao et al., 2009; Liu et al., 2013). The monitoring of microseismic events is a very effective tool in underground mines as they are associated with all forms of rock failures, that is, rock falls, rock bursts, bumps and outbursts (rock falls: loose rock falling due to their own weight; rock bursts: hazardous violent rock failures; bumps: non-hazardous violent rock failures; outbursts: rapid ejection of rock mass due to sudden release of gas) (Obert and Duvall, 1967; Osterwald, 1970; Cook, 1976). Microseismic emissions cannot be avoided in underground mines (where excessive stress is leading to rock failure) and they are often followed by formation of additional fault planes, which are parallel to the working faces (Cook, 1976). Apart from rock bursts assessment, microseismicity monitoring has also helped in mitigating additional mine-related problems, such as differentiating the signal characteristics between mine shocks and gas outbursts in coalmines (Ding et al., 2016). It has been effective in identification of susceptible zones to water inrush and gas emission in the coalmines as carried out in Taoyuan coal mine (Jian et al., 2011) and Velenje coal mine (Si et al., 2015), and also helped in studying the roof damage in shallow longwall face of Don Sheng Coalfield (III-2) (Yanfang, 2011). Apart from underground mining, microseismic monitoring has also helped in studying slope stability in opencast coal mine (Vinoth and Kumar, 2014). Microseismicity has helped in keeping an eye on the illegal mining in China (Guo et al., 2011) and also proved to be a very efficient tool in locating the rock fall sites in old abandoned mines of Korea, which was considered as serious geohazard (Waltham et al., 2011). Thus, the microseismic technique is a robust tool, which is helpful in several studies related to rock strata dynamics and its corresponding stress conditions.

For microseismic source parameters acquisition, two different instruments were used for the present study: (a) NIRM microseismic monitoring system (NMMS), (b) Integrated Seismic System (ISS) (Srinivasan et al., 2005).

a. NIRM Microseismic Monitoring System (NMMS): The NIRM Microseismic Monitoring System was developed by National Institute of Rock Mechanics (India) and was capable of recording the ERR in real time (Sivakumar et al., 2008). The instrument was designed to operate in both automated as well as manual mode. It worked on MS-DOS (version 6.0) platform and had 16-channel PC-based data acquisition system. The data acquisition was carried out using advanced digital grade SM4 (14 Hz) and SM6 (4.5 Hz) borehole sensors, which were installed at various depths (35–65 m) along the overlying strata of the panel using boreholes of very small diameter. The aforesaid sensors were connected to the amplifier and frequency modulator. A data was collected from sensors, which were bunched together in groups (each containing six sensors) and are connected to a junction box. The signal from those junction boxes were transmitted to the Central Lab, situated at a distance of 3km through ten pairs of twisted copper cable. In the Central Lab, the input data is passed through Signal Conditioning Unit, which consisted of Carrier Filter, Demodulators and Narrow Band Gain Signal Filters (70–150 Hz). The output data from the Signal Conditioning Unit is then fed to the industrial PC after amplification and normalization using a proper factor. The PC consists of an intelligent data acquisition having 16-channel 512MB data acquisition card and processing unit. The analogue signals were converted to digital format through a 12-bit A/D convertor having sampling frequency of 4KHz/channel. The digital signals were then made noise-free and fed into HDD for calculating the ERR. The block diagram of NMMS instrumentation setup is shown in Figure 10.10 (Srinivasan et al., 2005; Mondal et al., 2017a, 2017b; Mondal and Roy, 2019; Mondal et al., 2020).

b. Integrated Seismic System (ISS): The ISS was developed by ISSI (South Africa), and was a high resolution (24-bit), digital and intelligent seismic monitoring and control unit. The hardware components of ISS consisted of seismic sensors, data acquisition unit, communication units and central site computer, whereas software consisted of in-built plugin related to Run Time System (RTS), Seismological Processing (XMTS) and Visualization/Interpretation of Seismicity (XDS). This instrument was installed in parallel with NMMS for getting additional advantage regarding the small fractures created during the mining operations. The ISS used MS-9 borehole

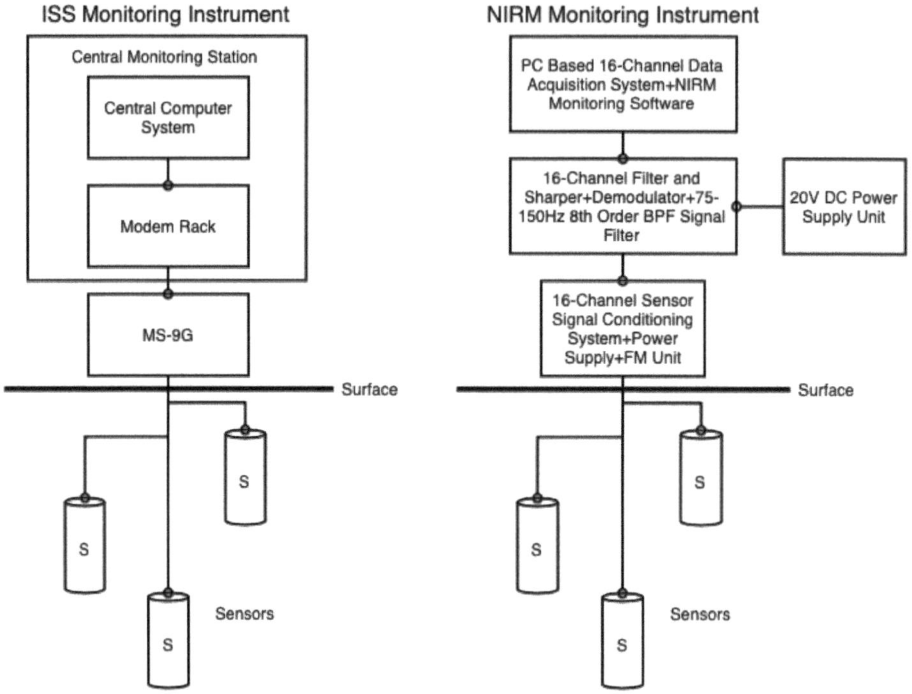

Figure 10.10 Block diagram of ISS and NIRM monitoring instrument. (Sivakumar et al., 2008; Srinivasan et al., 2005.)

geophones installed in the roof strata along with SM4 and SM6 sensors using boreholes between depths 36 and 67 m.

The Data Acquisition Unit helped in triggering all the active sensors for minimizing the unwanted noise and transmitting the acquired data to the Central Computer in real time after automated decimation. The Analog/Digital converter used in Integrated Seismic System had the following specifications: (a) Dynamic Range: 112 dB; (b) Signal to Distortion Ratio: 100 dB; (c) Resolution: 24 bit; (d) Active Channels: 9; (e) Operating Bandwidth: 800–1000 Hz; (f) Sampling Rate/Channel: 2–24 K/s. The important parameters recorded from the aforesaid instrument included microseismic source location (in x, y, z), magnitude, apparent volume, apparent stress, energy index, and so on. The block diagram of ISS instrumentation has been shown in Figure 10.10 (Srinivasan et al., 2005; Mondal et al., 2017a, 2017b; Mondal and Roy, 2019; Mondal et al., 2020). The borehole sensors used with the aforesaid instruments have been shown in Figure 10.11 (SM4, SM6, and MS-9).

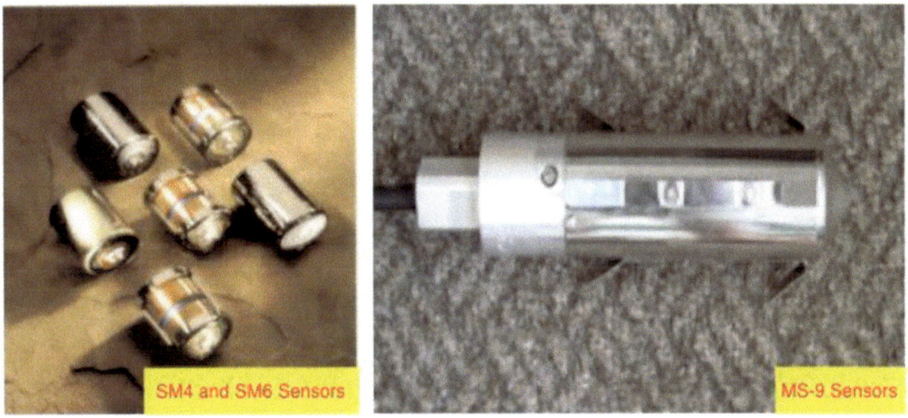

Figure 10.11 Borehole sensors used in the present study.

10.3.4 Correlation integral (c), correlation fractal dimension (*d*) and Seismic *b*-value

Fractal dimension (*d*) using correlation integral (*c*) is very widespread technique and has been used in several studies. A fractal set is considered to have self-similar events, which are scale invariant and seismic events are considered to obey fractal properties with regard to time, space, and magnitude (Kagan and Knopoff, 1978, 1980, 1981). This self-similarity can be explained with the help of Sierpinski gasket (Mondal and Roy, 2019) having a set of equilateral triangle, where each miniature triangle is a mini replica of the primary triangle. The concept of fractal is used to describe the complexity in structure through the ratio of change in dimension to change in scale. In our case the fractal dimension has been used for quantitative scale invariant measurement of microseismic clustering with respect to space and time for getting an idea about the seismicity of the region (Legrand et al., 1996; Nakaya and Hashimoto, 2002). The fractal structures are associated with earthquake seismicity as well as with the micro-fracturing of rocks (Mogi, 1963; Scholz, 1968; Hirata, 1987; Hirata et al., 1987). Previous studies on earthquake have shown that occurrence of seismic events formed clusters in both space and time prior to large earthquakes, which resulted in lowering of fractal dimension (Ouchi and Uekawa, 1986). Box counting method is considered to be the most generalized method for calculating fractal dimension, which is used for analysing the fault system and same has been done previously

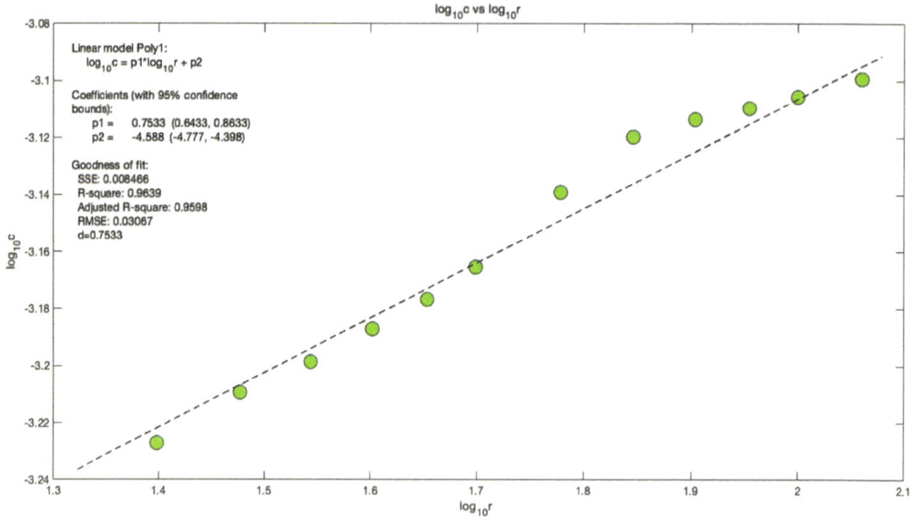

Figure 10.12 $\log_{10}c$ versus $\log_{10}r$ plot. (Mondal and Roy, 2019; Mondal et al., 2020.)

for studying the fault system of Japan (Hirata, 1989), Upper Silesian coal basin-Poland (Idziak and Teper, 1996), and Guemero state-Mexico (Brown et al., 1998). Though, box counting is a robust method but has its own limitations, where fractal values become highly unstable in case of limited number of seismic event (Greenside et al., 1982). Therefore, correlation integral method has been used for computing the fractal dimension in present case study, which gives an idea about the fraction of points lying between a given scale range (r). The fractal dimension (D) is calculated through as equations 10.2 and 10.3, which is basically the slope of the curve plotted between correlation integral (c) and scale range (r) on logarithmic scale (Figure 10.12) (Grassberger and Procaccia, 1983).

$$c(r) \sim r^d \tag{10.2}$$

$$d = \lim_{r \to 0} \frac{log_{10}c(r)}{log_{10}r} \tag{10.3}$$

The correlation integral in 2-D and 3-D space is calculated using equations 10.4, 10.5, and 10.6 (Grassberger and Procaccia, 1983; Henderson et al., 1999; Roy and Mondal, 2009; Roy and Gupta, 2015), where h is the Heaviside Step function (equation 10.7) (Al-Shameri, 2012), r is the scaling distance, n is the total number of event combinations in the fractal set, x_i and x_j represents the position of the two events separated by distance r_{ij}.

$$c(r) = \lim_{n\to\infty} \frac{1}{n^2}\{number\, of\, pairs\ (x_i, x_j)\ with\ r_{ij} < r\} \tag{10.4}$$

$$c_q(r) = \frac{1}{n}\sum_{j=1}^{n}\left\{\left(\frac{1}{n}\right)\sum_{i=1}^{n} h\left(r - |x_i - x_j|\right)^{q-1}\right\}^{1/(q-1)} \tag{10.5}$$

$$c(r) = \frac{1}{n(n-1)}\sum_{j=1}^{n}\sum_{i=j+1}^{n} h(r - r_{ij}) \tag{10.6}$$

$$h = \left\{\begin{array}{l} 1,\ for\ 0 \le (r - r_{ij}) \\ 0,\ for\ 0 > (r - r_{ij}) \end{array}\right\} \tag{10.7}$$

The inter-event distance between the microseismic events has been calculated with the help of P-wave travel time equation given as equation 10.8, where coordinates of the sensors and P-wave travel time are given as x_n, y_n, z_n, and t_n, and the coordinates of the microseismic events, their occurrence time and velocity of the generated P wave (Eq. 10.9) is given as x, y, z, t and v, respectively. E, ξ, and ρ are the modulus of elasticity, Poisson's ratio, and density of the medium (Hirata and Imoto, 1991; Roy and Gupta, 2015; Ming-Ruo et al., 2003; Tie and Hongguang, 2010).

$$(x - x_n)^2 + (y - y_n)^2 + (z - z_n)^2 = v^2(t - t_n)^2 \tag{10.8}$$

$$v = \sqrt[2]{\frac{E(1-\xi)}{(1+\xi)(1-2\xi)\rho}} \tag{10.9}$$

The selection of the most suitable window length for calculation of fractal dimension is very crucial and the best window length was considered to be between 50 and 75 (Roy and Gupta, 2015). Higher window length will lead to missing of shorter patters (like local stress build ups and fracturing), whereas smaller window length will result in reducing the linearity region in $\log_{10}(c)$ versus $\log_{10}(r)$ curve further resulting in

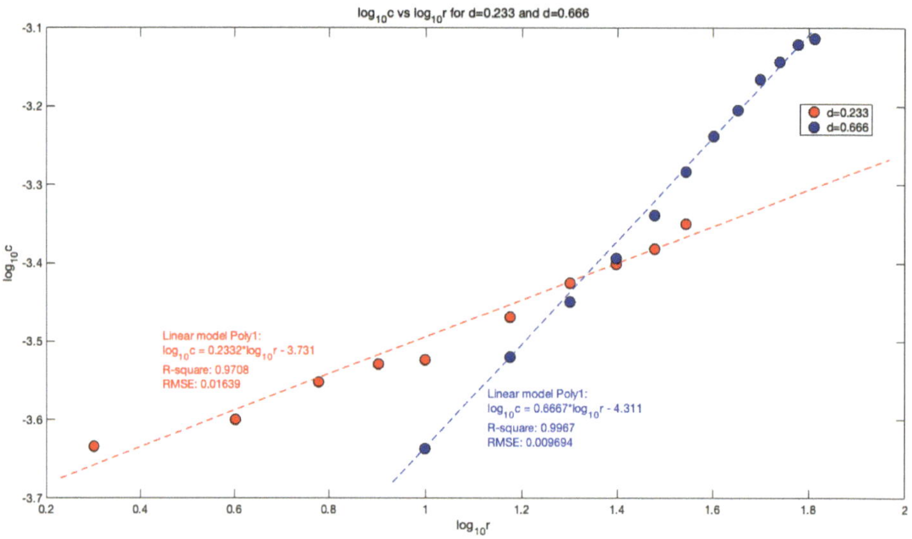

Figure 10.13 $\log_{10}(c)$ versus $\log_{10}(r)$ plot for $d = 0.233$ and $d = 0.666$. (Mondal and Roy, 2019; Mondal et al., 2020.)

making the fractal values very unstable and inaccurate (Roy and Mondal, 2012). Lower fractal values correspond to clustering of microseismic events, and it has been explained with the help of $\log_{10}(c)$ versus $\log_{10}(r)$ curve for $d = 0.233$ and $d = 0.666$ (Figure 10.13). During stress accumulation, the microseismic events are clustered in space and time resulting in elevation of correlation integral for lower scale ranges and vice versa. Thus, from Figure 10.13, it can be clearly seen that $\log10(c)_{\mathrm{d}=0.233}$ is greater than $\log10(c)_{\mathrm{d}=0.666}$ (for $\log_{10}(r) < 1.3$), whereas on the other hand $\log10(c)_{\mathrm{d}=0.233}$ is less than $\log10(c)_{\mathrm{d}=0.666}$ (for $\log_{10}(r) > 1.3$).

In parallel to fractal dimension, the mine-induced microseismic data was also used for calculating another important parameter known as the seismic *b*-value. The distribution of seismic events associated with earthquakes in space, time, and magnitude showed non-uniformity and scale-invariability, which seemed to be self-similar (refer to Figure 10.14), and the magnitude of the seismic events obeyed the power law relation. The aforesaid theory of earthquake magnitude and estimation of *b*-value was introduced by C. Richter in 1935, which proved to be very helpful is getting an idea about the occurrence of larger as well as smaller magnitude earthquakes in a region (Wang et al., 2009). Thus, self-similarity relation for n_e number of events (earthquakes) having magnitude greater

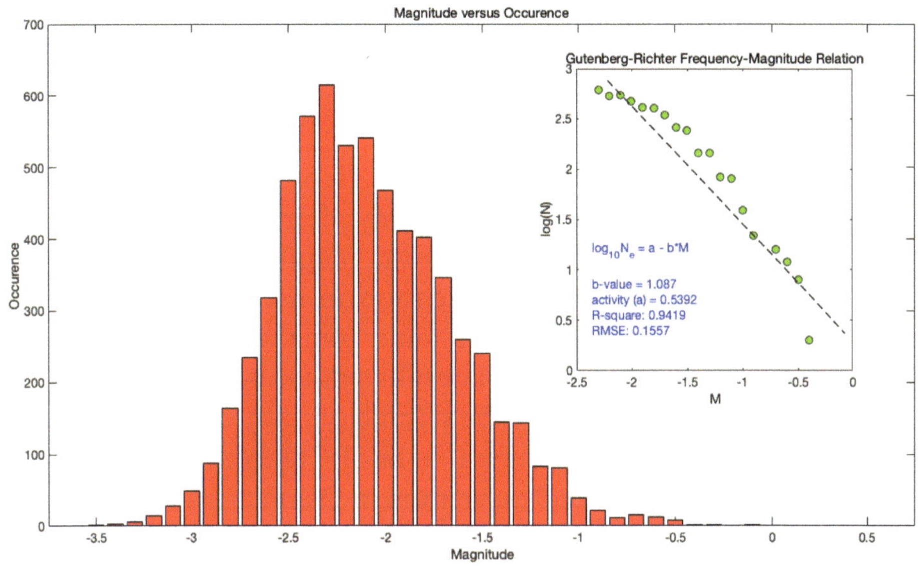

Figure 10.14 Distribution of earthquake magnitude with respect to their occurrence. (Mondal and Roy, 2019.)

than M (also known as the threshold magnitude) can be represented as equation 10.10, which is also known as the Gutenberg-Richter Magnitude Frequency Relation (Ishimoto and Iida, 1939).

$$log_{10}N_e(m > M) = a - bM \tag{10.10}$$

In the aforesaid equation, a (also known as the Microseismic Activity Parameter) and b are real positive constant representing the seismic activity and abundance of larger magnitude events over smaller ones in an area. The larger occurrence of smaller magnitude events is represented by high b-value (in equation 10.10, usually b is close to 1) and vice versa. The seismic b-value can also be estimated through maximum likelihood method (Eq. 10.11), which provides more accurate estimation than the least square method (Ustu, 1965; Aki, 1965), where M_{avg} and M_{min} are the average and minimum magnitude.

$$b = \frac{log_{10}e}{\left(M_{avg} - M_{min}\right)} = \frac{0.43}{\left(M_{avg} - M_{min}\right)} \tag{10.11}$$

The decline in seismic b-value and fractal dimension has been observed before several earthquakes (Ouchi and Uekawa, 1986), and similar results were expected in case of rock fracturing in underground mines. The precursory signatures of fractal dimension and *b*-value has been used for earthquake forecasting, which is a widespread method (Caneva and Smirnov, 2004). The study carried out for Tokachi-oki Earthquake (M = 8.0), which occurred in Hokkaido (Japan) in 2003 showed that fractal values started decreasing gradually after 1998 and reached minimum in 2002 (i.e., one year prior to the main shock) (Murase, 2004). The temporal study of fractal dimension in northwestern Himalayan region helped in finding the zones undergoing accumulation of stress (Roy and Mondal, 2009). Similar results related to lowering of fractal dimension was also seen during the seismicity study around the injection wells drilled in Geysers geothermal area of California (USA) (Henderson et al., 1999), before rock deformation in Oshima granite while carrying out microseismicity study in laboratory (Hirata et al., 1987) and also during the study of rock bursts hypocenters in mines (Xie and Pariseau, 1993). The lowering in seismic *b*-value was seen before main earthquake shock and swarms during Caracas earthquake in 1967 and Hengill Triple Junction, Iceland. Similar results were also seen during laboratory testing of microseismic events related to rock failure on various rock types such as Oshima granite, San Marcos gabbro, Westerly granite and Darley Dale sandstone (Scholz, 1968; Weeks et al., 1978; Hirata et al., 1987; Meredith et al., 1990).

10.3.5 A brief discussion on overlying strata of underground mines

Roof strata plays a very important role in underground mining. Longwall roof is broadly classified into three zones (refer Figure 10.15), caved zone, fractured zone, and continuous deformation zone (Mondal et al., 2020).

a. Caved Zone: It is the roof just above the mining area, which is allowed to cave over the mine floor after the face advance. The thickness of caved zone is usually about 2–8 times the mining height and depends on the mining height (h_{im}), sagging of lowest uncaved strata (s) and bulking factor (k) (refer Eq. 10.12; Mondal et al., 2020).

$$H = s + h_{im}(k - 1) \tag{10.12}$$

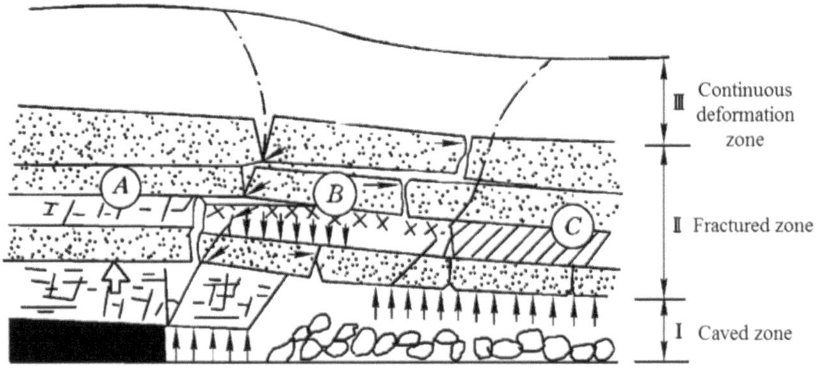

Figure 10.15 Caved zone, fractured zone, and continuous deformation zone. (Shu et al., 2019.)

The bulking factor (which is the ratio of change in volume of excavated strata with respect to volume in its in-situ state) acts as the key deciding factor in estimation of caving height. The bulking factor for various rocks and minerals, which are found in the sedimentary deposits of coalfields has been summarized in Table 10.2 (Chiang, 1980; Peng, 2019).

b. Fractured Zone: The fractured zone is situated just above the caved zone. It is sandwiched between caved zone and continuous deformation zone, and mainly consisted of broken blocks of rock mass (also known as the force-transmitting beams) having vertical/sub-vertical fractures with horizontal cracks (occurred due to bed separation). The thickness of this zone is approximately 28–42 times the mining height. Thus, the combined height of caved and fractured zone is about 30–50 times the mining height, also commonly known as destressed zone (Dahl and Vonschonfeldt, 1976; Mondal et al., 2020).

c. Continuous deformation zone: The upper-most layer is known as the continuous deformation zone. The downward gravitational pull leads to rock deformation in this zone without causing any major fractures and hence this zone acts as a continuous medium.

The presence of aforesaid three zones in the roof strata has been confirmed through laboratory experiments (Kuzyniazouv, 1954). Based on another classification carried out on the basis of roof caving, the overlying strata is divided into two major subgroups: [a] immediate roof

Table 10.2 Low fractal values obtained before minor and major falls

Sl. No.	*Fall No.*	*WN*	*Date*	*Time*	*FP (m)*	*Lowest D*	*WN*	*FT (Hrs.)*
			August and September, 2001					
1	LF1	1	28/08/2001	12:00	20.00	0.720	1	0
2	LF2	1	29/08/2001	06:20	21.00	0.720	1	0
3	LF3	1	30/08/2001	07:25	27.70	0.720	1	0
4	LF4	1	30/08/2001	17:20	27.70	0.720	1	0
5	LF5	2	31/08/2001	16:30	31.30	0.720	1	13.25
6	LF6	13	08/09/2001	03:00	44.35	0.176	4	148.25
7	RF1	14	08/09/2001	20:33	44.35	0.111	14	4.95
8	RF2	14	09/09/2001	03:18	45.07	0.111	14	11.58
9	LF7	20	13/09/2001	05:10	54.37	0.183	20	0
10	LF8	20	13/09/2001	05:35	54.37	0.183	20	0
11	LF9	20	13/09/2001	12:45	54.37	0.183	20	0
12	RF3	20	14/09/2001	00:00	Not Known	0.183	20	5.91
13	RF4	29	16/09/2001	21:16	59.90	0.160	27	16.00
14	LF10	30	17/09/2001	04:00	62.32	0.160	27	22.78
15	RF5	30	17/09/2001	13:22	62.32	0.160	27	32.06
16	LF11	30	17/09/2001	17:00	62.32	0.160	27	35.73
18	LF12	32	20/09/2001	13:00	63.55	0.766	31	57.41
19	LF13	32	20/09/2001	14:25	63.55	0.766	31	58.83
20	LF14	33	25/09/2001	04:10	Not Known	0.766	31	168.41

(overburden strata lying just above the roofline, which is allowed to cave in the gob area immediately after the face advance) and [b] main roof (portion of overburden strata lying at the lower portion of the fractured zone and above the immediate roof) (refer Figure 10.15). As discussed earlier, caving is a very crucial phenomenon in longwall mining and regular caving of the overhanging strata above the goaf is highly desired for releasing the load from the powered supports. The cavablity of the overlying strata is highly dependent on the geology of its constituent rock mass. Based its cavablity, the immediate roof is classified into several subgroups: [a] unstable immediate roof: Soft rock of mainly consisting of carbonaceous shale, sandy shale. Caving occurs immediately after the face advance creating smaller size fragments, [b] medium stable immediate roof: Comparatively harder rock consisting of hard shale, sandy shale, weak sandstone and the joints/fracture created are very less well defined. Caving occurs shortly after the face advance and fragments produced are larger in size, and [c] stable immediate roof: Very hard roof consisting of massive sandstone, conglomerate, and so forth. Caving is very irregular and overhanging roof can stay in the intact position upto 5–8hrs. Thus, the area under the overhanging immediate roof can reach up to 70,000–80,000ft^2 (6500–7440m^2), which results in introduction of tremendous pressure on the powered supports. Artificial/induced caving methods are highly recommended for such roof (Peng and Chiang, 1982, 1984; Mondal et al., 2017a, 2017b).

10.4 FRACTAL BEHAVIOR OF LONGWALL PANEL

10.4.1 Stress monitoring through correlation integral and correlation fractal dimension

The accumulation of stress in overlying strata is represented by clustering of microseismic events in both space and time. The fractal behavior of roof strata during different (high and low) stress condition has been discussed through event patterns corresponding to two fractal values $d = 0.032$ and $d = 1.017$, respectively (Figure 10.16). The event separation has been calculated for every possible combinations (total combination = $^nC_2$; where n is the total events in a window), and the values were arranged in ascending order (refer Figure 10.16a). Thus, from Figure 10.16b it can be seen that

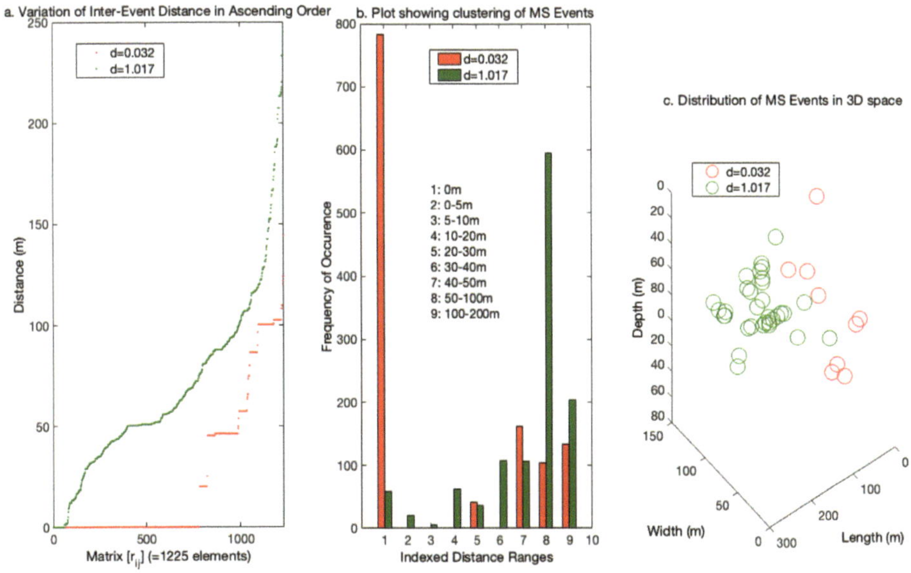

Figure 10.16 Variation of fractal dimension with event clustering.

higher frequency of occurrence is obtained lower values of r_{ij} for $d = 0.032$ (highest occurrence for $r_{ij} = 0$) (refer Figure 10.16b). On the other hand, for $d = 1.017$, higher occurrence is seen for greater values r_{ij} (highest occurrence frequency obtained for scale distance 50–100m; refer Figure 10.16b). The above results became much clearer after studying the spatial distribution of microseismic events (Figure 10.16c), which showed that events were more clustered for $d = 0.032$ (also concentrated at one place) and events corresponding to $d = 1.017$ are highly dispersed in space. Thus, fractal values acted as indicator for identification of stressed zones in roof strata.

10.4.2 Validation of the microseismic raw data

Since estimation of correlation integral, viz., fractal dimension depends on the spatial distribution of microseismic events, therefore, it's very necessary to pick only those events created within the study area, that is, roof strata of the panel under investigation. Addition of unwanted dataset/noise will result in inaccurate estimation of correlation integral, which will further contaminate the fractal values. Therefore, manual filters were applied for removing the noise, which included both unwanted events recorded outside the study area and also events containing error values. Thus, the filtered dataset of microseismic events generated from the panel roof from September 2001 to February 2002 has been shown in Figure 10.17

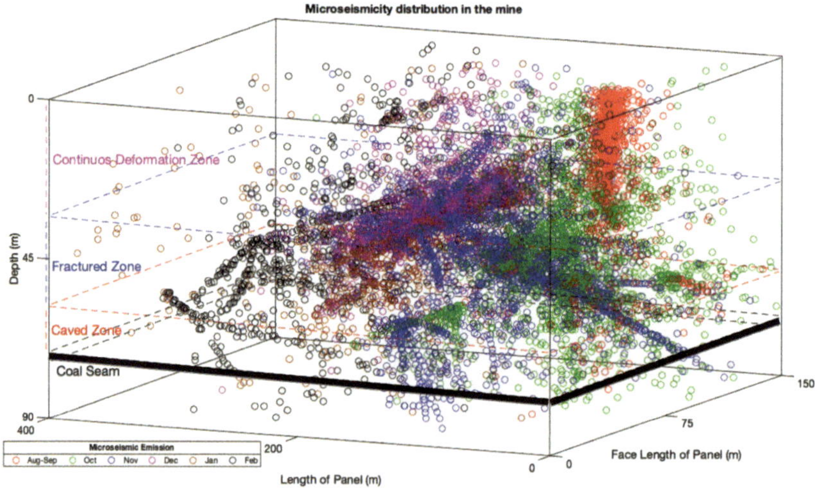

Figure 10.17 Microseismic emission during September 2001–February 2002. (Mondal et al., 2020.)

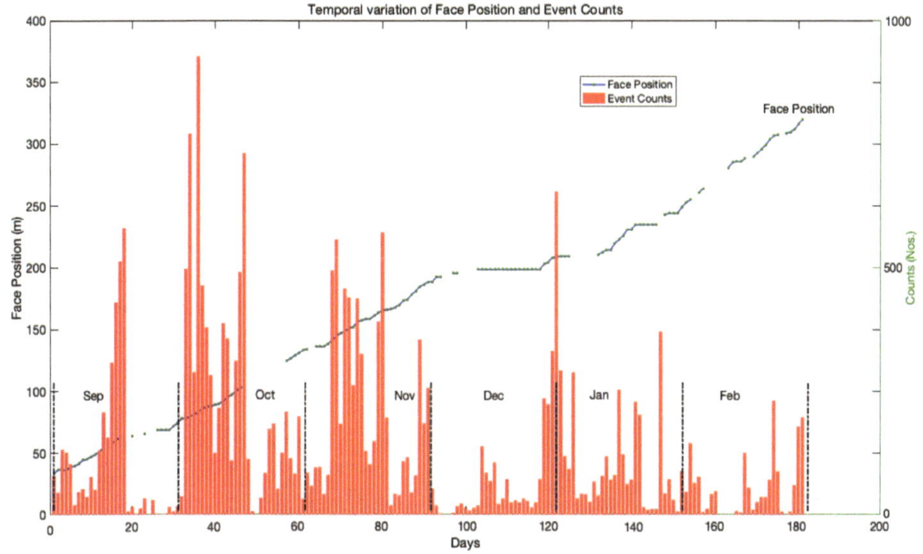

Figure 10.18 Temporal variation of event counts and face position. (Mondal and Roy, 2019.)

(Mondal et al., 2017a; Mondal and Roy, 2019; Mondal et al., 2020). The temporal variation of face position and event counts/day for the entire investigation period has been shown in Figure 10.18. It (Figure 10.18) showed that microseismic emission was high during October 2001–November 2001, which resulted due to stress readjustment in the virgin strata.

10.4.3 Temporal variation of fractal dimension and its gradient during the investigation period

a. From 28 August 2001 to 30 September 2001: The microseismic monitoring of the longwall panel started on 28 August 2001, when the face position was 20m. The total face shift for the month was around 51.6m as face shifted from 20m (as on 28 August 2001) to 71.6m (as on 30 September 2001). The temporal variation of fractal dimension and its gradient from 28 August 2001 to 30 September 2001 has been shown in Figure 10.19 and Figure 10.20, respectively. Primary readjustment of stress in the overlying strata due to initiation of mining activities has resulted in high microseismic emission and abrupt fluctuation in fractal variation. During aforesaid time period, a total of 14 minor falls (local falls) and five major rock bursts (roof fall) were reported

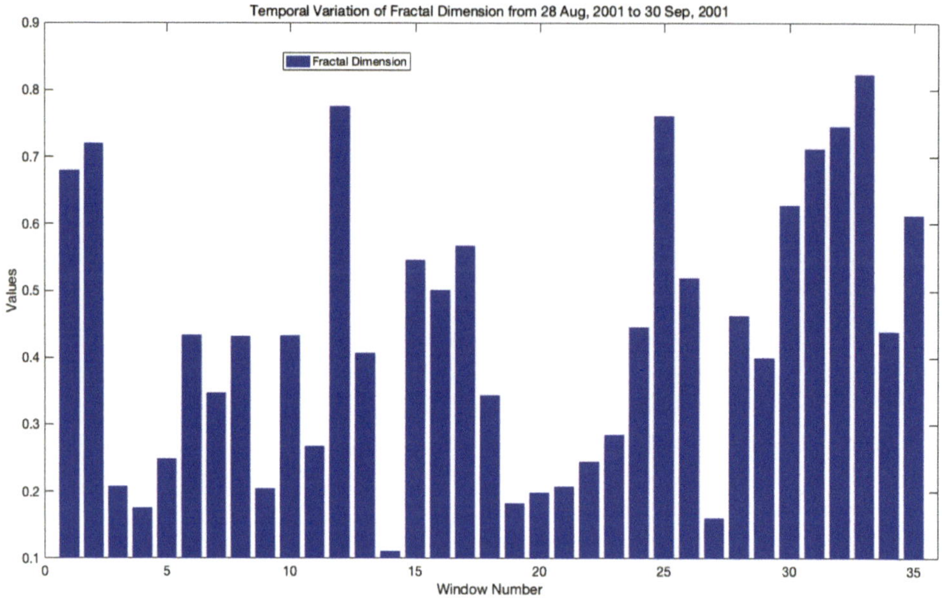

Figure 10.19 Fractal dimension between 28 August 2001 to 30 September 2001. (Mondal and Roy, 2019.)

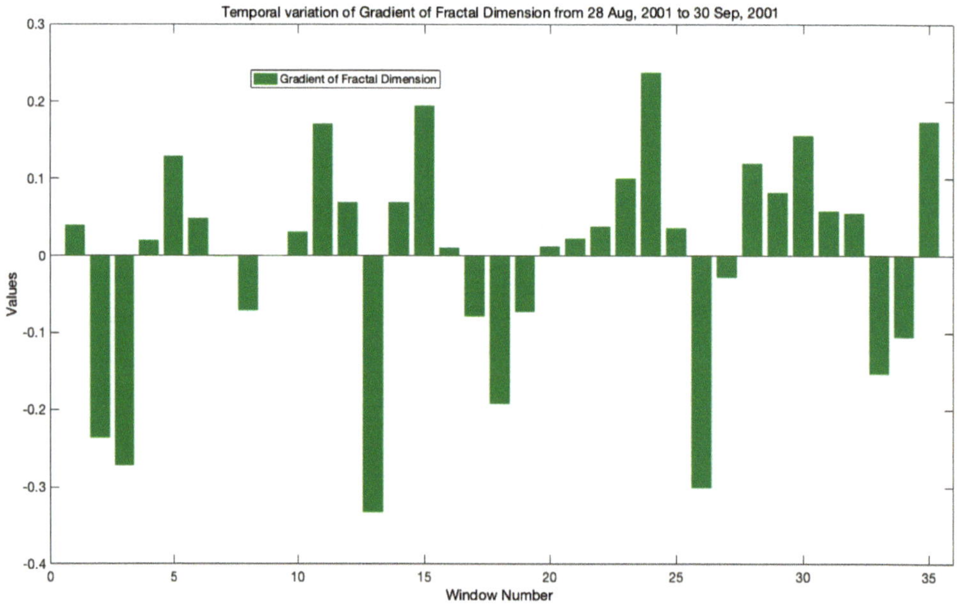

Figure 10.20 Gradient of fractal dimension between 28 August 2001 to 30 September 2001.

from the panel. The readjustment of strata has resulted in continuous stress accumulation and release in the roof strata leading to abrupt fluctuations in fractal values and its gradient (refer Figures 10.19 and 10.20). Lowering in the fractal dimension and high negative gradient was obtained before major as well as minor falls. The lower fractal values along with details related to window number, time to fall (*FT*) as obtained before both major and minor falls have been summarized in Table 10.2.

b. From 1 October 2001 to 31 October 2001: The face position on the first day of the month was 74.05m (1 October 2001), which shifted to 133.23m on 31 October 2001. Thus, the total face shift reported was around 75m, which showed that the rate of coal excavation increased in the month of October. The faster coal excavation resulted in leaving higher volume of goaf behind the powered supports leading to severe rate roof stress readjustment and higher emission of microseismic activity, which further increased the frequency of occurrence of major as well as minor falls. A total of 18 major falls and nine minor falls were reported in the month of October. The temporal variation of fractal dimension and its gradient from 1 October 2001 to 11 October 2001 has been shown in Figures 10.21 and 10.22, respectively. Initially, the stress

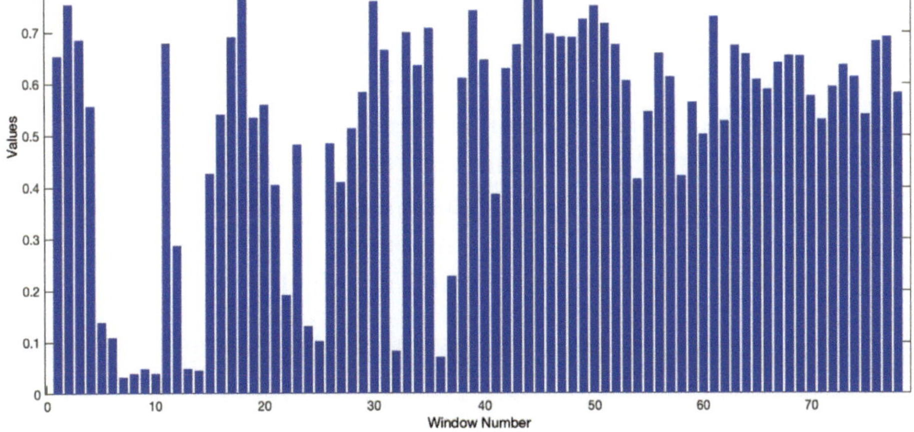

Figure 10.21 Fractal dimension between 1 October 2001 to 11 October 2001. (Mondal and Roy, 2019.)

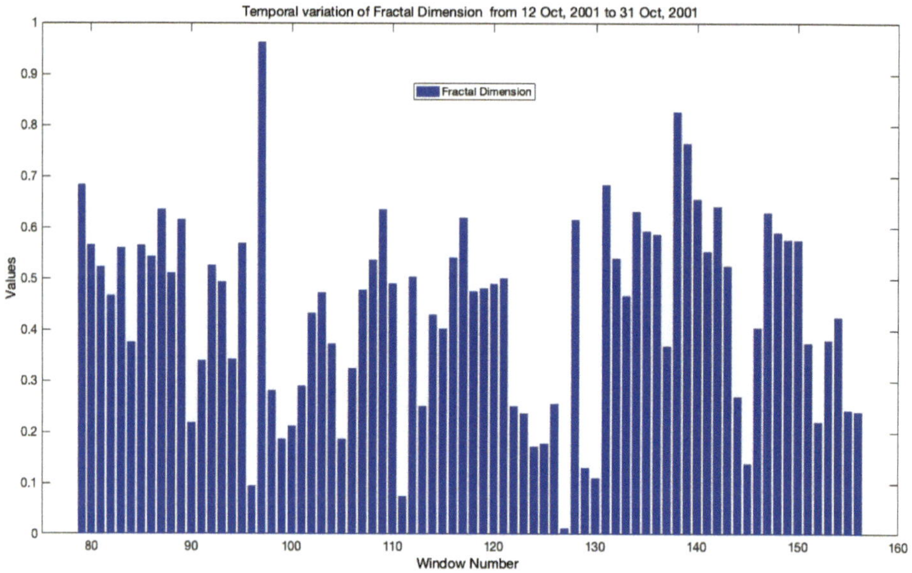

Figure 10.22 Fractal dimension between 12 October 2001 to 31 October 2001. (Mondal and Roy, 2019.)

adjustment in roof has resulted in highly fluctuating fractal dimension and its gradient values, as seen during the first half of the month (refer Figures 10.21 and 10.22), after that strata readjustment has little bit stabilized and relatively less fluctuations have been seen in the aforesaid parameters. The temporal variation of fractal dimension and its gradient for later half of the month, that is, from 12 October 2001 to 31 October 2001 has been shown in Figures 10.23 and 10.24, respectively. In the latter half of the month, the fluctuation in fractal dimension has increased again, which can be justified from the rapid change in the fractal gradient and the number of rock bursts (both major and minor) has also increased during this time. The lower fractal values obtained before both major and minor falls in the month of October have been summarized in Table 10.3.

c. From 1 November 2001 to 30 November 2001: November was another high productive month after October and total face shift was around 54.25m (from face position 134.05m on 1 November 2001 to 188.3m on 30 November 2001). Massive coal excavation has resulted in high

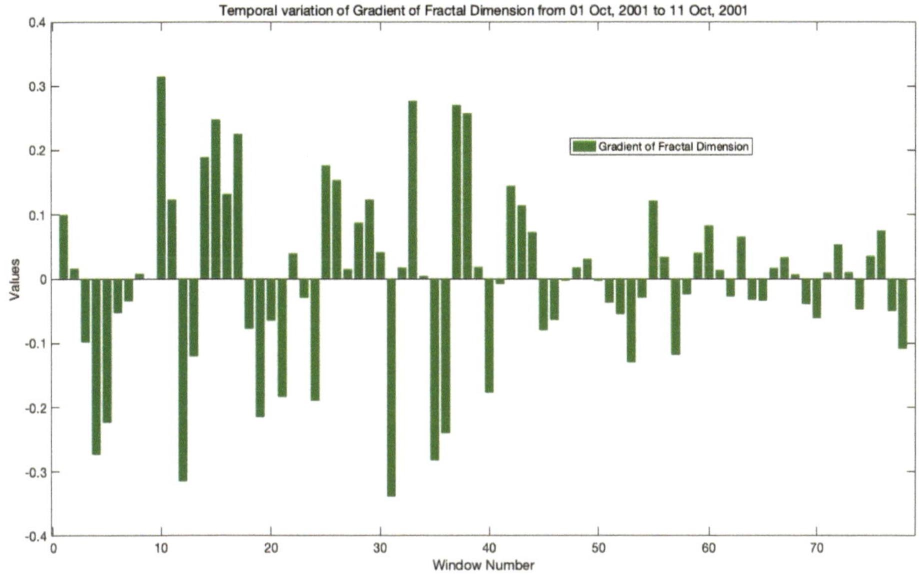

Figure 10.23 Gradient of fractal dimension between 1 October 2001 to 11 October 2001.

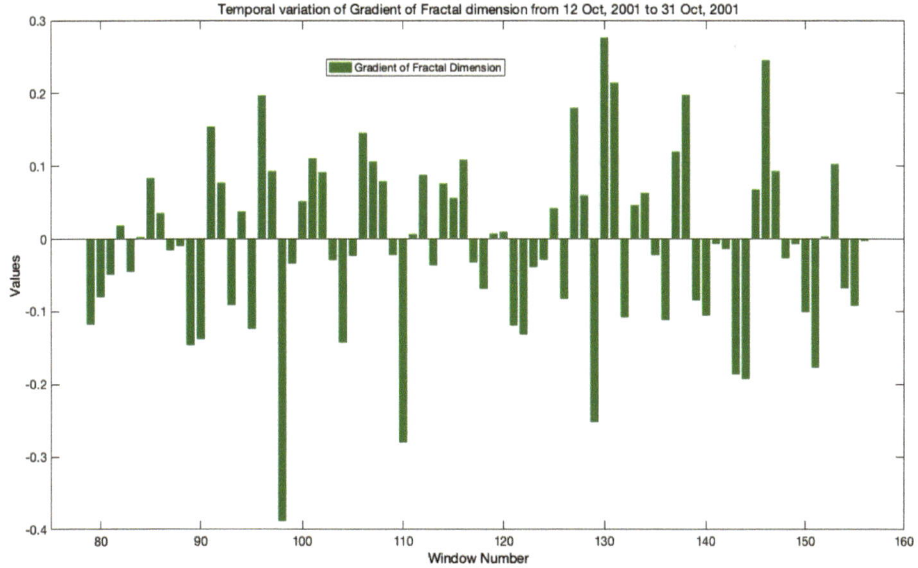

Figure 10.24 Gradient of fractal dimension between 12 October 2001 to 31 October 2001.

Table 10.3 Low fractal values obtained before minor and major falls

Sl. No.	*Fall No.*	*WN*	*Date*	*Time*	*FP (m)*	*Lowest D*	*WN*	*FT (Hrs.)*
				October, 2001				
1	LF15	1	02/10/2001	10:30	77.70	0.613	35	39.50
2	RF6	1	02/10/2001	18:24	77.70	0.613	35	47.33
3	RF7	2	03/10/2001	03:41	77.70	0.613	35	56.66
4	LF16	2	03/10/2001	09:00	77.70	0.613	35	62.00
5	LF17	2	03/10/2001	11:30	77.70	0.613	35	64.50
6	RF8	15	04/10/2001	03:55	79.25	0.032	7	6.39
7	RF9	16	04/10/2001	05:02	79.25	0.032	7	8.50
8	LF18	30	05/10/2001	17:15	80.62	0.102	25	21.66
9	RF10	45	06/10/2001	17:15	83.30	0.070	36	4.66
10	RF11	46	06/10/2001	17:22	83.30	0.070	36	4.75
11	LF19	67	09/10/2001	02:43	87.45	0.415	54	35.33
12	RF12	74	10/10/2001	21:31	88.75	0.530	71	30.16
13	LF20	75	11/10/2001	09:30	89.30	0.530	71	42.38
14	RF13	80	12/10/2001	00:17	92.25	0.405	80	0
15	LF21	90	13/10/2001	13:40	94.42	0.405	90	0.83
16	LF22	96	14/10/2001	17:00	96.80	0.095	96	1.58
17	LF23	96	14/10/2001	21:05	96.80	0.095	96	5.66
18	RF14	107	16/10/2001	11:15	102.20	0.187	105	5.66
19	RF15	107	16/10/2001	16:53	102.20	0.187	105	11.30
20	RF16	118	17/10/2001	10:32	103.55	0.075	111	16.12
21	RF17	123	17/10/2001	14:16	103.55	0.383	123	0.03
22	RF18	126	17/10/2001	20:39	103.55	0.172	124	4.83
23	RF19	131	22/10/2001	21:00	Not Known	0.012	127	118.09
24	RF20	147	28/10/2001	03:36	126.30	0.139	145	20.17
25	RF21	149	29/10/2001	05:47	129.00	0.590	148	19.16
26	RF22	149	29/10/2001	21:48	129.00	0.590	148	34.33
27	RF23	153	30/10/2001	04:45	130.62	0.220	152	3.05

microseismic emission from the panel roof, and a total of about 26 major falls and 6 minor falls was reported from the longwall mine. The temporal variation of fractal dimension and its gradient from 1 November 2001 to 14 November 2001 has been shown in Figures 10.25 and 10.26, respectively. High stress accumulation in the roof can be inferred from lower values of fractal dimension and also from the rapid decline in its gradient, which has resulted in occurrence of only major falls. The temporal variation of fractal dimension for the second phase of the month, that is, from 15 November 2001 to 30 November 2001 is shown in Figures 10.27 and 10.28, respectively, which also

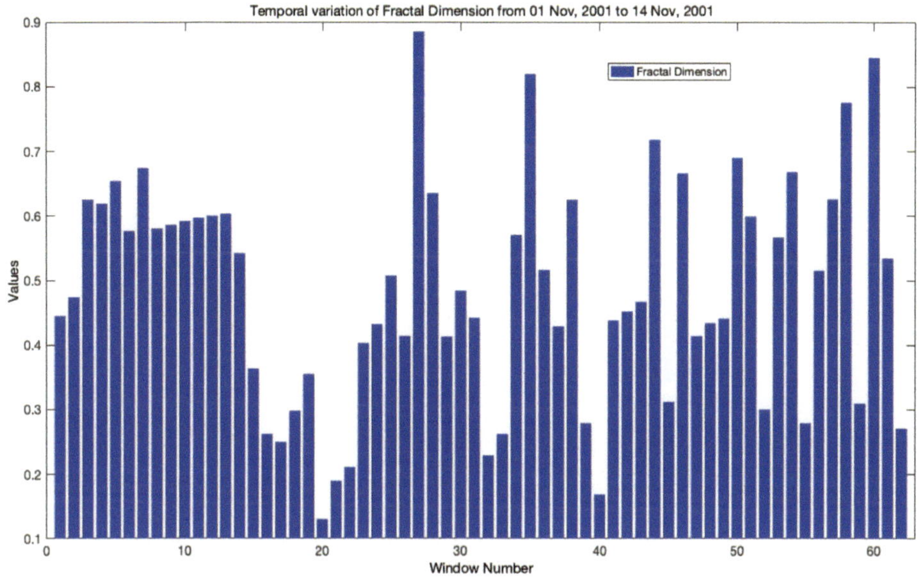

Figure 10.25 Fractal dimension between 1 November 2001 and 14 November 2001. (Mondal and Roy, 2019.)

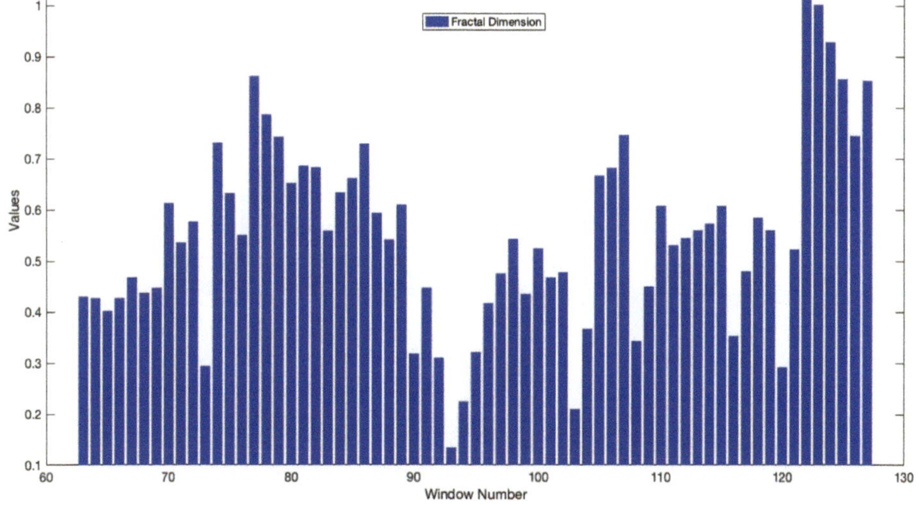

Figure 10.26 Fractal dimension between 15 November 2001 and 30 November 2001. (Mondal and Roy, 2019.)

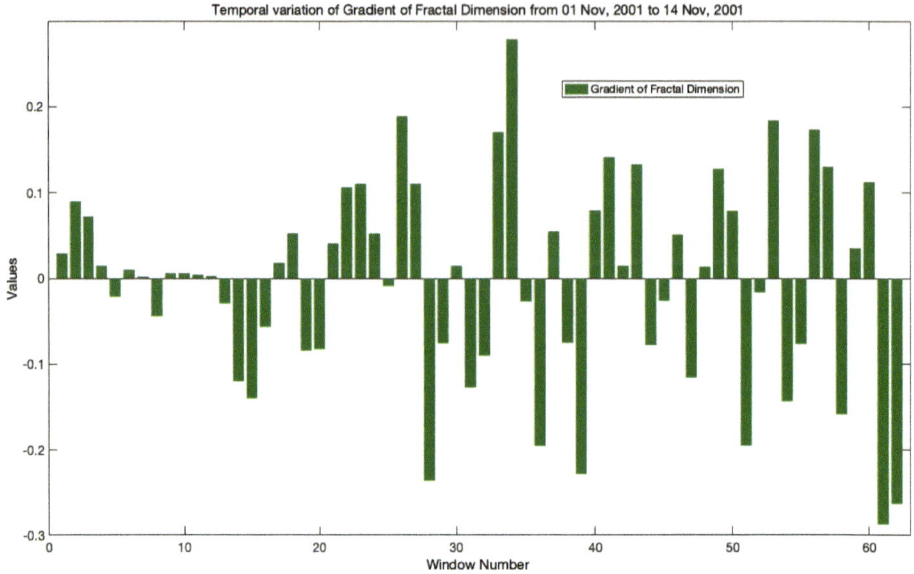

Figure 10.27 Gradient of fractal dimension between 1 November 2001 and 14 November 2001.

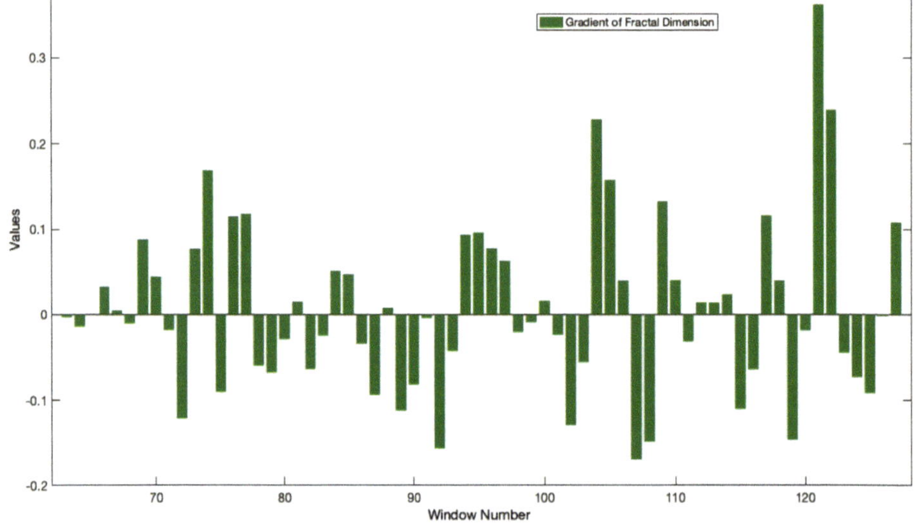

Figure 10.28 Gradient of fractal dimension between 15 November 2001 and 30 November 2001.

Table 10.4 Low fractal values obtained before minor and major falls

Sl. No.	*Fall No.*	*WN*	*Date*	*Time*	*FP (m)*	*Lowest D*	*WN*	*FT (Hrs.)*
November, 2001								
1	RF24	1	01/11/2001	10:37	134.05	0.240	156	14.16
2	RF25	8	06/11/2001	20:41	138.27	0.474	2	89.49
3	RF26	12	07/11/2001	05:07	141.45	0.585	9	5.83
4	RF27	12	07/11/2001	06:30	141.45	0.585	9	7.11
5	RF28	20	08/11/2001	01:01	144.50	0.130	20	0
6	RF29	26	08/11/2001	11:35	144.50	0.190	21	10.36
7	RF30	26	08/11/2001	11:45	144.50	0.190	21	10.46
8	RF31	27	08/11/2001	13:47	144.50	0.190	21	12.46
9	RF32	41	10/11/2001	16:35	148.60	0.169	40	4.00
10	RF33	48	11/11/2001	09:37	150.42	0.312	45	2.50
11	RF34	49	11/11/2001	16:19	150.42	0.312	45	11.20
12	RF35	53	12/11/2001	03:49	151.90	0.300	52	1.75
13	RF36	60	13/11/2001	10:18	155.65	0.279	55	23.58
14	RF37	82	18/11/2001	08:12	163.15	0.271	62	107.10
15	LF24	82	18/11/2001	08:20	163.15	0.271	62	107.31
16	RF38	86	18/11/2001	20:15	163.15	0.559	83	12.36
17	RF39	86	18/11/2001	20:21	163.15	0.559	83	12.46
18	RF40	92	19/11/2001	08:29	165.55	0.318	90	3.66
19	RF41	92	19/11/2001	10:37	165.55	0.318	90	5.83
20	RF42	104	23/11/2001	11:40	169.72	0.134	92	101.00
21	RF43	107	25/11/2001	02:53	174.45	0.682	106	4.75
22	RF44	107	25/11/2001	03:25	174.45	0.682	106	5.20
23	LF25	108	25/11/2001	10:30	174.45	0.343	108	4.70
24	LF26	108	25/11/2001	15:30	174.45	0.343	108	9.70
25	RF45	110	27/11/2001	12:05	180.92	0.450	109	24.58
26	RF46	110	27/11/2001	20:04	180.92	0.450	109	33.55
27	RF47	112	28/11/2001	04:00	184.55	0.531	111	5.00
28	RF48	116	28/11/2001	10:33	184.55	0.353	116	1.12
29	LF27	118	28/11/2001	22:20	184.55	0.480	117	14.90
30	LF28	120	29/11/2001	11:05	186.05	0.292	120	2.06
31	RF49	122	30/11/2001	09:18	188.30	0.523	121	16.30
32	LF29	127	30/11/2001	21:20	188.30	0.745	126	0.83

showed extreme variation in both fractal dimension and its gradient. The lower fractal dimension values as obtained before both major and minor falls in the month of November are given in Table 10.4.

d. From 1 December 2001 to 31 December 2001: The face position on 1 December 2001 was 188.30m, which moved to 208.3m on 31 December 2001 (total face shift: 20m). The productivity of December

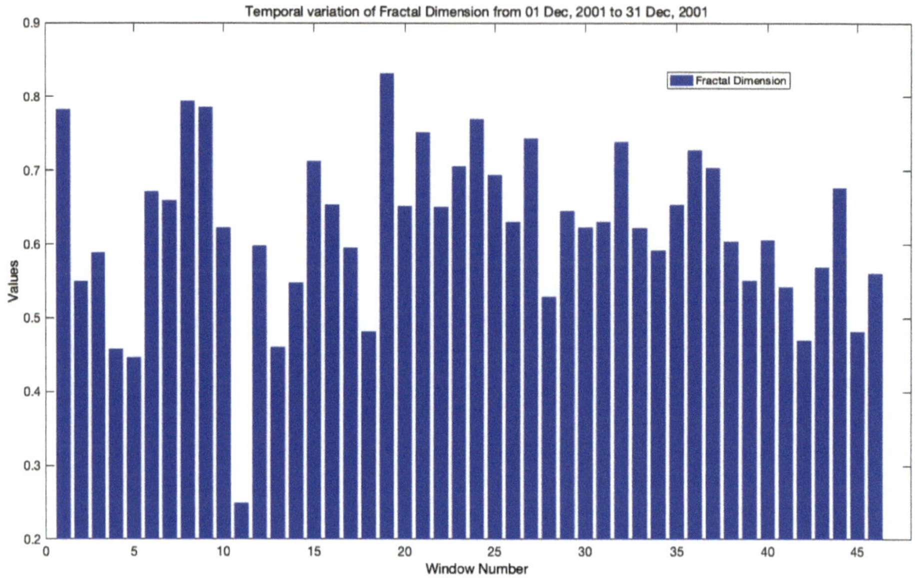

Figure 10.29 Fractal dimension between 1 December 2001 and 31 December 2001. (Mondal and Roy, 2019.)

was very low and coal excavation hampered severely due to shearer breakdown from 14 December 2001 to 27 December 2001. Thus, shearer break down has resulted in a large number zero/non-production day. The temporal variation of fractal dimension and its gradient for the entire month (i.e., from 1 December 2001 to 31 December 2001) has been shown in Figures 10.29 and 10.30, respectively. Thus, aforesaid figures showed the occurrence of highly fluctuated fractal values only during the beginning of month and later it has become steady and very few (ten) minor falls were recorded in the month. The low fractal dimension values obtained before the minor falls during the month of December has been summarized in Table 10.5.

e. From 1 January 2001 to 31 January 2001: The face position in the beginning of the month (i.e., on 1 January 2002) was 208.95m, which shifted to 252.6m on 31 January 2002. Thus, January was a highly productive month and the total face shift reported was around 46.95m. As initiation of production resulted in high microseismic emission, which again decreased drastically as production stopped on 02/01/2002 due to shearer motor break down till 10/01/2001. The

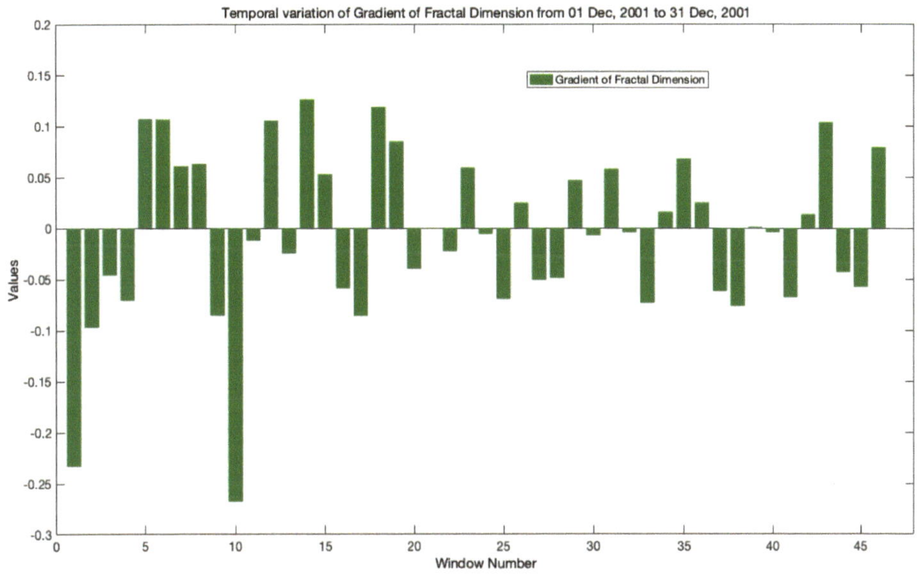

Figure 10.30 Gradient of fractal dimension between 1 December 2001 and 31 December 2001.

re-initiation of mining activities has resulted in continuous stress re-settling of roof strata and a large number of major as well as minor falls was reported during the period (18 major and 9 minor falls). The temporal variation of fractal dimension and its gradient from 1 January 2002 to 31 January 2002 has been shown in Figures 10.31 and 10.32.

Table 10.5 Low fractal values obtained before minor and major falls

Sl. No.	*Fall No.*	*WN*	*Date*	*Time*	*FP (m)*	*Lowest D*	*WN*	*FT (Hrs.)*
December, 2001								
1	LF30	2	02/12/2001	21:15	192.25	0.207	2	0
2	LF31	4	13/12/2001	07:00	198.27	0.458	4	0
3	LF32	4	13/12/2001	07:40	198.27	0.458	4	0.48
4	LF33	17	27/12/2001	15:40	198.27	0.250	11	235.00
5	LF34	17	27/12/2001	20:00	198.27	0.250	11	239.25
6	LF35	19	28/12/2001	15:00	202.50	0.482	18	18.78
7	LF36	20	28/12/2001	16:00	202.50	0.652	20	0.26
8	LF37	30	30/12/2001	18:15	207.90	0.529	28	16.75
9	LF38	35	31/12/2001	02:55	208.30	0.592	34	3.36
10	LF39	36	31/12/2001	05:00	208.30	0.592	34	5.45

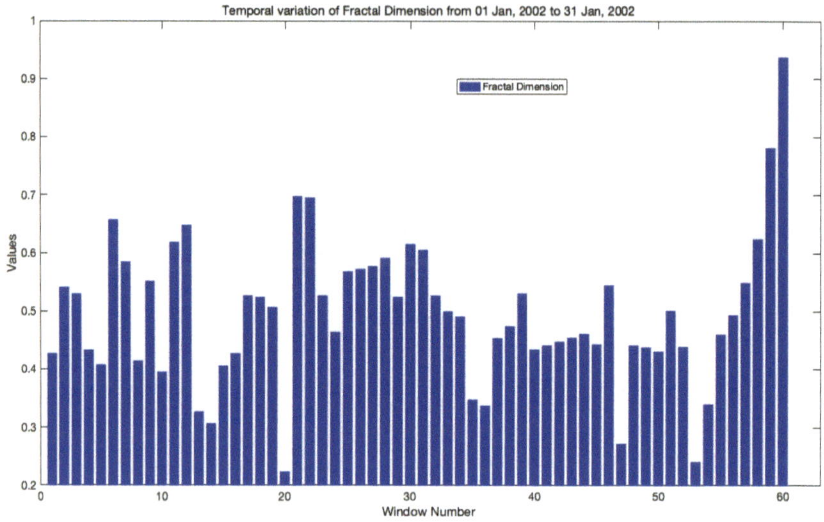

Figure 10.31 Fractal dimension between 1 January 2002 and 31 January 2002. (Mondal and Roy, 2019.)

As rock bursts was seen to occur regularly, therefore, the variation of fractal values was not as high as seen during the previous months, which was also confirmed from the variation of fractal gradient. The low fractal dimension values obtained before both major and minor falls in January has been summarized in Table 10.6.

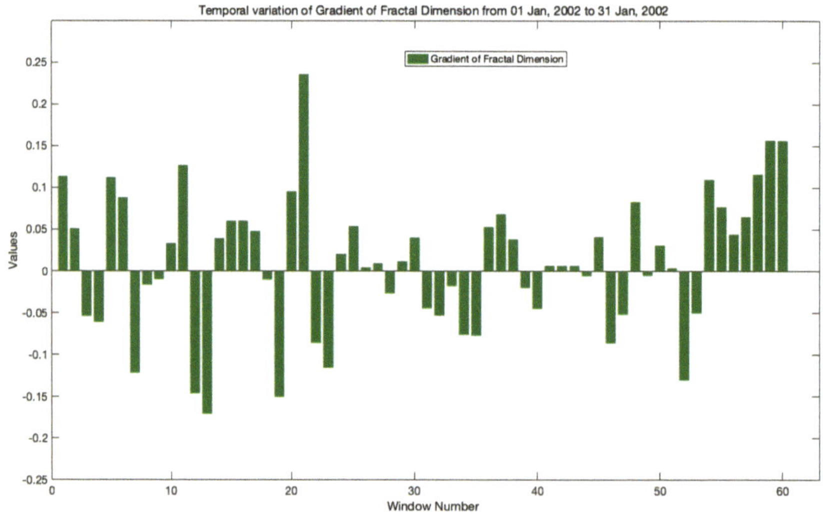

Figure 10.32 Gradient of fractal dimension between 1 January 2002 and 31 January 2002.

Table 10.6 Low fractal values obtained before minor and major falls

Sl. No.	*Fall No.*	*WN*	*Date*	*Time*	*FP (m)*	*Lowest D*	*WN*	*FT (Hrs.)*
				January, 2002				
1	RF50	21	11/01/2002	13:36	212.30	0.224	20	29.00
2	RF51	25	13/01/2002	01:00	214.15	0.464	24	8.58
3	RF52	30	15/01/2002	13:15	222.35	0.524	29	10.41
4	RF53	31	15/01/2002	17:10	222.35	0.524	29	14.32
5	RF54	33	15/01/2002	21:24	222.35	0.524	29	17.55
6	LF40	33	15/01/2002	23:07	222.35	0.524	29	19.30
7	RF55	33	16/01/2002	05:02	225.47	0.524	29	26.16
9	LF41	36	18/01/2002	00:30	230.75	0.337	36	1.38
10	LF42	36	18/01/2002	00:45	230.75	0.337	36	1.91
11	RF56	37	18/01/2002	12:36	230.75	0.337	36	13.83
12	RF57	37	18/01/2002	12:39	230.75	0.337	36	13.88
13	LF43	38	19/01/2002	01:05	234.50	0.337	36	26.20
14	LF44	38	19/01/2002	10:30	234.50	0.337	36	36.11
16	RF58	41	19/01/2002	17:40	234.50	0.435	40	0
17	LF45	41	19/01/2002	18:10	234.50	0.435	40	0.33
18	RF59	41	19/01/2002	21:00	234.50	0.435	40	3.21
19	RF60	45	20/01/2002	07:31	234.70	0.444	45	1.03
20	RF61	45	20/01/2002	12:55	234.70	0.444	45	6.41
21	RF62	45	20/01/2002	15:00	234.70	0.444	45	8.51
22	RF63	46	23/01/2001	19:53	234.70	0.545	46	68.50
23	RF64	48	25/01/2002	05:38	Not Known	0.272	47	4.20
24	RF65	48	25/01/2002	07:31	Not Known	0.272	47	6.08
25	RF66	52	25/01/2002	20:58	Not Known	0.431	50	2.45
26	LF46	57	30/01/2002	21:00	248.96	0.241	53	119.06
27	LF47	58	30/01/2002	22:00	248.96	0.241	53	120.06
28	RF67	60	31/01/2002	07:40	252.60	0.524	59	6.35
29	LF48	60	31/01/2002	21:00	252.60	0.524	59	20.66

f. From 1 February 2001 to 28 February 2001: The face position at the beginning of February (i.e., on 1 February 2002) was 254.8m, which shifted to 319.7m on 28 February 2002. Thus, the total face shift was around 64.9m. As the production progressed with face advance, the values of fractal dimension were seen to decrease and similar pattern continued till the end of the month, which showed continuous stress build-up in the roof strata. A total of 9 major and 16 minor falls were reported in the month. The temporal variation of fractal dimension and its gradient has been shown in Figures 10.33 and 10.34, respectively. The sharp decrease in the fractal values and its gradient was

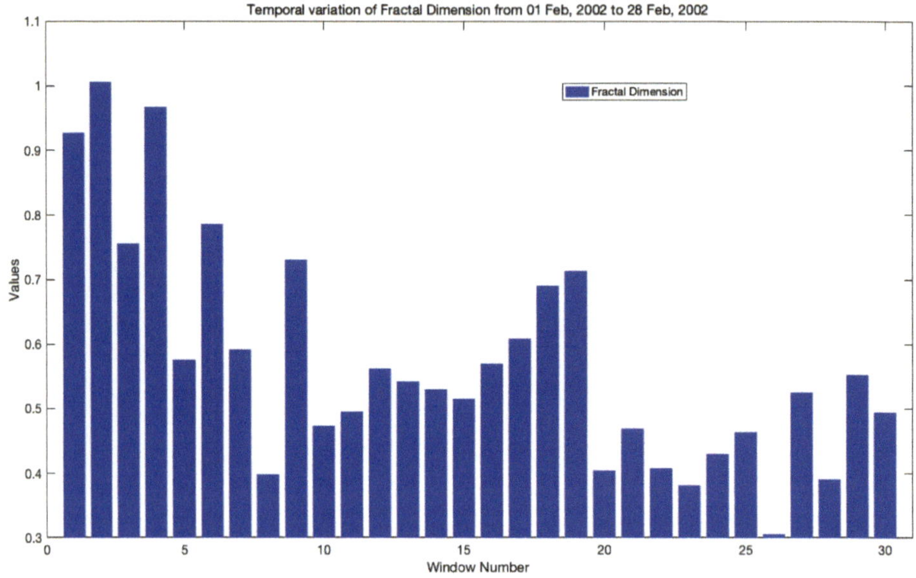

Figure 10.33 Fractal dimension between 1 February 2002 and 28 February 2002. (Mondal and Roy, 2019.)

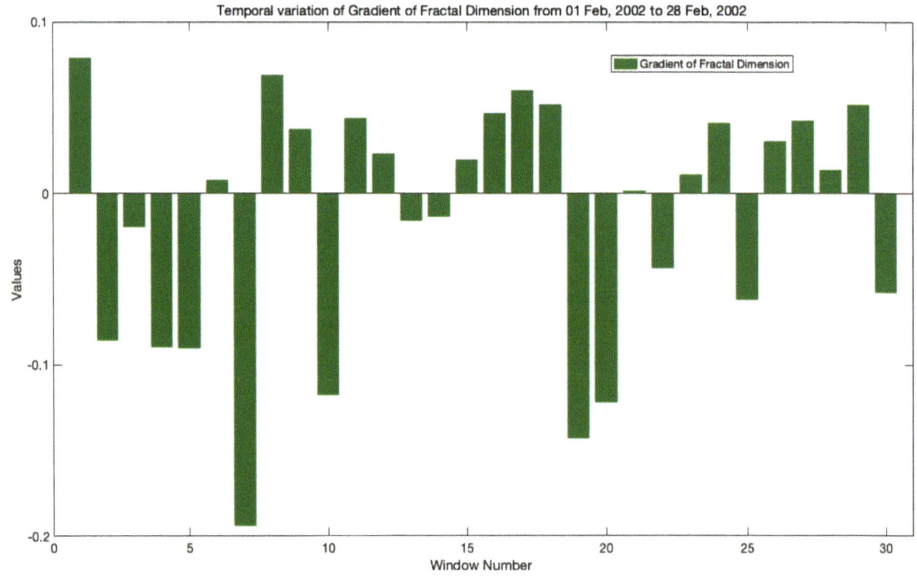

Figure 10.34 Gradient of fractal dimension between 1 February 2002 and 28 February 2002.

Table 10.7 Low fractal values obtained before minor and major falls

Sl. No.	*Fall No.*	*WN*	*Date*	*Time*	*FP (m)*	*Lowest D*	*WN*	*FT (Hrs.)*
			February, 2002					
1	LF49	2	01/02/2002	08:10	254.80	0.927	1	0
2	RF68	2	01/02/2002	17:41	254.80	0.927	1	0.12
3	LF50	3	01/02/2002	21:10	254.80	0.927	1	3.58
4	RF69	5	03/02/2002	07:24	260.75	0.576	5	5.50
5	LF51	5	03/02/2002	18:00	260.75	0.576	5	16.16
6	LF52	5	04/02/2002	12:00	263.90	0.576	5	22.16
7	LF53	8	12/02/2002	18:15	285.65	0.411	8	0
9	RF70	10	14/02/2002	18:26	288.10	0.473	10	0
10	RF71	12	17/02/2002	07:52	292.40	0.495	11	43.17
11	RF72	12	17/02/2002	22:45	292.40	0.495	11	58.10
12	LF54	12	18/02/2002	08:25	295.90	0.495	11	67.67
13	LF55	13	19/02/2002	09:30	298.80	0.495	11	93.84
14	RF73	13	19/02/2002	20:42	298.80	0.495	11	104.05
16	LF56	15	20/02/2002	19:00	303.40	0.515	15	0
17	LF57	15	20/02/2002	19:35	303.40	0.515	15	0.58
18	LF58	15	20/02/2002	20:25	303.40	0.515	15	1.41
19	LF59	15	20/02/2002	22:20	303.40	0.515	15	2.36
20	LF60	15	20/02/2002	23:25	303.40	0.515	15	3.41
21	RF74	20	21/02/2002	22:17	306.50	0.501	20	0
22	LF61	20	22/02/2002	04:35	307.40	0.501	20	0.66
23	LF62	20	22/02/2002	05:40	307.40	0.501	20	1.75
24	LF63	21	25/02/2002	01:25	309.40	0.501	20	69.58
25	LF64	22	27/02/2002	03:30	316.20	0.408	22	20.41
26	RF75	29	28/02/2002	12:38	319.70	0.306	26	13.05
27	RF76	29	28/02/2002	12:55	319.70	0.306	26	13.16

seen before the rock bursts. The low fractal dimension values obtained before both major and minor falls in February has been summarized in Table 10.7.

10.5 STUDY OF SEISMIC *B*-VALUE IN LONGWALL MINE

10.5.1 Estimation of seismic *b*-value and its significance

The concept of earthquake magnitude was first introduced by Richter in 1935, and it was seen that distribution of earthquakes was non-uniform with regard to space, time, and magnitude. It was also observed that the distribution of earthquakes with respect to its magnitude obeys a power law relation, and are self-similar and scale invariant. Theoretically, it limits the event size in terms of earthquake magnitude and the power law is given as equation 10.10., where N_e represents the discrete frequency for events having magnitudes within M and $M + \delta M$, and constants a and b represents the overall microseismic activity and magnitude distribution of microseismic events. The equation 10.10 is also known as the Gutenberg-Richter Frequency Magnitude relation. Previous studies have shown that the lowering of b-value was considered as a significant precursor signature for earthquake monitoring. The variation in the b-value before and after main shock of earthquake was first studied by Suyehiro in 1966 (Suyehiro, 1966), and since then it (b-value) has been used in several earthquake studies (Moatti et al., 2013; Tormann et al., 2014, 2015; Wang et al., 2016). The factors influencing the seismic b-value includes: (a) the tectonic activity (Tsapanos, 1990; Khan et al., 2011), (b) geology, stress distribution, type of faulting, and confining pressure (Schorlemmer and Weimer, 2005; Schorlemmer et al., 2005; Tormann et al., 2014, 2015), (c) friction (Shaw, 1995; Wang, 1996, 1997), (d) Self-similarity of geological structure and structural heterogeneity (King, 1983), (e) hydrothermal activity (Wang, 1988), and (f) dilatancy generation (Rikitake, 1979).

The b-value has also been very successful in studying the seismicity associated with mines. The seismicity study in the Bobrek mine showed that lower b-value and its negative trend was observed before strong seismic activities (Mutke et al., 2016). Another study carried out in the Fankou lead-zinc mine for monitoring the variation of b-value during rock mass excavation showed sharp decrease in b-value at the time of disaster (Chun-lai et al., 2009).

10.5.2 Temporal variation of *b*-value in longwall mine

a. From 28 August 2001 to 30 September 2001: The variation in the b-value for September has been shown in Figure 10.35. It can be seen that strata dynamics of virgin rock due to the initiation of coal excavation has resulted in highly fluctuating b-values. It can be seen that lower b-value has been observed during both major as well as minor falls. The magnitude versus their occurrence for September has been shown in Figure 10.36, which showed that the highest occurrence is obtained for events corresponding to magnitude $M = -2.1$. The

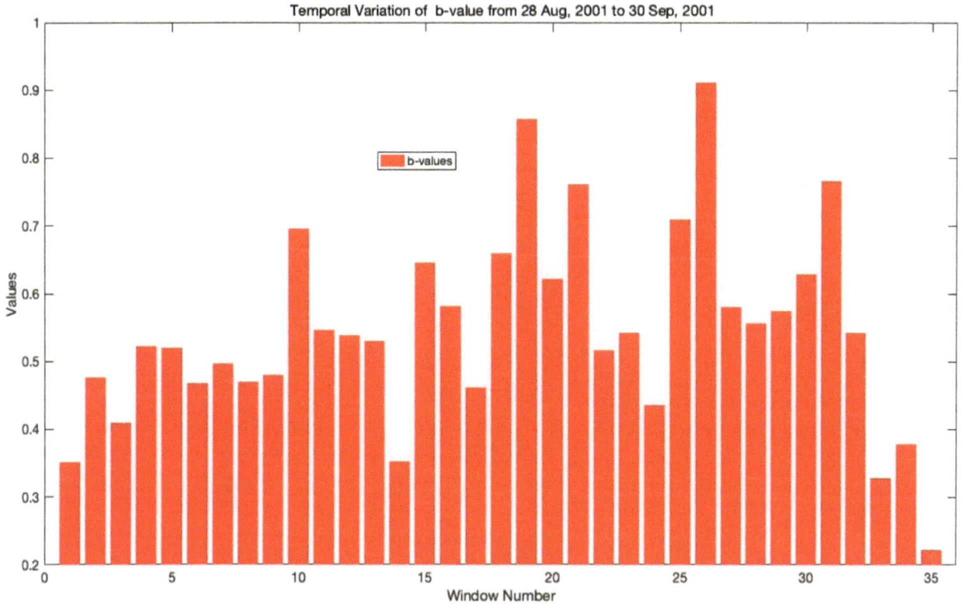

Figure 10.35 Seismic *b*-value between 28 August 2001 and 30 September 2001. (Mondal and Roy, 2019.)

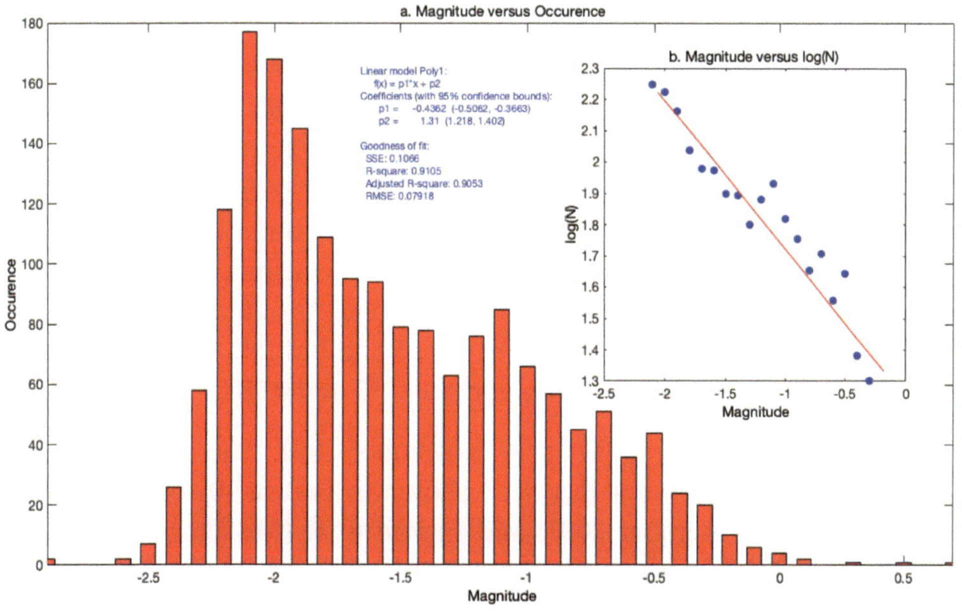

Figure 10.36 Seismic activity and magnitude distribution for September 2001. (Mondal and Roy, 2019.)

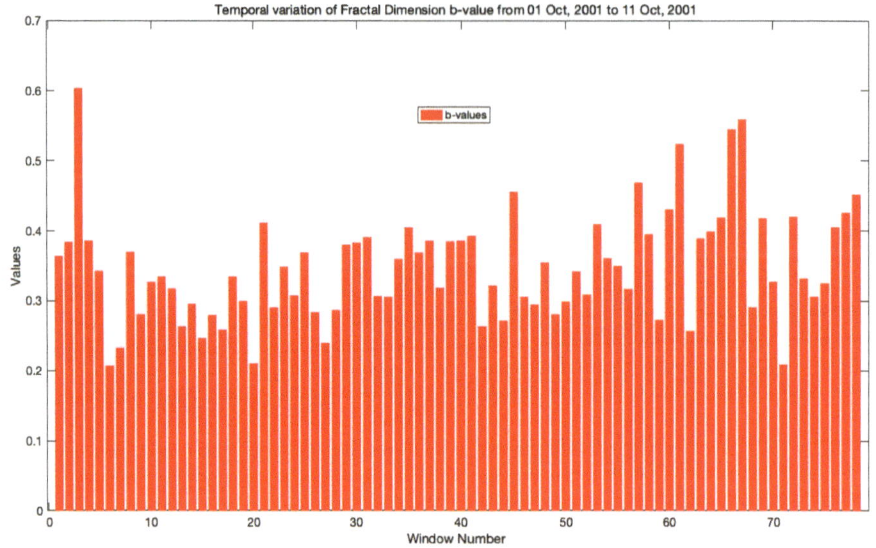

Figure 10.37 Seismic *b*-value between 1 October 2001 and 11 October 2001. (Mondal and Roy, 2019.)

constants *a* and *b* obtained from Gutenberg-Richter relation (refer Figure 10.36) for the month was 1.31 and 0.436.

b. From 1 October 2001 to 31 October 2001: The variation in the *b*-value for the month of October has been in Figure 10.37 (1 October 2001 to 11 October 2001) and Figure 10.38 (12 October 2001 to 31 October 2001), respectively. Due to the increase in rate of coal excavation, the variation of *b*-value has become more fluctuating than the previous month. The creation of large vacant space (goaf) resulted in intense fracturing of rock leading to emission of higher magnitude events and lowering of the *b*-value. The event magnitude versus their occurrence for the month is shown in Figure 10.39 and highest occurrence is obtained for events corresponding to $M = -1.2$. The constants *a* and *b* obtained from the month as per Gutenberg-Richter relation was 1.72 and 0.988, respectively.

c. From 1 November 2001 to 30 November 2001: The variation in the *b*-value for November has been shown in Figure 10.40 (1 November 2001 to 14 November 2001) and Figure 10.41 (14 October 2001 to 30 October 2001), respectively. The complexity in the variation of *b*-value for the month is much more fluctuating than the previous month and it can also be seen that major falls are mostly reported during this period, whereas very few minor falls were reported in the latter half

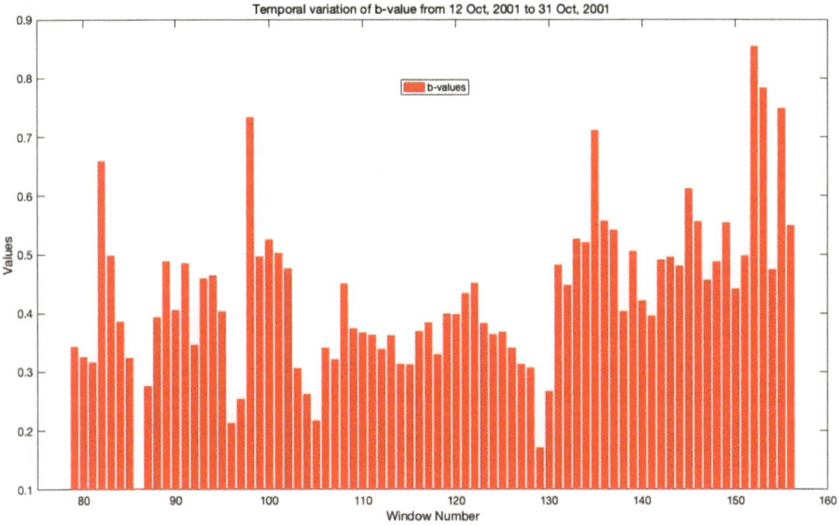

Figure 10.38 Seismic *b*-value between 12 October 2001 and 31 October 2001. (Mondal and Roy, 2019.)

of the month (i.e., from 15 November 2001 to 30 November 2001). The surface blasting has also resulted in lowering the *b*-value, which were carried out to release the stress from the roof strata. The magnitude versus their occurrence for November is shown in Figure 10.42, and the highest occurrence was seen for events corresponding to $M = -2.3$.

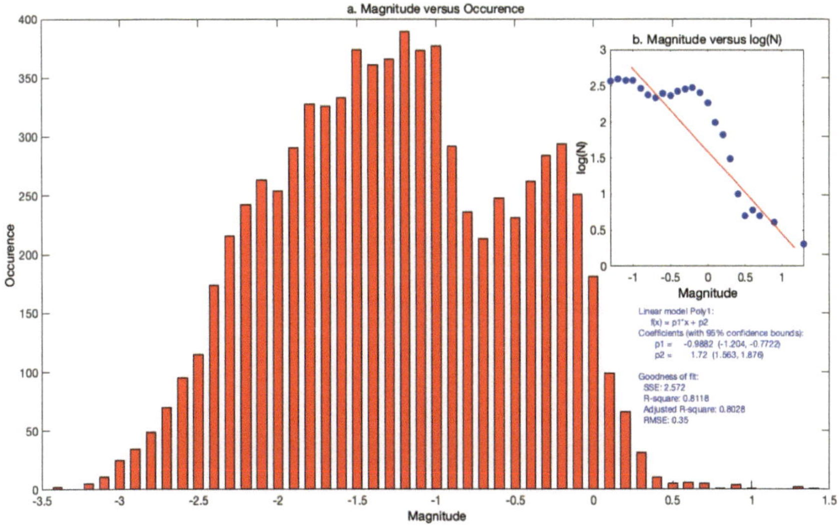

Figure 10.39 Seismic activity and magnitude distribution for October 2001. (Mondal and Roy, 2019.)

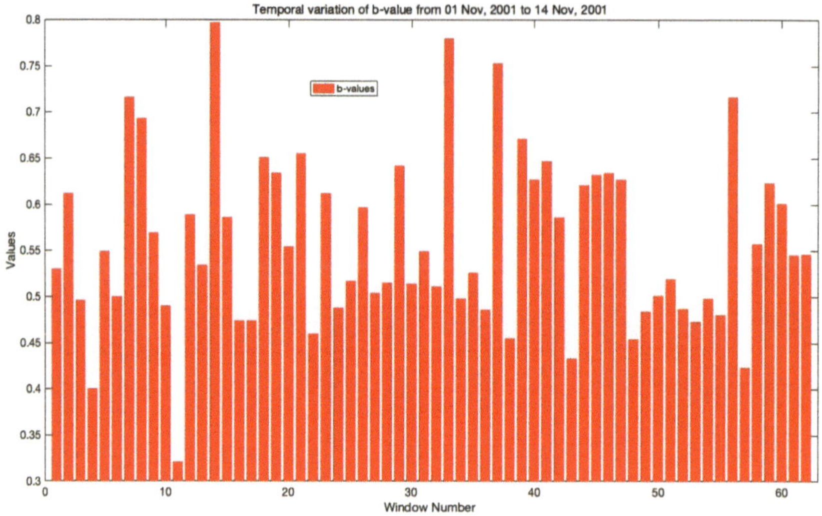

Figure 10.40 Seismic *b*-value between 1 November 2001 to 14 October 2001. (Mondal and Roy, 2019.)

The constants *a* and *b* obtained from Gutenberg-Richter relation for the month was 0.539 and 1.087, respectively.

d. From 1 December 2001 to 31 December 2001: The variation of *b*-value during December has been shown in Figure 10.43. The variation of seismic *b*-value is smoother as compared to previous months

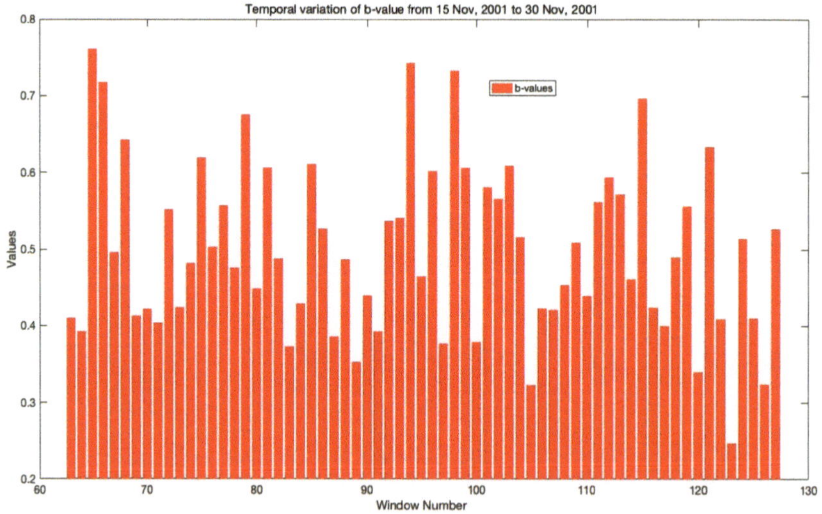

Figure 10.41 Seismic *b*-value between 15 November 2001 and 30 November 2001. (Mondal and Roy, 2019.)

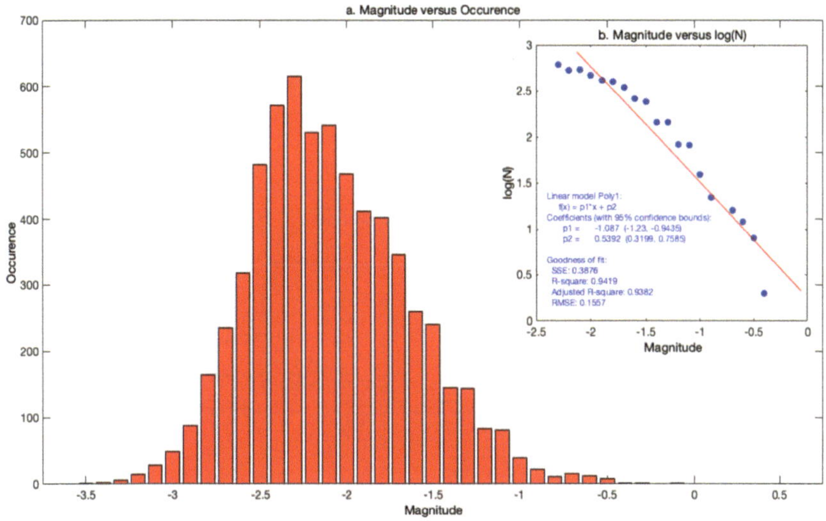

Figure 10.42 Seismic activity and magnitude distribution for November 2001. (Mondal and Roy, 2019.)

as mining activities were stopped for few days due to instrumental break down, which also resulted in occurrence of only minor falls. However, lower *b*-value was obtained during minor falls and surface blasting. The event magnitude versus their occurrence for December has been shown in Figure 10.44, and the highest occurrence is

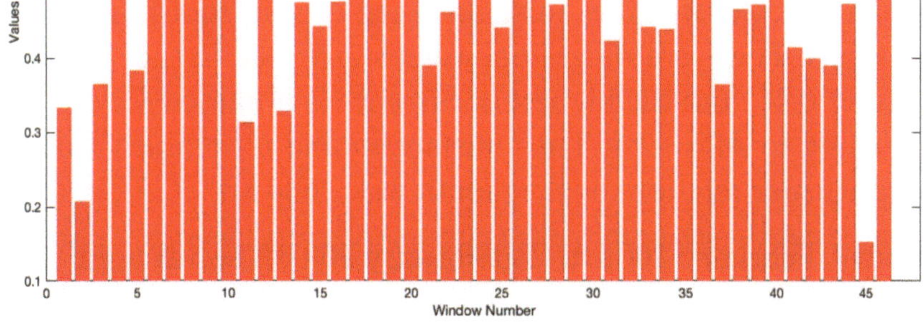

Figure 10.43 Seismic *b*-value between 1 December 2001 and 31 December 2001. (Mondal and Roy, 2019.)

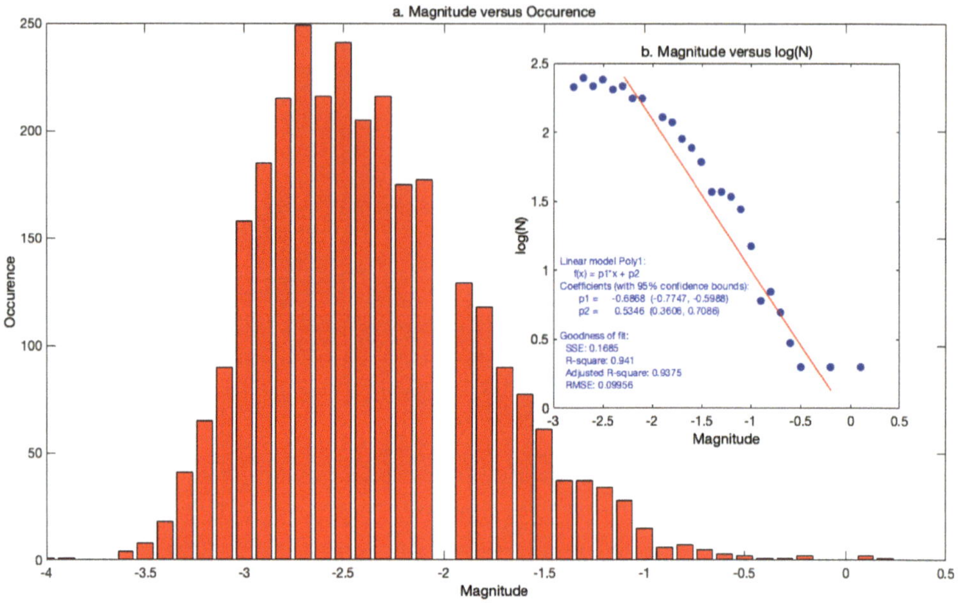

Figure 10.44 Seismic activity and magnitude distribution for December 2001. (Mondal and Roy, 2019.)

obtained for events corresponding to $M = -2.7$. The constants a and b as per Gutenberg-Richter relation for the month was 0.534 and 0.686, respectively.

e. From 1 January 2001 to 31 January 2001: The b-value variation for the month of January has been shown in Figure 10.45. Previously, in the month of December, the temporal variation of the b-value was smooth due to the absence of mining activities, which has again become highly fluctuating due to re-initiation of coal excavation and again here also lower b-values were obtained during roof falls and surface blastings. The event magnitude versus their occurrence was shown in Figure 10.46, and the highest occurrence is obtained for events corresponding to $M = -2.4$. The constants a and b obtained from Gutenberg-Richter relation for the month was 0.684 and 0.733, respectively.

f. From 1 February 2001 to 28 February 2001: The variation in b-value for February has been shown in Figure 10.47. The occurrence of higher magnitude events has resulted in lowering the b-value. The stress distribution in the overlying strata has readjusted a lot due to which large

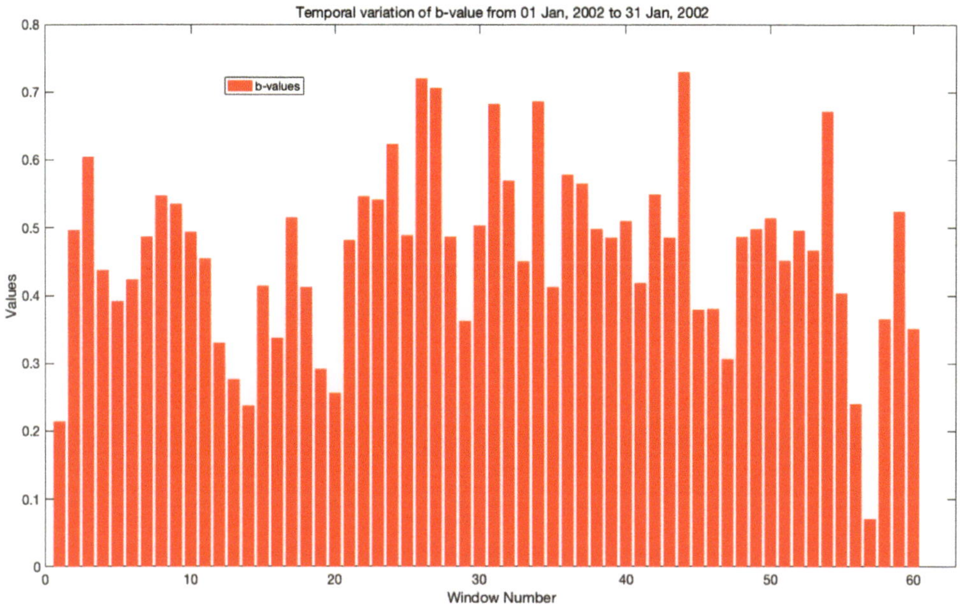

Figure 10.45 Seismic *b*-value between 1 January 2002 and 31 January 2002. (Mondal and Roy, 2019.)

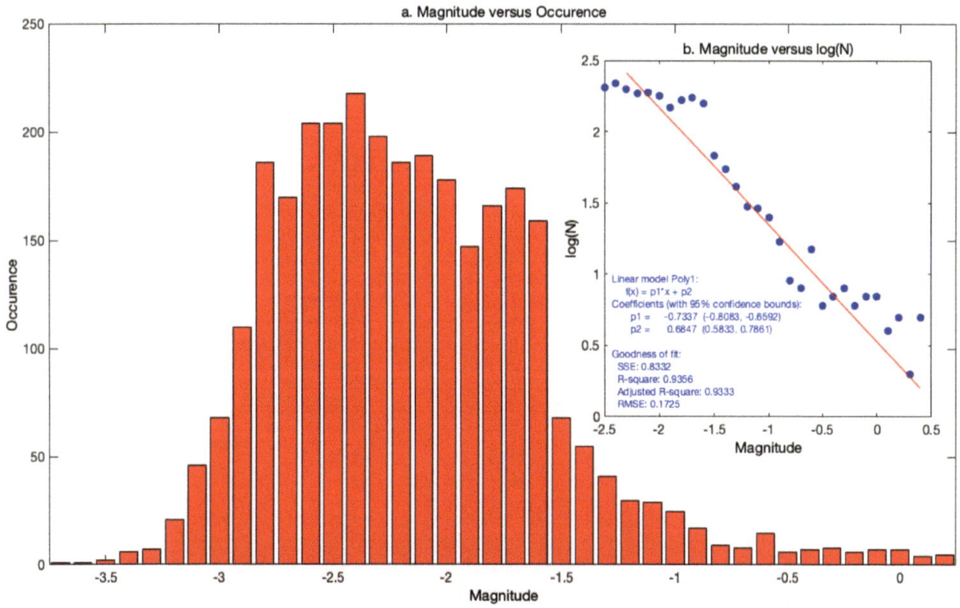

Figure 10.46 Seismic activity and magnitude distribution for January 2002. (Mondal and Roy, 2019.)

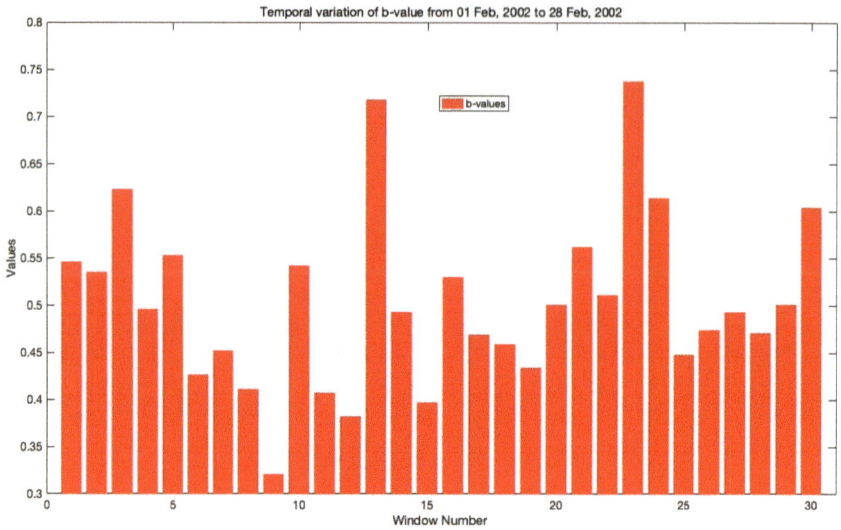

Figure 10.47 Seismic *b*-value between 1 February 2002 and 28 February 2002. (Mondal and Roy, 2019.)

number of rock bursts have been reported very frequently. The event magnitude versus their occurrence has been shown in Figure 10.48, and the highest occurrence is obtained for events corresponding to $M = -2.0$. The constants a and b as per the Gutenberg-Richter relation for the month was 0.048 and 1.069, respectively.

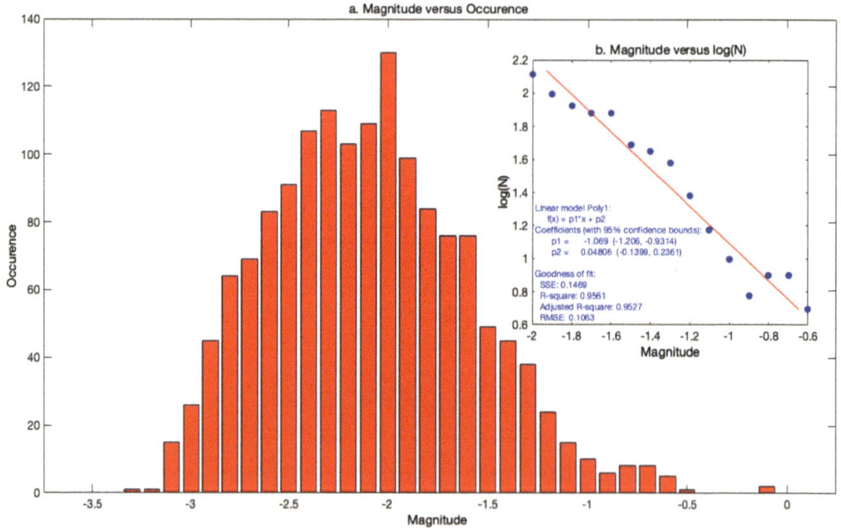

Figure 10.48 Seismic activity and magnitude distribution for February 2002. (Mondal and Roy, 2019.)

10.6 ROOF FALL FORECASTING IN THE LONGWALL MINE

The geology of immediate roof showed that the working panel was overlain by thick hard sandstone roof, which reduced the roof cavability and further leading to occurrence of around 98 major roof falls during observation period. These roof falls basically occurred due to the sudden release of the accumulated heavy stress in roof through weaker section of rock mass and are highly responsible fatal accidents in underground mine. Thus, forecasting of such massive roof falls is very necessary for ensuring safety in vicinity of the working areas of underground mines. The lowering of fractal dimension prior to major roof falls has helped in getting a precursory signature for roof fall forecasting in space and time, which has been explained through a case study of MS-ISS-RF10.

10.6.1 Case study of MS-ISS-RF10 (reported on 06/10/2001-17:15hrs)

The MS-ISS-RF10 was reported on 6 September 2001 at 17:15hrs. The position of the longwall face on the day of the fall was around 83.30m from the beginning. The temporal variation of the fractal dimension before the aforesaid roof fall has been shown in Figure 10.49 (Table 10.8).

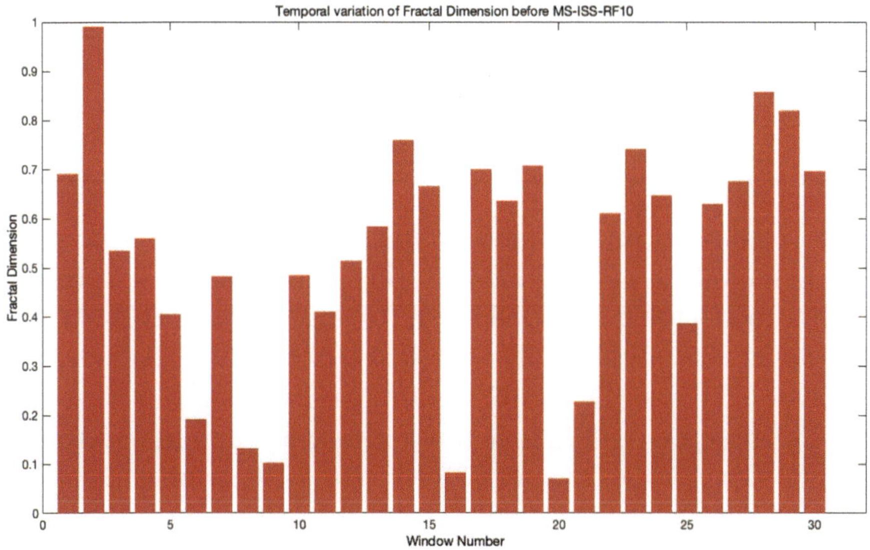

Figure 10.49 Temporal variation of fractal dimension before roof-fall MS-ISS-RF10.

Table 10.8 Fractal dimension values obtained before roof-fall MS-ISS-RF10

Sl. No.	*Event Interval*	*Fractal Dimension*
1	04/10/2001 (05:41hrs)–04/10/2001 (07:57hrs)	0.690
2	04/10/2001 (07:58hrs)–04/10/2001 (12:27hrs)	0.990
3	04/10/2001 (12:28hrs)–04/10/2001 (13:51hrs)	0.534
4	04/10/2001 (13:53hrs)–04/10/2001 (15:05hrs)	0.559
5	04/10/2001 (15:06hrs)–04/10/2001 (16:58hrs)	0.404
6	04/10/2001 (17:00hrs)–04/10/2001 (18:13hrs)	0.191
7	04/10/2001 (18:14hrs)–04/10/2001 (19:34hrs)	0.482
8	04/10/2001 (19:35hrs)–04/10/2001 (21:03hrs)	0.131
9	04/10/2001 (21:03hrs)–04/10/2001 (23:12hrs)	0.102
10	04/10/2001 (23:14hrs)–04/10/2001 (23:58hrs)	0.484
11	04/10/2001 (23:58hrs)–04/10/2001 (00:19hrs)	0.409
12	05/10/2001 (00:23hrs)–05/10/2001 (02:41hrs)	0.513
13	05/10/2001 (02:43hrs)–05/10/2001 (05:13hrs)	0.583
14	05/10/2001 (05:17hrs)–05/10/2001 (14:27hrs)	0.759
15	05/10/2001 (14:30hrs)–05/10/2001 (18:48hrs)	0.665
16	05/10/2001 (18:49hrs)–05/10/2001 (23:31hrs)	0.082
17	05/10/2001 (23:31hrs)–06/10/2001 (01:20hrs)	0.699
18	06/10/2001 (01:21hrs)–06/10/2001 (03:20hrs)	0.635
19	06/10/2001 (03:24hrs)–06/10/2001 (08:01hrs)	0.707
20	06/10/2001 (08:03hrs)–06/10/2001 (12:30hrs)	0.070
21	06/10/2001 (12:32hrs)–06/10/2001 (13:19hrs)	0.227
22	06/10/2001 (13:19hrs)–06/10/2001 (13:45hrs)	0.61
23	06/10/2001 (13:46hrs)–06/10/2001 (14:14hrs)	0.741
24	06/10/2001 (14:14hrs)–06/10/2001 (14:58hrs)	0.646
25	06/10/2001 (14:59hrs)–06/10/2001 (15:33hrs)	0.386
26	06/10/2001 (15:34hrs)–06/10/2001 (15:50hrs)	0.629
27	06/10/2001 (15:50hrs)–06/10/2001 (16:10hrs)	0.675
28	06/10/2001 (16:11hrs)–06/10/2001 (16:34hrs)	0.857
29	06/10/2001 (16:34hrs)–06/10/2001 (16:55hrs)	0.819
30	06/10/2001 (16:55hrs)–06/10/2001 (17:15hrs)	0.696

Abrupt decrement in the fractal dimension occurred before roof fall, where lower values were obtained during time intervals 4 October 2001 (17:00hrs) to 4 October 2001 (18:13hrs) [D = 0.191], 4 October 2001 (19:35hrs) to 4 October 2001 (21:03hrs) [D = 0.131], 4 October 2001 (21:03hrs) to 4 October 2001 (23:12hrs) [D = 0.102], 5 October 2001 (18:49hrs) to 5 October 2001 (23:31hrs) [D = 0.082], 6 October 2001 (08:03hrs) to 6 October 2001 (12:30hrs) [D = 0.070], 6 October 2001 (12:32hrs) to 6 October 2001 (13:19hrs) [D = 0.227] and 6 October 2001 (14:59hrs) to 6 October 2001 (15:33hrs) [D = 0.386]. The microseismic events corresponding to low fractal dimension values have been plotted in 3-D space (Figure 10.50). Few local falls

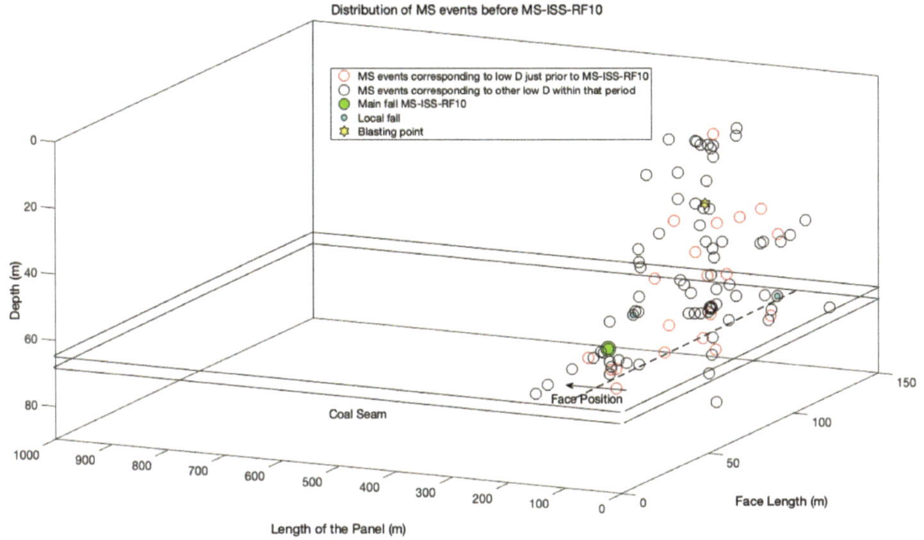

Figure 10.50 Distribution of microseismic events in 3-D space corresponding to lower fractal values obtained before MS-ISS-RF10.

were also reported in the mine on 4 and 5 October 2001 above the panel (in the goaf area). But still the excess stress above the powered supports along the right side of the face failed to cave. Therefore, surface blasting was carried out on 6 October 2001 (06:45hrs) using 13 holes and 1325kg charge size to release the stress, which released in the form of a major fall (MS-ISS-RF10) at the predicted place, where stress was accumulated after 11 hours on the same day (i.e., 6 October 2001 at 17:15hrs).

10.7 FRACTAL BEHAVIOR OF THE LONGWALL OVERLYING STRATA

The monthly fractal variation of microseismicity along the face of the mine has been shown in Figure 10.51. Based on the overall trend of fractal variation, whole observation period has been divided into several zones (namely 'A', 'B', 'C', 'D', 'E', 'F' and 'G'). The fractal trends in the aforesaid zones have been discussed next.

a. Zone 'A': Zone 'A' mainly consisted of fractal values obtained of microseismic emission during September. Initially, in September, since the overlying strata was in virgin state, therefore, initiation of mining has induced enormous change in the stress distribution of

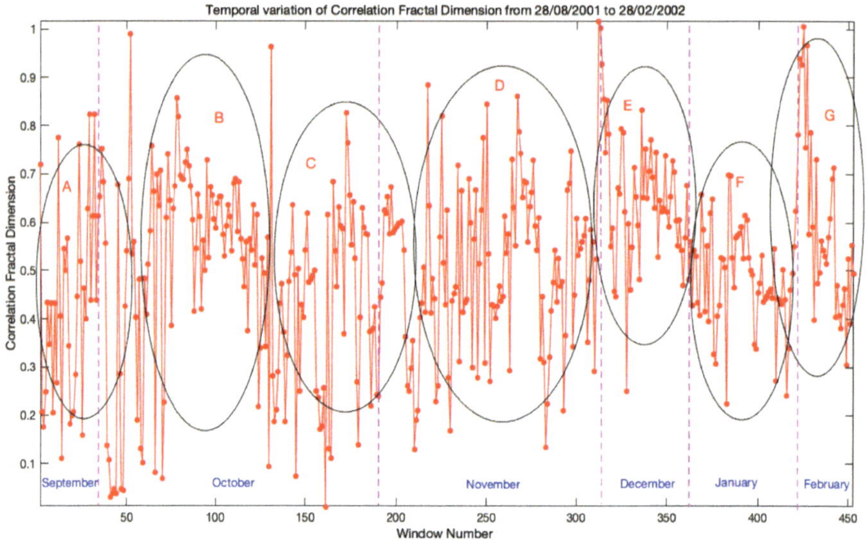

Figure 10.51 Temporal variation of fractal dimension from 28 August 2001 to 28 February 2002.

the immediate roof. The area of overhanging strata behind the powered supports also increased gradually with face advance leading to increase in microseismic activity. The increasing stress resulted in several rock bursts (both major and minor falls, as reported on 6, 08, 13, 16, 17, and 20 September) and rock spalling near the face of the panel (reported on 12, 15, and 25 September). Thus, accumulation of initial stress can be considered as a prime reason towards decrease in fractal values for that period. Later on, mining was closed due to shearer breakdown (from 26 to 29 September), which resulted in reduction in microseismic activity further leading to high fractal values.

b. Zone 'B' and 'C': The Zone 'B' mostly consisted of the fractal variation obtained during the first half of October. As after three days, mining was again started, which resulted in inducing alteration in the stress distribution along the immediate roof and correspondingly high microseismic activity has been observed during this period. The increasing volume of goaf further resulted in massive rock falls. The observation carried out during this period showed that stress could not hold for long time in roof strata due to the presence of unstable roof in that zone. Thus, periodic caving (in the form of major and minor fall; on 2, 3, 4, 5, 6, 9, 11, 13, 16, and 17 October) and rock spalling (7, 8, 10, and 14 October) was reported during this time, which has resulted in timely release of accumulated stress from the roof strata leading to average increase in the fractal values during this period. Zone 'C' consisted of fractal values calculated during latter half of the month. The delay in the caving has resulted in accumulation of stress in the roof, which was mainly due to the presence of stable roof type and can also be inferred from the use of surface blasting techniques during this period. Hence, lower fractal dimension values were obtained. The regular caving (on 28, 29, and 30 October) was again seen at the end of the month, which again led to increase in fractal values during this period and similar pattern continued during the first week of November.

c. Zone 'D': The Zone 'D' consisted of fractal values calculated from microseismic data recorded in November, which showed a general elevation due to the presence of unstable roof above the panel, and the same can be inferred from the frequent roof falls (on 6, 7, 8, 10, 11, and 13 November). After 15 November, a relatively lower trend in the fractal value is seen due to the presence of stable roof as surface blastings were carried out to release the excess accumulated stress. Also, delayed roof falls have been reported during this period (18, 19, 23, 25 November). At the end of the month, fractal values have shown rise due to occurrence of regular caving, mostly due to the presence of less compact sandstone roof in that area (roof bursts in this period were reported on 27, 28, 29, and 30 November).

d. Zone 'E': Zone 'E' consisted of fractal values that obtained from microseismic data recorded in the month of December. Mining activities were closed for most of the days in December due to variable machinery breakdowns. Moreover, minor falls were reported during the entire month (on 2, 13, 27, 28, 30, and 31 December). Therefore, stress couldn't be built to threshold limits leading to an average variation in fractal dimension for that period.

e. Zone 'F': Mining again started in the month of January and Zone 'E' consisted of fractal values calculated from microseismic data recorded during the month of January. Irregular caving was reported during this time and roof falls were reported in relatively larger time intervals (on 11, 13, 15, 18, 19, 20, 25, 30, and 31 January), which resulted in accumulation of stress in roof for longer time leading to higher emission of microseismic events from the weaker points. Thus, lower fractal values were obtained during this period.

f. Zone 'G': Zone 'G' consisted of fractal values in the month of February. During this period, an initial increase in the fractal dimension can be seen as stress was released from the roof strata at nearly regular intervals (on 1, 3, and 4 February). Then it took six days for the next roof fall leading to stress accumulated in the immediate roof, which could not be released and hence resulting in lowering of fractal values during this time (on 11, 12, 14, and 17 February). At the end of the month, roof fall was reported on regular intervals but then also further decrease in the fractal values have been seen, which resulted due to the presence of stable immediate roof above the panel leading to high stress build-up and more concentration of microseismic events.

10.8 RESULTS AND DISCUSSION

The present work was carried out for studying the heterogeneity in the overburden rock through analysis of fractal dimension and *b*-value over passive mine microseismicity, which proved its potential in mitigating several safety-related issues. The valuable results drawn from the present study has been mentioned next.

1. The study helped in understanding the relationship between the microseismic event distribution (in 3-D space) and fractal dimension, which showed that correlation integral had a linear relation with scale range (on logarithmic scale).
2. Clustering of microseismic events results in higher correlation integral values for lower scale ranges, which leads to lower fractal values. Therefore, lower fractal dimension values are obtained at the time of stress build-up in the rock strata and vice versa.
3. High fluctuations have been seen in the temporal variation of fractal dimension and its gradient, which showed that continuous stress was building-up in roof strata and released at certain intervals.
4. The temporal variation of fractal and its gradient showed a valid justification with mining activities and related seismicity. Higher

production resulted in large microseismic activity with fluctuating tendency of fractal values and its gradient (e.g., November 2001), whereas during lower production (such as non-productive working days) led to reduction in microseismic activity with relative stable variation in fractal dimension and its gradient (e.g., December, 2001).

5. Accumulation of stress resulted in lower fractal values and negative gradient, which was seen prior to roof falls.
6. The magnitude of mine microseismic events obeyed the Gutenberg-Richter Frequency-Magnitude relation and hence the *b*-value study helped in getting magnitude distribution and seismic activity for the respective production months.
7. The temporal variation of *b*-value, as calculated from maximum-likelihood method over uniform event windows helped in understanding the magnitude variation of microseismic events emitted before and after the rock bursts.
8. The nature of *b*-value fluctuations was similar to that of fractal dimension. During productive months, high-microseismic activity resulted in highly fluctuating tendency of *b*-value (e.g., November 2001), whereas during non-productive working days, the variation in the aforesaid parameter was relatively stable (e.g., December 2001).
9. Variation of the fractal dimension and *b*-value with the emitted seismicity has been successfully explained by characterizing the dynamics in various zones.

10.9 SCOPE OF THE WORK

Thus, it can be said that the present study has been very helpful in mitigating various issues, which were related to the safety of the mine like identification of stressed zones, spatio-temporal forecasting of rock bursts, impact of surface blasting, microseismic event magnitude study, and so on. As the study used the mine-induced microseismicity emitted during the creation of micro-fractures in the roof strata, therefore, similar study can also be implemented in various other forms of underground mining, such as room and pillar mining. Thus, the present study can be very helpful in enhancing safety in underground mines, where working panel is overlain by adverse rock type.

ACKNOWLEDGMENT

The authors would like to acknowledge the National Institute of Mine Safety (India) and South Eastern Coalfield Limited (India) for generating valuable real-time microseismic data from the roof strata of the longwall mine. The authors would like to extend their sincere gratitude to Mr. R.N. Jha (Director T/RD&T, CMPDI), Mr. A.K. Rana (Director T/P&D, CMPDI), Mr. S.K. Gomasta (Director T/CRD, CMPDI), Mr. B.A. Reddy (General Manager ICT, CMPDI), Lakshmi Narayan Rajavolu (ICT, CMPDI), Sandeep Kumar (ICT, CMPDI), Awanish Kumar (ICT, CMPDI), Abhijit Bose (ICT, CMPDI), Nilesh Kumar (ICT, CMPDI), Vinay Kumar (ICT, CMPDI), Ritesh Kumar (ICT, CMPDI), Saurabh Kishore (ICT, CMPDI) and other executives of CMPDI for their constant support, motivation, and encouragement for completion of the work.

Funding Agencies

The authors are grateful to acknowledge the Ministry of Earth Science (Govt. of India) for partly sponsoring the work by Project No. MOES/P.O.(Seismo)/1(364)/2019.

Conflict of Interest

The authors would like to state that there was NO CONFLICT OF INTEREST while carrying out the study described here.

REFERENCES

Aki, K. 1965. Maximum likelihood estimate of b in the formula logN= a-bM and its confidence limits. *Bull. Earthq. Res. Inst.* 43, 237–239. https://www.doi.org/10.15083/0000033631

Al-Shameri, W.F.H. 2012. Correlation dimension of an attractor generated by an orbit of general two-dimensional iterated quadratic map. *Int. J. Contem. Math. Sci.* 7(9), 413–424. https://www.m-hikari.com/ijcms/ijcms-2012/9-12-2012/alshameriIJCMS9-12-2012.pdf

Arabasz, W.J., Nava, S.J., and Phelps, T.W. 1996. Mining Seismicity in the Wasatch Plateau and Book Cliffs Coal Mining Districts, Utah, USA. In: Proceedings of International Symposium on Rockburst and Seismicity in Mines (Krakow, Poland), 111–116. https://www.osti.gov/etdeweb/biblio/613594

Baltz, R., and Hucke, A. 2008. Rockburst prevention in German coal industry. In: Proceedings of 27th International Conference on Ground Control in Mining (SME-ICGCM), 46–50. https://www.onemine.org/document/abstract.cfm?docid=150262&title=Rockburst-Prevention-in-the-German-Coal-Industry

Bischoff, M., Cete, A., Fritschen, R., and Meier, T. 2010. Coal mining induced seismicity in the Ruhr area, Germany. *Pure Appl. Geophys.* 167, 63–75. https://www.doi.org/10.1007/s00024-009-0001-8

BP Statistical Review of World Energy. 2016. https://www.oilproduction.net/files/especial-BP/bp-statistical-review-of-world-energy-2016-full-report.pdf

BP Statistical Review of World Energy. 2017. https://www.oilproduction.net/files/especial-BP/bp-statistical-review-of-world-energy-2017-full-report.pdf

Brown, F.A., Guzmán, A.R., Yépez, E., Navarro, A.R., and Miller, C.P. 1998. Fractal geometry and seismicity in the Mexican subduction zone. *Geofís. Int.* 37(1), 29–38. https://www.redalyc.org/pdf/568/56837103.pdf

Cai, M., Kaiser, P.K., and Martin, C.D. 2001. Quantification of rock mass damage in underground excavations from microseismic event monitoring. *Int. J. Rock Mech. Min. Sci.* 38(8), 1135–1145. https://www.doi.org/10.1016/S1365-1609(01)00068-5

Caneva, A., and Smirnov, V. 2004. Using the fractal dimension of earthquake distributions and the slope of the recurrence curve to forecast earthquakes in Colombia. *Earth Sci. Res. J.* 8(1), 3–9. https://www.researchgate.net/publication/228595496

Chiang, H.S. 1980. *Fully Mechanized Longwall Coal Mining.* Shandong Scientific and Technical Publishing Co.

Christman, R.C., Haslbeck, J., Sedlik, B., Murray, W., and Wilson, W. 1980. Activities, Effects and Impacts of the Coal Fuel Cycle for a 1,000-MWe Electric Power Generating Plant. In: U.S. Nuclear Regulatory Commission. https://catalog.hathitrust.org/Record/003117666

Chun-lai, W., Ai-xiang, W., Xiao-hui, L., and Rui, L. 2009. Study on fractal characteristics of *b* value with microseismic activity in deep mining. *Procedia Earth Planet. Sci.* 1, 592–597.

Coal Mine Safety. 2010. National coal mining safety accidents report in China. State Administration (Coal Mine Safety). https://www.chinasafety.gov.cn

Cook, N.G.W. 1964. The application of seismic techniques to problems in rock mechanics. *Int. J. Rock Mech. Min. Sci. Geomech. Abstr.* 1(2), 169–179. https://www.doi.org/10.1016/0148-9062(64)90025-7

Cook, N.G.W. 1976. Seismicity associated with mining. *Eng. Geol.* 10(2–4), 99–122. https://www.doi.org/10.1016/0013-7952(76)90015-6

DGMS. 2011. Fatal Accidents in Indian Coal Mines. DGMS Strategic Plan 2011–15 (Directorate General of Mines Safety). https://www.scribd.com/document/59115207/Dgms-Strategic-Plan-2011

Dhanam, K., Kumar, P.S., Mysaiah, D., Prasad, P.P., and Seshunarayana, T. 2013. High resolution seismic imaging of the Sohagpur Gondwana Basin Central India: Evidence for syn-sedimentary subsidence and faulting. *J. Earth Syst. Sci.* 122(6), 1495–1505. https://www.doi.org/10.1007/s12040-013-0362-3

Ding, Y., Dou, L., Cai, W., Chen, J., Kong, Y., Su, Z., and Li, Z. 2016. Signal characteristics of coal and rock dynamics with micro-seismic monitoring technique. *Int. J. Min. Sci. Technol.* 26(4), 683–690. https://www.doi.org/10.1016/j.ijmst.2016.05.022

Dodson, J., Li, X., Sun, N., Atahan, P., Zhou, X., Liu, H., Keliang, Z., Songmei, H., and Yang, Z. 2014. Use of coal in the Bronze Age in China. *The Holocene.* 24(5), 525–530. https://www.doi.org/10.1177/0959683614523155

Ellenberger, J.L., Heasley, K.A., Swanson, P.L., and Mercier, J. 2001. Three dimensional microseismic monitoring of a Utah longwall. In: Proceedings of 38th US Symposium on Rock Mechanics (USRMS, American Rock Mechanics Association), 1321–1326. https://www.stacks.cdc.gov/view/cdc/9859

Gautam, S., Awatar, R., Tewari, R., and Goswami, S. 2016. Permian-Triassic palynofloral transition in Sohagpur coalfield, South Rewa Gondwana Basin, Madhya Pradesh, India. *Paleobot.* 65, 109–129.

Ge, M. 2005. Efficient mine microseismic monitoring. *Int. J. Coal Geol.* 64(1–2), 44–56. https://www.doi.org/10.1016/j.coal.2005.03.004

Geng, X. 2008. Construction of micro seismic monitoring system for detecting mining rock burst. *HeiLongJiang Science and Technology Information.* 31.

Ghose, A.K. 1988. Mechanized Longwall mining in India- Collating the Decadal Experience (1978–1987). *J. Min. Metal. Fuel.* 267–273.

Ghose, A.K. 2003. Why Longwall in India has not succeeded as in other developing countries like China. *IE(I)-J-MN.* 84, 1–4. https://www.researchgate.net/publication/284259715

Ghose, G.K., and Sivakumar, C. 2018. Application of underground microseismic monitoring for ground failure and secure longwall coal mining operation: A case study in an Indian mine. *J. Appl. Geophys.* 150, 21–39. https://doi.org/10.1016/j.jappgeo.2018.01.004

Gibowicz, S. J., and Kijko, A. 1994. Chapter 13 – Prediction and prevention of large seismic events. In: Dmowska, R. (Eds.), *An Introduction to Mining Seismology*, pp. 336–346. London: Academia Press Inc. https://www.doi.org/10.1016/B978-0-12-282120-2.50017-0

Grassberger, P., and Procaccia, I. 1983. Measuring the strangeness of strange attractors. *Physica D.* 9, 189–208. https://doi.org/10.1016/0167-2789 (83)90298-1

Greenside, H.S., Wolf, A., Swift, J., and Pignataro, T. 1982. Impracticality of a box-counting algorithm for calculating the dimensionality of strange attractors. *Phys. Rev. A.* 25(6), 3453–3459. https://www.doi.org/10.1103/PhysRevA.25.3453

Guo, X., Tang, C., Nan, S., Zhang, J., and Wang, W. 2011. Application of mine micro-seismic monitoring system on preventing against illegal mining. *Procedia Eng.* 26, 2206–2212. https://www.doi.org/10.1016/j.proeng. 2011.11.2426

Harris, J., Kirsch, P., Shi, M., Li, J., Gagrani, A., Krishna, E.S., Tabish, A., Arora, D., Kothandaraman, K., and Cliff, D. 2014. Comparative analysis of coal fatalities in Australia, South Africa, India, China and USA, 2006-2010. In: 4th Coal Operators Conference, University of Wollongong (The Australasian

Institute of Mining and Metallurgy & Mine Managers Association of Australia), 399–407. https://ro.uow.edu.au/cgi/viewcontent.cgi?article=2197&context=coal

Hayes, P. 2000. Moonee Colliery: Renewing the economic viability of a mine using microseismic and hydraulic fracturing techniques in massive roof conditions. In: Proceedings of 19th International Conference on Ground Control in Mining (SME ICGCM), 38–44. https://www.onemine.org/document/abstract.cfm?docid=149856&title=Moonee-Colliery-Renewing-The-Economic-Viability-Of-A-Mine-Using-Microseismic-And-Hydraulic-Fracturing-Techniques-In-Massive-Roof-Conditions

Henderson, J.R., Barton, D.J., and Foulger, G.R. 1999. Fractal clustering of induced seismicity in the geysers geothermal area, California. *Geophys. J. Int.* 139(2), 317–324. https://www.doi.org/10.1046/j.1365-246x.1999.00939.x

Hirata, T. 1987. Omori's Power Law aftershock sequences of microfracturing in rock fracture experiment. *J. Geophys. Res.* 92, 6215–6221. https://www.doi.org/10.1029/JB092iB07p06215

Hirata, T. 1989. Fractal dimension of fault systems in Japan: Fractal structure in rock fracture geometry at various scales. *Pure Appl. Geophys.* 131(1), 157–170. https://www.doi.org/10.1007/BF00874485

Hirata, T., and Imoto, M. 1991. Multifractal analysis of spatial distribution of micro earthquakes in the Kanto region. *Geophys. J. Int.* 107(1), 155–162. https://www.doi.org/10.1111/j.1365-246X.1991.tb01163.x

Hirata, T., Satoh, T., and Ito, K. 1987. Fractal structure of spatial distribution of microfracturing in rock. *Geophys. J. Int.* 90(2), 369–374. https://www.doi.org/10.1111/j.1365-246X.1987.tb00732.x

Idziak, A., and Teper, L. 1996. Fractal dimension of faults network in the upper Silesian coal basin (Poland): Preliminary studies. *Pure Appl. Geophys.* 147(2), 239–247. https://www.doi.org/10.1007/BF00877480

Ishimoto, M., and Iida, K. 1939. Observations sur les seismes enregistres par le microseismographe construit dernierement (1). *Bull. Earthq. Res. Inst.* 17, 443–478.

Jasper, A., Agnihotri, D., Tewari, R., Spiekermann, R., Pires, E.F., Da Rosa, Á.A.S., and Uhl, D. 2016. Fires in the mire: Repeated fire events in Early Permian 'peat forming' vegetation of India. *Geol. J.* 52, 955–969. https://www.doi.org/10.1002/gj.2860

Jian, S., Lianguo, W., Zhansheng, W., Huaqiang, H., and Yifeng, S. 2011. Determining areas in an inclined coal seam floor prone to water-inrush by micro-seismicmonitoring. *Min. Sci. Technol.* 21(2), 165–168. https://www.doi.org/10.1016/j.mstc.2011.02.009

Jiang, F.X., Ye, G.X., Wang, C.W., Zhang, D.Y., and Guan, Y.Q. 2008. Application of high-precision microseismic monitoring technique to water inrush monitoring in coal mine. *Chinese J. Rock Mech. Eng.* 27(9), 1932–1938. https://www.rockmech.org/EN/Y2008/V27/I09/1932

Jie, P., and Tai, M. 1988. AE phenomenon of gas outbursts. *Safety in Coal Mine.* 7, 1–7.

Kagan, Y., and Knopoff, L. 1978. Statistical study of the occurrence of shallow earthquakes. *Geophys. J. Int.* 55(1), 67–86. https://www.doi.org/10.1111/j.1365-246X.1978.tb04748.x

Kagan, Y.Y., and Knopoff, L. 1980. Spatial distribution of earthquakes: The two-point correlation function. *Geophys. J. Int.* 62, 303–320. https://www.doi.org/10.1111/j.1365-246X.1980.tb04857.x

Kagan, Y.Y., and Knopoff, L. 1981. Stochastic synthesis of earthquake catalogs. *J. Geophys. Res.* 86, 2853–2862. https://doi.org/10.1029/JB086iB04p02853

Khan, P.K., Ghosh, M., Chakraborty, P.P., and Mukherjee, D. 2011. Seismic *b*-value and the assessment of ambient stress in Northeast India. *Pure Appl. Geophys.* 168(10), 1693–1706. https://www.doi.org/10.1007/s00024-010-0194-x

King, G. 1983. The accommodation of large strains in the upper lithosphere of the earth and other solids by self-similar fault systems: the geometrical origin of *b*-value. *Pure Appl. Geophys.* 121(5–6), 761–815. https://www.doi.org/10.1007/BF02590182

Kuzyniazouv, G.N. 1954. The mechanics of interactions between surroundings strata and supports on the longwall faces in slightly inclined coal seam. In: *Collog. on Investigation into the Suitability of Mine Pressures on Mechanized Supports*, p. 40. Moscow: Coal Mining Technical Publishing Co.

Legrand, D., Cisternas, A., and Dorbath, L. 1996. Multifractal analysis of the 1992 Erzincan aftershock sequence. *Geophys. Res. Lett.* 23(9), 933–936. https://www.doi.org/10.1029/96GL00725

Li, S.L. 2009. Discussion on microseismic monitoring technology and its application to underground projects. *Chinese J. Undergr. Space Eng.* 5, 122–128. https://en.cnki.com.cn/Article_en/CJFDTOTAL-BASE200901023.htm

Li, T., Cai, M.F., and Cai, M., 2007. A review of mining-induced seismicity in China. *Int. J. Rock Mech. Min. Sci.* 44(8), 1149–1171. https://www.doi.org/10.1016/j.ijrmms.2007.06.002

Linden, W.V.D. 2000. World best longwall mining practices, what is applicable to India. In: Ghose, A.K., and Dhar, B.B. (Eds.), *Chapter 3 – Mining Challenges of the 21st Century. (Institute of Engineers, India)*, pp. 23–38. New Delhi: A.P.H. Publishing Corporation. ISBN: 81-7648-158-0. https://www.google.co.in/books/edition/Mining_Challenges_of_the_21st_Century/C5XyR0jHe8QC?hl=en&gbpv=1

Liu, J., Feng, X., Li, Y., Xu, S., and Sheng, Y. 2013. Studies on temporal and spatial variation of microseismic activities in a deep metal mine. *Int. J. Rock Mech. Min. Sci.* 60, 171–179. https://www.doi.org/10.1016/j.ijrmms.2012.12.022

Luo, X., and Hatherly, P. 1998. Application of microseismic monitoring to characterize geomechanical conditions in longwall mining. *Explor. Geophys.* 29(4), 489–493. https://www.doi.org/10.1071/EG998489

MacNeill, P. 2008. Overview of the project undertaken at the NSW Department of Primary Industries as a vacation industrial work experience project. International Mining Fatality Database. https://www.resourcesandenergy.nsw.gov.au/__data/assets/pdf_file/0009/182484/International-Mining-Fatality-Database-project-report.pdf

Mandal, A., and Sengupta, D. 2000. The analysis of fatal accidents in Indian coal mines. *Calcutta Statistical Association Bull.* 50(1–2), 95–120. https://www.doi.org/10.1177/0008068320000109

Masabattula, V. 2010a. Reduction of risk from roof and side falls in Indian Coal Mines. Technology. https://www.slideshare.net/venkoos/reduction-of-risk-from-roof-and-side-fall-in-indian-coal-mines

Masabattula, V. 2010b. Longwall mining in India- Experiences and Expectations. https://www.slideshare.net/venkoos/longwall-exp-indiaexpectations

Meredith, P.G., Main, I.G., and Jones, C. 1990. Temporal variations in seismicity during quasi-static and dynamic rock failure. *Tectonophysics.* 175(1–3), 249–268. https://doi.org/10.1016/0040-1951(90)90141-T

Ming-Ruo, J., Chun'an, T., and Cuornin, Z. 2003. Numerical test of influence of mesoscopic heterogeneity on macroscopic behavior of rock failure and seismic sequence types. *Chinese J. Geophys.* 46(4), 659–666. https://www.doi.org/10.1002/cjg2.414

Ming-Xiao, W., Tao, Z., Miao-Rong, X., Bin, Z., and Ming-Qiu, J. 2011. Analysis of national coal-mining accident data in China, 2001-2008. *Public Health Rep.* 126(2), 270–275. https://www.doi.org/10.1177/003335491112600218

Moatti, A., Amin-Nasseri, M.R., and Zafarani, H. 2013. Pattern recognition on seismic data for earthquake prediction purpose. In: International Conference on Environment, Energy, Ecosystems and Development, 141–146. https://www.researchgate.net/publication/305103043

Mogi, K. 1963. Magnitude-frequency relation for elastic shocks accompanying fractures of various materials and some related problems in earthquakes (2nd paper). *Bull. Earthq. Res. Inst.* 40, 831–853. https://core.ac.uk/display/15891341

Mondal, D., Ghosh, S., Naveen, P., Kumar, M., Majumder, A., and Panda, A.K. 2021. An integral study on the geochemical, geophysical and geomechanical characteristics of the organic deposits (Coal and CBM) of eastern Sohagpur coalfield, India. *Gondwana Res.* 96, 122–141. https://www.doi.org/10.1016/j.gr.2021.03.016

Mondal, D., and Roy, P.N.S. 2019. Fractal and seismic b-value study during dynamic roof displacements (roof fall and surface blasting) for enhancing safety in the longwall coal mines. *Eng. Geol.* 253, 184–204. https://www.doi.org/10.1016/j.enggeo.2019.03.018

Mondal, D., Roy, P.N.S., and Behera, P.K. 2017a. Use of correlation fractal dimension signatures for understanding the overlying strata dynamics in Longwall Coal Mines. *Int. J. Rock Mech. Min. Sci.* 91. 210–221. https://doi.org/10.1016/j.ijrmms.2016.11.019

Mondal, D., Roy, P.N.S., and Behera, P.K. 2017b. Application of Correlation Integral and Fractal Dimension in Longwall Mine Safety. In: 79th EAGE Conference and Exhibition 2017 (Paris). https://www.doi.org/10.3997/2214-4609.201701417

Mondal, D., Roy, P.N.S., and Kumar, M. 2020. Monitoring the strata behavior in the destressed zone of a shallow Indian longwall panel with hard sandstone cover using mine-microseismicity and borehole televiewer data. *Eng. Geol.* 271, 105593 (1–25). https://www.doi.org/10.1016/j.enggeo.2020.105593

Murase, K. 2004. A characteristic change in fractal dimension prior to the 2003 Tokachi-oki Earthquake (MJ= 8.0), Hokkaido, Northern Japan. *Earth Planet. Sp.* 56(3), 401–405. https://www.doi.org/10.1186/BF03353072

Mutke, G., Pierzyna, A., and Baranski, A. 2016. b-value as a criterion for the evaluation of rockburst hazard in coal mines. In: 3rd International Symposium on Mine Safety Science and Engineering, 1–5. https://www.isms2016.proceedings.mcgill.ca/article/view/146

Nakaya, S., and Hashimoto, T. 2002. Temporal variation of multifractal properties of seismicity in the region affected by the mainshock of the October 6, 2000 Western Tottori Prefecture, Japan, earthquake (M = 7.3). *Geophys. Res. Lett.* 29, 133. https://www.doi.org/10.1029/2001GL014216

Obert, L., and Duvall, W.I. 1967. *Rock Mechanics and the Design of Structures in Rock.* New York: Wiley. https://www.semanticscholar.org/paper/Rock-mechanics-and-the-design-of-structures-in-rock-Obert-Duvall/c97a8286f9aee573dff7d3f9c8e75dfe36292f1b

Oraee, K., Shahriar, K., and Bakhtavar, E. 2009. Roof falls: An inherent risk in underground coal mining. In: Proceedings of 28th International Conference on Ground Control in Mining (ICGCM). https://www.researchgate.net/publication/44174300

Osterwald, F.W. 1970. Comments on rock bursts, outbursts and earthquake prediction. *Bull. Seismol. Soc. Am.* 60(6), 2083–2085. https://www.doi.org/10.1785/BSSA0600062083

Ouchi, T., and Uekawa, T. 1986. Statistical analysis of the spatial distribution of earthquakes-variation of the spatial distribution of earthquakes before and after large earthquakes. *Phys. Earth Planet. Inter.* 44(3), 211–225. https://www.doi.org/10.1016/0031-9201(86)90071-3

Pareek, H.S. 1987. Petrographic, chemical and trace-elemental composition of the coal of Sohagpur coalfield, Madhya Pradesh, India. *Int. J. Coal Geol.* 9(2), 187–207. https://www.doi.org/10.1016/0166-5162(87)90045-0

Peng, S. 2019. *Longwall Mining.* London: CRC Press. https://www.doi.org/10.1201/9780429260049

Peng, S.S., and Chiang, H.S. 1982. Roof stability in Longwall Coal Faces. In: Proceedings of 1st International Conference on Underground Stability (Vancouver, Canada), 295–335.

Peng, S.S., and Chiang, H.S. 1984. *Longwall Mining*. New York: John Wiley and Sons. https://www.osti.gov/scitech/biblio/6285667

Qin, S., Cheng, J., and Zhu, S. 2012. The application and prospect of microseismic technique in coalmine. *Procedia Environ. Sci.* 12, 218–224. https://www.doi.org/10.1016/j.proenv.2012.01.269

Raja Rao, C.S. 1983. Coalfields of India: Coal resources of Madhya Pradesh, Jammu and Kashmir. *Bull. Geol. Survey India (Series A)*. 3(45), 1–204.

Rikitake, T. 1979. Classification of earthquake precursors. *Tectonophysics*. 54(3–4), 293–309. https://www.doi.org/10.1016/0040-1951(79)90372-X

Rowell, G.A. 1989. Microseismic monitoring in coal and hard under deep cover. In: Proceedings of ISRM International Symposium (Pau, France), 321–328. https://www.onepetro.org/ISRMIS/proceedings-abstract/IS89/All-IS89/ISRM-IS-1989-041/45762

Roy, P.N.S., and Gupta, D.K. 2015. Fractal Analyzer: A Matlab application for multifractal seismicity analysis. *Seismol. Res. Lett.* 86(5), 1424–1431. https://www.doi.org/10.1785/0220150013

Roy, P.N.S., and Mondal, S.K. 2009. Fractal nature of earthquake occurrence in northwest Himalayan region. *J. Ind. Geophys. Union*. 13(2), 63–68. http://citeseerx.ist.psu.edu/viewdoc/download?doi=10.1.1.580.6218&rep=rep1&type=pdf

Roy, P.N.S., and Mondal, S.K. 2012. Multifractal analysis of earthquakes in Kumaun Himalaya and its surrounding region. *J. Earth Syst. Sci.* 121(4), 1033–1047. https://www.doi.org/10.1007/s12040-012-0208-4

Scholz, C.H. 1968. Microfractures, aftershocks, and seismicity. *Bull. Seismol. Soc. Am.* 58(3), 1117–1130. https://www.doi.org/10.1785/BSSA0580031117

Schorlemmer, D., and Wiemer, S. 2005. Earth science: Microseismicity data forecast rupture area. *Nature*. 434(7037), 1086–1086. https://www.doi.org/10.1038/4341086a

Schorlemmer, D., Wiemer, S., and Wyss, M. 2005. Variations in earthquake-size distribution across different stress regimes. *Nature*. 437(7058), 539–542. https://www.doi.org/10.1038/nature04094

Shahdol, 2011. Profile/Mineral Resources. Shahdol District Administration. District Shahdol, Madhya Pradesh (Government of India). https://www.shahdol.nic.in

Shaw, B.E. 1995. Frictional weakening and slip complexity in earthquake faults. *J. Geophys. Res. Solid Earth*. 100, 18239–18251. https://doi.org/10.1029/95JB01306

Shu, J., Jiang, L., Kong, P., Wang, P., and Zhang, P. 2019. Numerical modeling approach on mining-induced strata structural behavior by considering the fracture-weakening effect on rock mass. *Appl. Sci.* 9, 1832. https://www.doi.org/10.3390/app9091832

Si, G., Durucan, S., Jamnikar, S., Lazar, J., Abraham, K., Korre, A., Shi, J.Q., Zavšek, S., Mutke, G., and Lurka, A. 2015. Seismic monitoring and analysis of excessive gas emissions in heterogeneous coal seams. *Int. J. Coal Geol.* 149, 41–54. https://www.doi.org/10.1016/j.coal.2015.06.016

Sivakumar, C., Srinivasan, C., Singh, A.K., Sivaram, Y.V., Sudhakar, K., Willy, Y.A., and Srinivasulu, T. 1998. Prediction of roof falls in goaf area in Churcha West Coalmine SECL by using the microseismic monitoring method. Project Completion Report- Coal S&T Project #MT98

Sivakumar, C., Srinivasan, C., Willy, Y.A., Murthy, Ch.S.N. 2008. Real Time Microseismic Monitoring to Study Geomechanics of Underground Structures. In: 12th International Conference of International Association for Computer Methods and Advances in Geomechanics, 1972–1979. https://citeseerx.ist.psu.edu/viewdoc/download?doi=10.1.1.384.1650&rep=rep1&type=pdf

Smith, A. 1997. Provenance of coals from Roman Sites in England and Wales. *Britannia.* 28, 297–324. https://www.doi.org/10.2307/526770

Srinivasan, C., Sivakumar, C., and Gupta, R.N. 2005. Source parameters of seismic events in a coal mine in India. In: 27th Seismic Research Review: Ground-Based Nuclear Explosion Monitoring Technologies, 653–662. https://l2a.ucsd.edu/local/Meetings/2005_SRR/PAPERS/03-18.PDF

Suyehiro, S. 1966. Difference between aftershocks and foreshocks in the relationship of magnitude to frequency of occurrence for the great Chilean earthquake of 1960. *Bull. Seismol. Soc. Am.* 56(1), 185–200. https://www.doi.org/10.1785/BSSA0560010185

Swanson, P., Boltz, M.S., and Chambers, D. 2016. Seismic monitoring strategies for deep longwall coal mines. Department of Health and Human Services, National Institute for Occupational Safety and Health, NIOSH Publication. https://www.cdc.gov/niosh/mining/UserFiles/works/pdfs/2017-102.pdf

Swanson, P.L. 2001. Development of an automated PC-network-based seismic monitoring system. In: Proceedings of 5th International Symposium on Rock bursts and Seismicity in Mines (Institute of Mining and Metallurgy). https://www.cdc.gov/niosh/mining/UserFiles/works/pdfs/doaap.pdf

Tang, L.Z., and Xia, K.W. 2010. Seismological method for prediction of areal rock bursts in deep mine with seismic source mechanism and unstable failure theory. *J. Cent. South Univ. Technol.* 17(5), 947–953. https://www.doi.org/10.1007/s11771-010-0582-5

Tie, L., and Hongguang, J. 2010. Identification of water inrush process of unknown water body using microseismic monitoring technique. *Chinese J. Rock Mech. Eng.* 29(1), 134–139. http://en.cnki.com.cn/Article_en/CJFDTotal-YSLX201001016.htm

Tormann, T., Enescu, B., Woessner, J., and Wiemer, S. 2015. Randomness of megathrust earthquakes implied by rapid stress recovery after the Japan earthquake. *Nature Geosci.* 8(2), 152–158. https://www.doi.org/10.1038/ngeo2343

Tormann, T., Wiemer, S., and Mignan, A. 2014. Systematic survey of high-resolution b value imaging along Californian faults: inference on asperities. *J. Geophys. Res. Solid Earth.* 119, 2029–2054. https://www.doi.org/10.1002/2013JB010867

Tsapanos, T.M. 1990. b-values of two tectonic parts in the Circum-Pacific belt. *Pure Appl. Geophys.* 134(2), 229–242. https://www.doi.org/10.1007/BF00876999

Ustu, T. 1965. A method for determining the value of b in the formula logn=a-bm showing the magnitude-frequency relation for earthquake. *Geophys. Bull. Hokkaido Univ.* 13, 99–103. https://www.semanticscholar.org/paper/A-method-for-determining-the-value-of-b-in-a-log-Utsu/163f20fb00c77fc5cfa8410d729fe125b97503f4

Vinoth, S., and Kumar, L.A. 2014. Applying real time seismic monitoring technology for slope stability assessment- An Indian opencast coal mine perspective. *Int. J. Min. Sci. Technol.* 24(1), 75–80. https://www.doi.org/10.1016/j.ijmst.2013.12.013

Waltham, T., Park, H.D., Suh, J., Yu, M.H., Kwon, H.H., and Bang, K.M. 2011. Collapses of old mines in Korea. *Eng. Geol.* 118(1–2), 29–36. https://www.doi.org/10.1016/j.enggeo.2010.11.007

Wang, C., Wu, A., Liu, X., and Li, R. 2009. Study on fractal characteristics of b value with microseismic activity in deep mining. *Procedia Earth Planet. Sci.* 1, 592–597. https://www.doi.org/10.1016/j.proeps.2009.09.094

Wang, J.H. 1988. b values of shallow earthquakes in Taiwan. *Bull. Seismol. Soc. Am.* 78 (3), 1243–1254. https://www.doi.org/10.1785/BSSA0780031243

Wang, J.H. 1996. Velocity-weakening friction as a factor in controlling the frequency-magnitude relation of earthquakes. *Bull. Seismol. Soc. Am.* 86 (3), 701–713. https://www.doi.org/10.1785/BSSA0860030701

Wang, J.H. 1997. Effect of frictional healing on the scaling of seismicity. *Geophys. Res. Lett.* 24(20), 2527–2530. https://www.doi.org/10.1029/97GL02523

Wang, J.H., Chen, K.C., Leu, P.L., and Chang, C.H. 2016. Precursor times of abnormal b-values prior to mainshocks. *J. Seismol.* 20(3), 905–919. https://www.doi.org/10.1007/s10950-016-9567-7

Weeks, J., Lockner, D., and Byerlee, J. 1978. Change in b-values during movement on cut surfaces in granite. *Bull. Seismol. Soc. Am.* 68 (2), 333–341. https://www.doi.org/10.1785/BSSA0680020333

World Coal Association. 2014. Coal Facts. https://www.worldcoal.org/coal-facts-2014

World Coal Institute. 2009. The Coal Resource – A Comprehensive overview of Coal. (World Coal Institute). https://refman.energytransitionmodel.com/publications/1264

Xie, H., and Pariseau, W.G. 1993. Fractal character and mechanism of rock bursts. *Int. J. Rock Mech. Min. Sci. Geomech. Abs.* 30(4), 343–350. https://www.doi.org/10.1016/0148-9062(93)91718-X

Xu, N., Tang, C., Li, H., and Liang, Z. 2012. Application of microseismic monitoring technique in hydroelectric projects. In: Boroujeni, H.S., (Eds.), *Hydropower – Practice and Application*, IntechOpen. https://www.doi.org/10.5772/31625

Yanfang, R., 2011. Study on monitoring technology of the overlying strata spatial structures' failure in shallow seam long wall face. *Procedia Eng.* 26, 928–933. https://www.doi.org/10.1016/j.proeng.2011.11.2258

Yang, C., Luo, Z., Hu, G., and Liu, X. 2007. Application of a microseismic monitoring system in deep mining. *J. Univ. Sci. Technol. Beijing Miner. Metall. Mater.* 14, 6–8. https://www.doi.org/10.1016/S1005-8850(07)60002-4

Yerpude, R.R., and Walke, D.V. 2014. Investigation into the risk factors associated with Roof-falls in Underground Coal Mines. *Int. J. Eng. Res. Technol.* 3(10), 1249–1257. https://www.ijert.org/research/investigation-in-to-the-risk-factors-associated-with-roof-falls-in-underground-coal-mine-IJERTV3IS100670.pdf

Young, R.P., and Collins, D.S. 2001. Seismic studies of rock fracture at the Underground Research Laboratory, Canada. *Int. J. Rock Mech. Min. Sci.* 38(6), 787–799. https://www.doi.org/10.1016/S1365-1609(01)00043-0

Zhao G., Ma, J., Dong, L., Li, X., Chen, G., and Zhang, C., 2015. Classification of mine blasts and microseismic events using starting-up features in seismograms. *T. Nonferr. Metal. Soc. China.* 25(10), 3410–3420. https://www.doi.org/10.1016/S1003-6326(15)63976-0

Zhao, X.D., Li, Y.H., Liu, J.P., and Tian, J. 2009. Study on microseismic activity in potential rock burst zone during deep excavation in Hongtoushan Mine. *J. Northeastern Univ.* 30, 1330–1333. https://www.en.cnki.com.cn/Article_en/CJFDTOTAL-DBDX200909029.htm

Chapter 11

Source parameters and scaling relation of the Baikal rift's earthquakes

Anna A. Dobrynina, Vyacheslav V. Cheptsov, Vladimir A. Sankov, Vladimir V. Chechelnitsky, and Rajib Biswas

11.1 INTRODUCTION

The stress-strain state of the medium is a product of active geodynamic processes occurring in the lithosphere and mantle of the Earth. Under different conditions, seismogenic ruptures of different morphological, genetic and geodynamic types are formed, which differ in characteristic types of movements in earthquake foci: strike-slip, normal and reverse faults. Different mechanisms of impact can, in addition to kinematics, also affect the level of stresses in the medium in different geodynamic regions. These differences in different conditions contribute to the fact that in different zones the source parameters of earthquakes and their relations differ significantly from each other.

Despite a significant number of determinations of the source parameters of individual strong earthquakes and aftershock sequences in different regions of the world, the comparison of the obtained values and their relations has not been sufficiently studied. This especially concerns the values of the stress drop in the earthquake sources, their geometrical dimensions and the ratio "seismic moment – seismic energy".

In present study, the task is to determine the source parameters of earthquakes (the seismic moment, the moment magnitude, the source dimension – radius, the stress drop and the displacement amplitude in the source) in the Baikal rift system (BRS), to compare with ones in different geodynamic regions of the world and to identify patterns typical for

DOI: 10.1201/9781003177692-11

zones of different types, to obtain correlations between the local magnitudes and the source parameters of earthquakes.

The first spectral determination of the seismic moment was made by K. Aki (Aki, 1966) for the Niigata earthquake on 16 June 1964 ($M = 7.5$), according to the records of surface G waves. Later, J. Brune proposed a model of the source of an earthquake, according to which, the spectrum of displacements of the transverse S wave in the far-field zone is linked to the seismic moment, and the source dimension by considerable simple relations (Brune, 1970; 1971). Hanks and Wyss extended this model, using the spectra of the teleseismic longitudinal P waves (Hanks and Wyss, 1972). Thatcher and Hanks (Thatcher and Hanks, 1973) worked out, in detail, the methodology towards the calculation of the source parameters based on the spectra of the transverse S waves for regional earthquakes ($M_L = 2.0–7.0$). At present, the method of determining the source parameters based on the spectra of the displacement of the body and surface waves is extensively used for this study of the strong and weak earthquakes, recorded by the teleseismic, regional, and local networks of seismic stations, and also for laboratory studies.

Herein, we aim to determine the source parameters of the earthquakes of the BRS, one of the most seismically active regions of Russia via spectral signatures of seismic waves. Despite the high seismic potential of the region, the determination of the source parameters of earthquakes (such as the seismic moment, the moment magnitude, the source dimension, the stress drop) has rarely been addressed for this region; owing to the absence of a digital data recording system. Till now, only four studies were carried out in this direction. For instance, using data of the temporary network with magnetic recordings, the source parameters of weak events located in the Barguzin depression were obtained (Danzig, 1981). Data of the temporary local seismic station network in the North Muya region were used for the determination of the scaling relation “seismic moment – local magnitude” (Kochetkov et al., 1985). An express method of the estimation of the seismic moment and the corner frequency using amplitudes and periods of seismic waves on the seismogram, as proposed by S.S. Arefiev (Arefiev, 1979) was adapted for Baikal’s earthquakes by A.V. Klyuchevskii (Klyuchevskii, 1986). Following this, values of the seismic moments and the source dimensions of the events that occurred in the BRS during 1968–1994 were obtained (using data of an analog seismic station network worked in the studied area till 1998) (Klyuchevskii and Demjanovich, 2002). The highly sensitive digital instrumentation of the seismic stations of the region in 1998–2003 appreciably improved the conditions of the recording of seismic events in the BRS (Masalsky et al., 2007); thereby enabling it to estimate the source parameters. Thus, the seismic moments, the moment magnitude, the source dimension, the stress drop, and the displacement amplitude were estimated for 62 earthquakes of the Baikal rift by using Brune's model (Dobrynina, 2009). There, the estimation of the source parameters was carried out without reckoning the site corrections and the regional attenuation model. As such, we have incorporated the local site effect for each seismic station and the regional frequency dependence quality factor for the S waves $Q_S(f)$ in this work.

11.2 STUDY AREA

The BRS, the second largest continental rift system in the world, is in North Eurasia. In a neotectonic sense, the BRS is confined to the boundary of the North Eurasian and the Amurian lithospheric plates (Figure 11.1). According to the assessed data by the GPS geodesy, the elongation in the central part of BRS is normal to main rift structures; and the relative velocity of the Amurian plate's to the North Eurasian plate in this place is 3.4 ± 0.7 mm/year (Sankov et al., 2009). On the southwestern wing, a shortening in the north-eastern direction dominates whereas, the north-eastern wing has not been studied by the GPS geodesy.

The high contemporary mobility in between the North Eurasian and the Amurian lithospheric plates determines a high level of the seismic process taking place there. Since 1950, more than 10^5 earthquakes, with magnitude $M \geq 1.4$, occurred in this area; 13 of them were strong events with magnitude $Ms \geq 6.0$. The epicentral field analysis (Sankov et al., 2012) shows that the majority of earthquakes are confined to the rift system (Figure 11.2). The main part of events trends in the rift depression with less extent towards, inter-basin spur and ridges and on the rift's arm. As such, the belts surrounding apart blocks tend to have less seismicity. In general, the orientation of these belts agrees with a strike of the main rift structures, and also the epicentral zones oriented across the

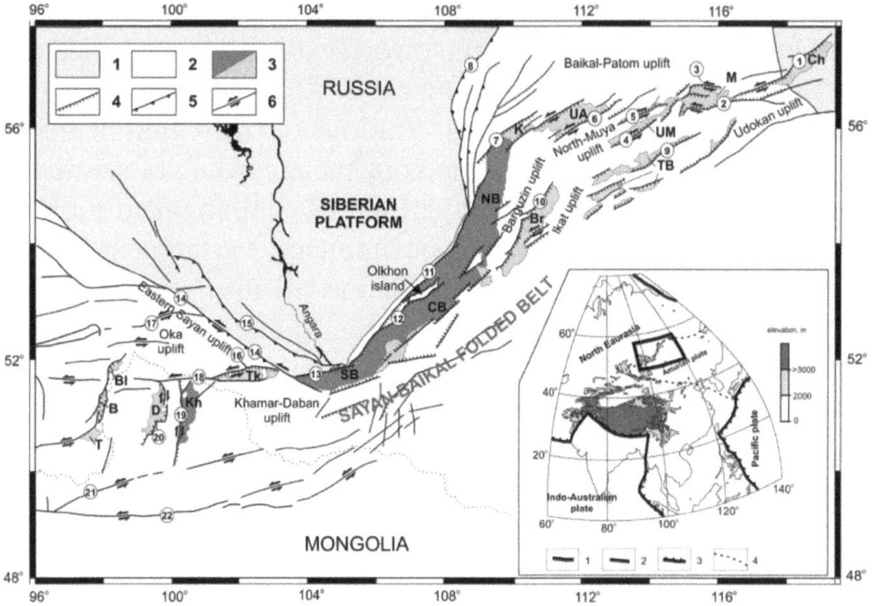

Figure 11.1 Neotectonic scheme of the Baikal rift system and surroundings (Dobrynina et al., 2016). 1 – the Siberian platform; 2 – the Sayan-Baikal folded area; 3 – the Cenozoic rift basins: Ch – Chara, M – Muya, UM – Upper Muya; TB – Tsipa-Baunt; UA – Upper Angara; K – Kitchera; NB – North Baikal; Br – Barguzin; CB – Central Baikal; SB – South Baikal; Tk – Tunka; Kh – Khubsugul; D – Darkhat; Bl – Belinskaya; B – Busingol; T – Terekhol; 4–6 – faults: 4 – normal; 5 – thrust and reverse; 6 – strike-slip. Numbers within circles denote the main faults: 1 – Kodar; 2 – South-Muya; 3 – North-Muya; 4 – Upper Muya; 5 – Muyakansky; 6 – Upper Angara; 7 – Kitchera; 8 – Akitkan; 9 – Tsipa-Baunty; 10 – Barguzin; 11 – Primorsky; 12 – Morskoy; 13 – Obruchev; 14 – Main Sayan Fault; 15 – Peredovoy; 16 – Tunka; 17 – Okino-Zhombolok; 18 – Baikal-Mondy; 19 – Khubsugul; 20 – Darkhat; 21 – Tsetserleg; 22 – Bolnai (Khangai). Inset locates the study region within Asia: 1 – the study zone; 2 – the main strike-slip faults; 3 – the subduction zones; and 4 – the Amurian plate boundary.

direction of the main structures exist (e.g., an area of the Barguzin ridge and others). At the same time, some large rifting faults (such as the Main Sayan fault, Northbaikalsky fault, the eastern part of the Tynkinsky fault) are characterized by weak seismicity. The plot of the seismic moment distribution based on data of events with magnitude $M \geq 2.5$ occurring in BRS during 1950–2010 shows that the maximum part of seismic energy is confined to the south-western and north-eastern wings. Apart from this, almost all strong earthquakes were located except events that occurred in the South and Middle Baikal basins. The localization of strong events

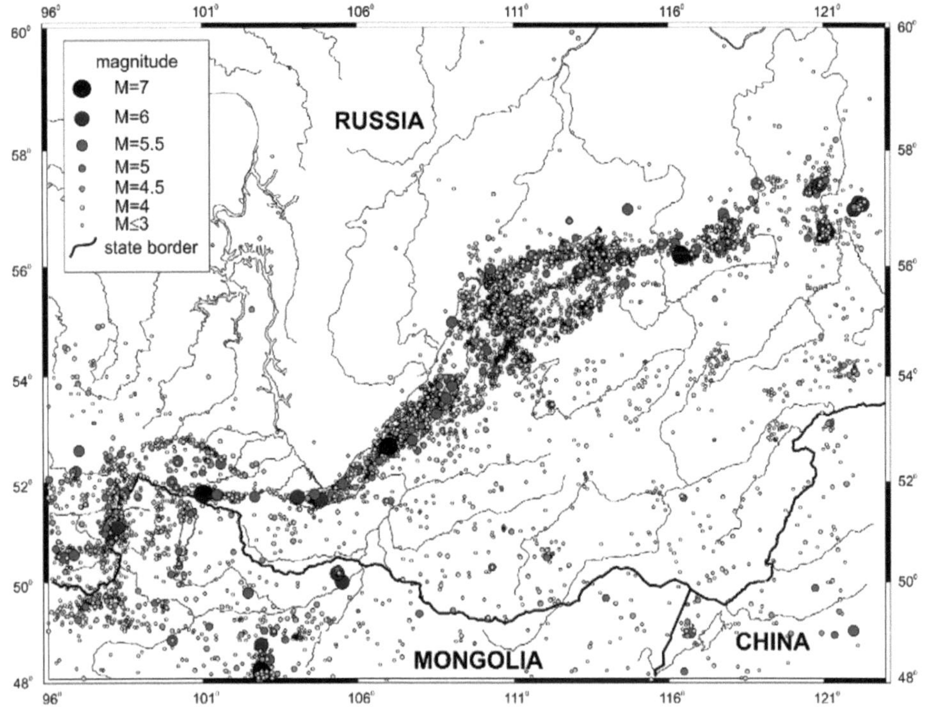

Figure 11.2 Seismicity of the Baikal rift system during 1950–2012. (Dobrynina et al., 2016.)

to the rift's wing may be a consequence of the difference in the stress state of the medium in the different parts of the rift system. The character of the seismic realization of the studied area is various. Despite the single earthquake, a large quantity of aftershock and foreshock-aftershock sequences and swarms are observed (Golenetsky, 1990; Radziminovich and Ochkovskaya, 2013). The precise depths determination in the BRS has not been addressed till date because of the sporadic distribution of seismic stations (the sparse seismic station network is located along the rift system) and the unknown velocity structure of the crust. The depth of the seismic active layer is ambiguously evaluated in different works. For the model of the homogeneous medium, the earthquakes hypocenters are located at depths 3–20 km (Golenetsky, 1990; Radziminovich, 2010), in the north-eastern part of the BRS – from the surface down to 30 km and deeper, down to the lower crust (Deverchere et al., 2001; Radziminovich, 2010). For the layered model, the earthquakes foci are located in the lower crust and the upper mantle at depths 32–46 km (Deverchere et al., 1991). Most earthquakes in the south-western part of the BRS (the South Baikal and Tunka basins) occur at depths 10–25 km (Radziminovich, 2010).

Earthquakes focal mechanism solutions are determined in BRS using the standard method of the first polarities of P waves (Honda, 1962). The database of source mechanisms contains about 600 solutions for individual events and 175 composites (group) solutions for the period 1950–2012 (Sankov and Dobrynina, 2018) with further inclusion of seismic moment tensor solutions obtained by international seismological agencies (ISC, NEIC, CMT).

In Figure 11.3, focal mechanism solutions of BRS earthquakes and their classification diagrams characterized by quantitative ratios of events with different types of slip in the source (strike-slip, dip-slip, or thrust) are presented. It is shown that, in the central part of BRS (regions II and III, Figure 11.2), the sources of "rift" type (the compression axis is vertical, the extension axis is horizontal, the nodal planes are oriented according to the general rift structures). Another picture is observed for the wings of BRS. In the south-western wing (region I), strong events (with magnitude $M \geq 6$) are characterized by the strike-slip displacement in the source; for example, the Mondy earthquake of 4 April 1950 with magnitude $Ms = 7.0$ can be referred to (Delouis et al., 2002). Also, for the south-western part of the BRS, the increasing portion of thrust events is observed as compared with the central part of BRS. For the north-eastern wing of the rift system (regions IV and V, Figure 11.2), the absence of thrusts is characteristic; here events with the dip-slip component in the focus dominate. But, at the same time, for strong earthquakes, the strike-slip component in the source prevails. The Muya earthquake on 27 June 1957 with a magnitude $Ms = 7.6$ (Doser, 1991) could be mentioned, for example. Thus, it can be asserted that the extension regime dominates in the central part of the rift system, while the manifestation of strike-slips is characteristic for latitudinal segments of the rift (in rift's wings).

11.3 METHOD

For calculating the seismic moment, the standard model of circular dislocation proposed by Brune was applied (Brune, 1970; 1971). In accordance with it, the spectrum of displacements in the far-field zone $\Omega(f)$ can be

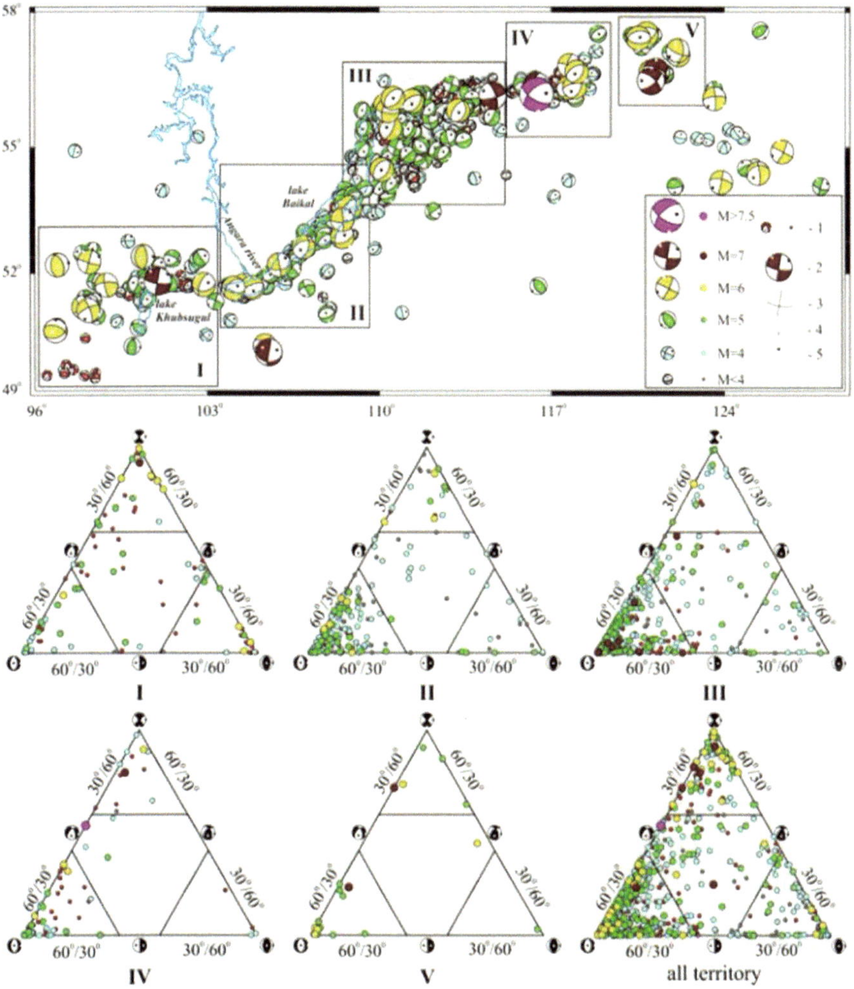

Figure 11.3 The focal mechanisms of the earthquakes of the BRS (lower hemisphere) for the period 1950–2010 (top) and the classification diagrams (bottom) (Sankov and Dobrynina, 2018). Different colors and sizes show the earthquakes of different magnitudes. Rectangles indicate the areas for which the diagrams are plotted, the area numbers are shown in Latin numerals. Legend: 1 – the composite focal mechanism solutions; 2 – the focal mechanism; 3 – the nodal planes; 4, 5 – the axes of the main stresses of the tension and the compression, respectively.

described by a constant amplitude Ω_0 up to a corner frequency f_C. It decreases as f^{-2} for frequencies, greater than f_C (Haskell, 1964; Aki, 1967):

$$\Omega(f) = \frac{\Omega_0}{1+\left(f/f_C\right)^2} \qquad (11.1).$$

For determining the spectral characteristics (Ω_0 and f_C), it is necessary to consider the influence of external factors on the recording of the earthquake. In general, the spectrum of displacements, recorded at a seismic station, can be represented in the form of the product of the source spectrum and a couple of filters which is given as follows:

$$S(f)=\Omega(f)\cdot I(f)\cdot G(f)\cdot P(f)\cdot A(f)\cdot FS. \tag{11.2}$$

In equation (11.2), $\Omega(f)$ is the source spectrum, $I(f)$ is the response of a recording device, $G(f)$ is the correction for attenuation of body waves as a result of the geometric spreading, $P(f)$ is the inelastic attenuation along the path "source-receiver", and FS is the influence of free surface ($FS = 2$) (Trifunac, 1972; Thatcher and Hanks, 1973; Aptekman et al., 1989). The inelastic attenuation along the way of the propagation of the seismic wave path was calculated by the formula (Douglas and Ryall, 1972):

$$\exp\left[\frac{\pi f \Delta}{Q_S V_S}\right], \tag{11.3}$$

Δ is the epicentral distance, Q_S is the quality factor for direct S waves.

The spectrum parameters Ω_0 and f_C were determined via the comparison of the theoretical and observed spectra. Further, they were corrected for the amplitude-frequency characteristic of the instrument, the geometric spreading, and inelastic attenuation on the path "source-receiver" by formulas (2) and (3).

The level of the spectral amplitude Ω_0 (cm · sec) and the seismic moment M_0 (dyne · cm = 10^7 N · m) are connected by the relationship:

$$M_0 = \frac{4\pi\rho\Delta V_S^3\Omega_0}{R_{\theta\varphi}}, \tag{11.4}$$

where ρ is the density, ρ = 2.7 g/cm^3 (Riznichenko, 1985), V_S is the velocity of the transverse waves (for the BRS V_S = 3.55 km/sec [Golenetskii and Novomeiskaya, 1975]) and $R_{\theta\varphi}$ is the S-wave radiation pattern coefficient. The mean seismic moment and the corner frequency of the earthquake were determined from the values of seismic moments and corner frequencies obtained at different stations, according to Archuleta et al., 1982:

$$\langle X_0 \rangle = anti\log\left[\frac{1}{N}\sum_{i=1}^{N}\log X_i\right], \quad (11.5)$$

where N is the number of stations, X_i is the seismic moment, or the corner frequency determined at the i^{-th} station. According to the classical statistical approach, the standard deviation of the logarithm was determined for the M_0 and f_C values (Archuleta et al., 1982):

$$\sigma(\log X) = \left[\frac{1}{N-1}\sum_{i=1}^{N}\left(\log X_i - \log\langle X \rangle\right)^2\right]^{1/2} \quad (11.6)$$

The source dimension (the radius of the equivalent circular dislocation) r (km), the stress drop $\Delta\sigma$ (dyne/cm^2 = 0.1 Pa) and the amplitude of the displacement along the fault D (cm) were obtained by the following equations:

$$r = \frac{2.34 V_S}{2\pi f_C} \quad (11.7)$$

$$\Delta\sigma = \frac{7 M_0}{16 r^3}, \quad (11.8)$$

$$D = \frac{M_0}{\mu S}, \quad (11.9)$$

where $S = \pi r^2$ is the area of the circular dislocation and $\mu = 3.2 \cdot 10^{11}$ dyne/cm^2 is the shear modulus (Riznichenko, 1985).

Attenuation and site corrections. For the calculation of the quality factor of S-waves $Q_S(f)$, the method of the predominant period was used (Dobrynina et al., 2016). Sixteen strong regional earthquakes with magnitudes m_b = 5.0–6.0 occurred in BRS during 2003–2006 were selected. For the determination of Q_S values, the records of the direct S waves were only used. The epicentral distances Δ changed within 20–169 km. The seismic traces were processed by Butterworth filter at six frequency ranges: 0.3 ± 0.2, 0.75 ± 0.25, 1.5 ± 0.5, 3 ± 1, 6 ± 2 and 12 ± 4 Hz. The use of a relatively narrow frequency range is due to the fact that, we tried to carry out measurements within the flat part of the frequency response so that there was no distortion of the seismic signal. In the calculation of Q_S values, the maximum displacement amplitude (AS) and the corresponding frequency were measured on the filtered traces and the displacement amplitudes for different earthquakes were recomputed for the same magnitude. Finally, the Q_S value is determined based on the slope of the least-square straight-line fit between the maximum amplitude A_S versus the epicentral distance Δ for each frequency range (Figure 11.4). Local earthquakes (Δ ≤ 200 km) are characterized by short-period radiation. The final estimations of the Q_S values are based

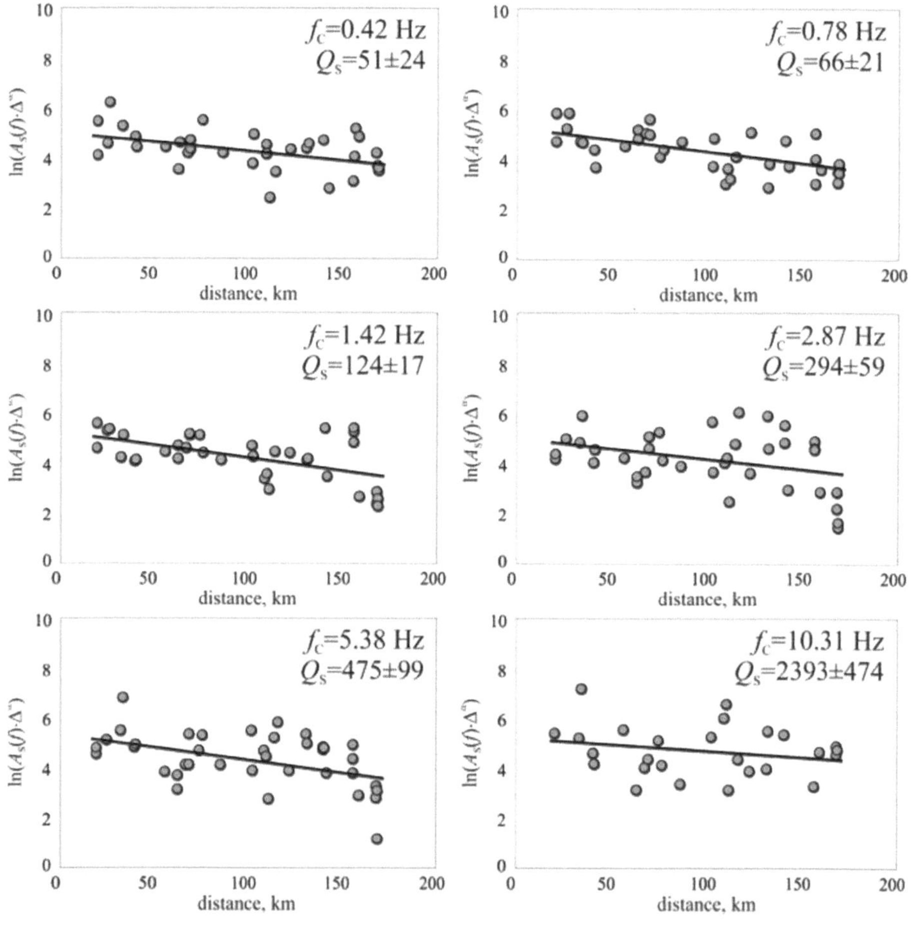

Figure 11.4 The example of the Q_S values calculation for the different frequency ranges (for each frequency range, Q_S value and central frequency are shown).

on more than 200 individual measurements. And the best fitting expression for the regional dependence of $Q_S(f)$ at low frequencies (0.1–8.0 Hz) is $(98 \pm 15) \cdot f^{(0.93 \pm 0.05)}$.

The estimation of the local site corrections was carried out using the method of H/V ratios of Nakamura (Nakamura, 1989). The technique of using the spectral ratios between the horizontal (H) and vertical (V) components (H/V ratio) of the microtremors for the medium response estimation for the first time was proposed by M. Nogoshi (Nogoshi and Igarashi, 1971) and later was popularized by Y. Nakamura (1989). Nakamura's method is effective for the calculation of both fundamental periods (Field and Jacob, 1993; Ohmachi et al., 1994) and the corresponding amplification factor (Lermo and Chavez-Garcia, 1994; Konno and Ohmachi, 1995). The comparison of the local site effect obtained by the H/V ratios method and one estimated by other methods show a satisfactory agreement (Field and Jacob, 1995; Bonilla et al., 1997; Parolai et al., 2004). The base of the H/V ratios method is an assumption about equality of the amplitude of the spectral ratio between the horizontal and vertical components of the microtremors measured at the ground surface R_{MB} and the amplification factors of a site A_S (Nakamura, 1989).

For the local site-effect estimations, the records of the microtremors were used. In order to escape the influence of random noise (traffic, earthquakes, weather, and others), a one-hour record length was processed. To account for daily, weekly, and seasonal variations, we processed the data of the permanent seismic registration for days, different days of the week (weekdays and weekend separately), and for the different seasons (summer and winter). Because of the apparatus characteristics (short-period instruments), the reliable site characteristics were obtained for the frequency range from 0.5 to 10 or 20 Hz (Table 11.1). As a result, the daily, weekly, and seasonal variations of the spectral characteristics of the microtremors were determined (Dobrynina et al., 2018). According to the obtained characteristics, the sites under study are divided into three groups:

1. the stable site – independent on the season (BOD, CRS, CIT, KMO, KPC, LSTR, MOY, NLYR, OGRR, ORL, SYVR, TUP, UKT, YLYR, YOA, ZAK) (Figure 11.5a);

2. the site with daily and/or seasonal variations (ARS, IRK, NIZ, SVKR, TLY, TRG) (Figure 11.5b);
3. the unstable site (KAB).

Analysis of the spectra horizontal and vertical components separately for the sites in the second group shows an increase in the amplitude of the horizontal oscillations in the comparison with the vertical ones for the warm season. This increase may be connected with the temporal changes

Table 11.1 The seismic stations of the Baikal rift system

	Coordinates, degrees				
Station/international code	*N°*	*E°*	*h, m*	*Type*	*f, Hz*
1	*2*	*3*	*4*	*5*	*6*
Arshan/ARS	51.920	102.423	970	B–11	0.5–20
Bodaibo/BOD	57.819	114.004	245	B–11	0.5–10
Chara/CRS	56.900	118.269	700	B–11	0.5–20
Chita/CIT	52.021	113.552	790	B–11	0.5–20
Irkutsk/IRK	52.243	104.271	467	B–10	0.5–10
Kabansk/KAB	52.050	106.654	468	B–10	0.5–10
Khapcheranga/KPC	49.704	112.379	953	B–11	0.5–20
Kumora/KMO	55.887	111.201	480	B–11	0.5–20
Listvyanka/LSTR	51.868	104.832	450	B–11	0.5–20
Mondy/MOY	51.667	100.993	1303	B–11	0.5–20
Nelyaty/NLYR	56.491	115.703	596	B–11	0.5–20
Nizhneangarsk/NIZ	55.775	109.541	487	B–10	0.5–10
Ongureny/OGRR	53.643	107.595	495	B–11	0.5–20
Orlik/ORL	52.539	99.810	1360	B–11	0.5–20
Severomuysk/SVKR	56.117	113.559	850	B–10	0.5–10
Suvo/SYVR	53.658	109.999	490	B–11	0.5–20
Talaya/TLY	51.681	103.644	579	B–11	0.5–20
Tupik/TUP	54.425	119.954	650	B–11	0.5–20
Tyrgan/TRG	52.760	106.348	718	B–11	0.5–20
Uakit/UKT	55.489	113.627	1140	B–11	0.5–20
Ulyunhan/YLYR	54.876	111.162	560	B–11	0.5–20
Uoyan/YOA	56.133	111.724	500	B–10	0.5–10
Zakamensk/ZAK	50.382	103.281	1200	B–11	0.5–20

Note: 1 – station's name and the international code; 2, 3 – the coordinates (the latitude and longitude); 4 – the elevation; 5 – the type of the receiver: B-10 is the "Baikal-10", B-11 is the "Baikal-11" (details in the text); 6 – the frequency band in Hz.

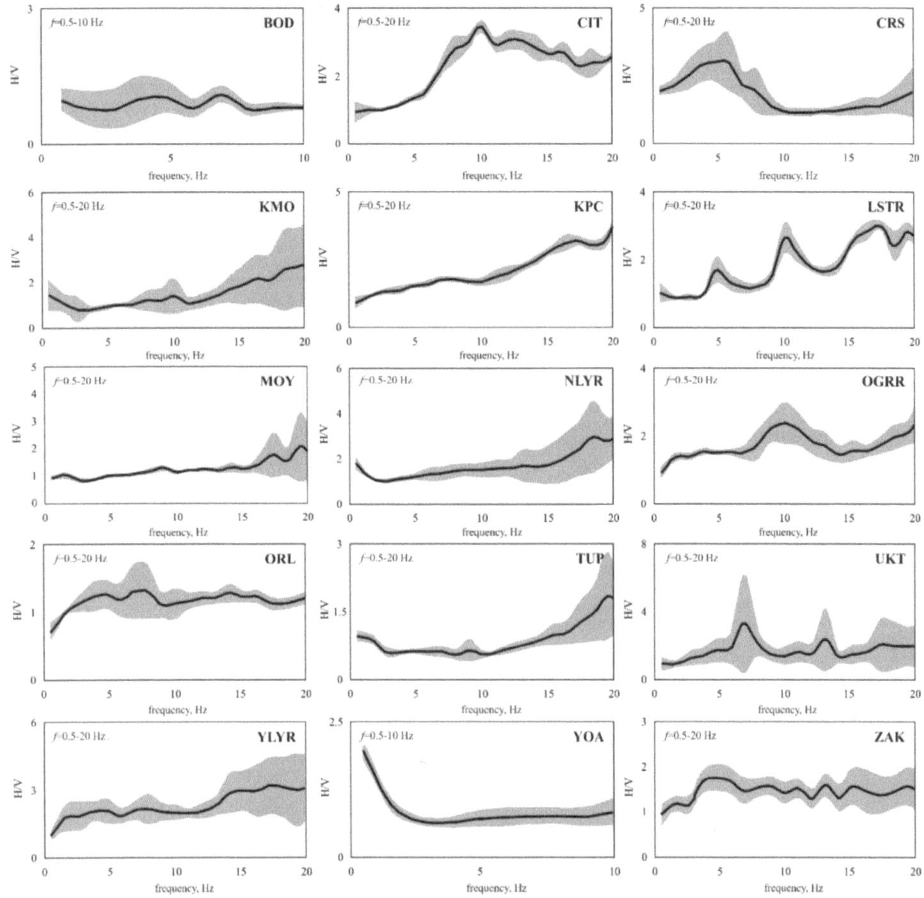

Figure 11.5a The stable H/V ratio curves 1 – the H/V curve; 2 – the variations of H/V curves; CIT, CRS etc. – the indications of the stations.

in the wave composition of the microtremors depending on the season and man-made conditions.

Most of the sites in the second group possess seasonal variations of the site characteristics that may be connected with the tidal and surf effect of Lake Baikal. These stations are situated on the coastline (NIZ, TRG, TLY, KAB, see Figure 11.5b). Variations of the local site effect for the seismic stations IRK and SVKR may be explained by the existing significant man-made noise (traffic, hydroelectric power stations, railway) (Dobrynina et al., 2018).

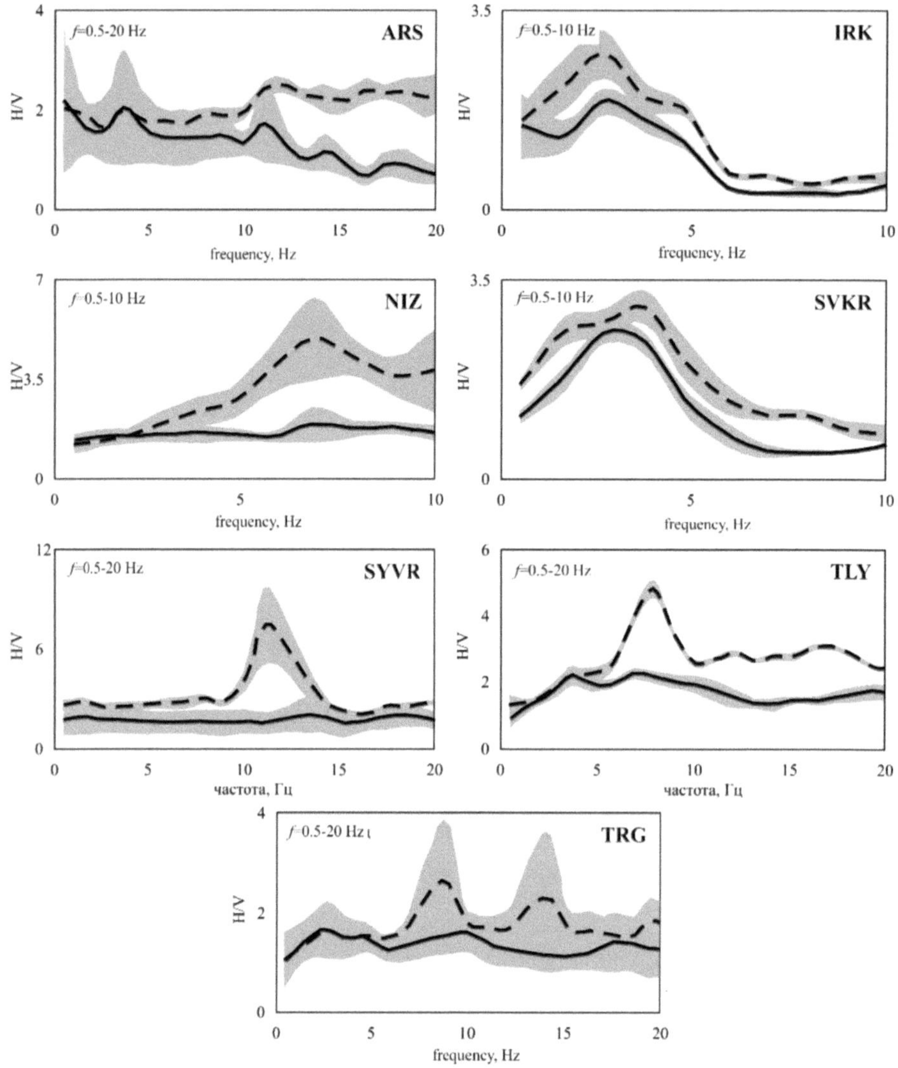

Figure 11.5b The curves of H/V ratio with the seasonal variations. 1 – the H/V curve in the winter period; 2 – the H/V curve in the summer period; 3 – the variations of the H/V curves; ARS, IRK etc. – the indications of the stations.

11.4 DATA

The earthquake waveforms data used in the present work were obtained by a permanent regional network of the digital seismic stations of Baikal Office of the Federal Research Center "Joint Geophysical Survey of Russian Academy of Science" (international code BAGSR). The network

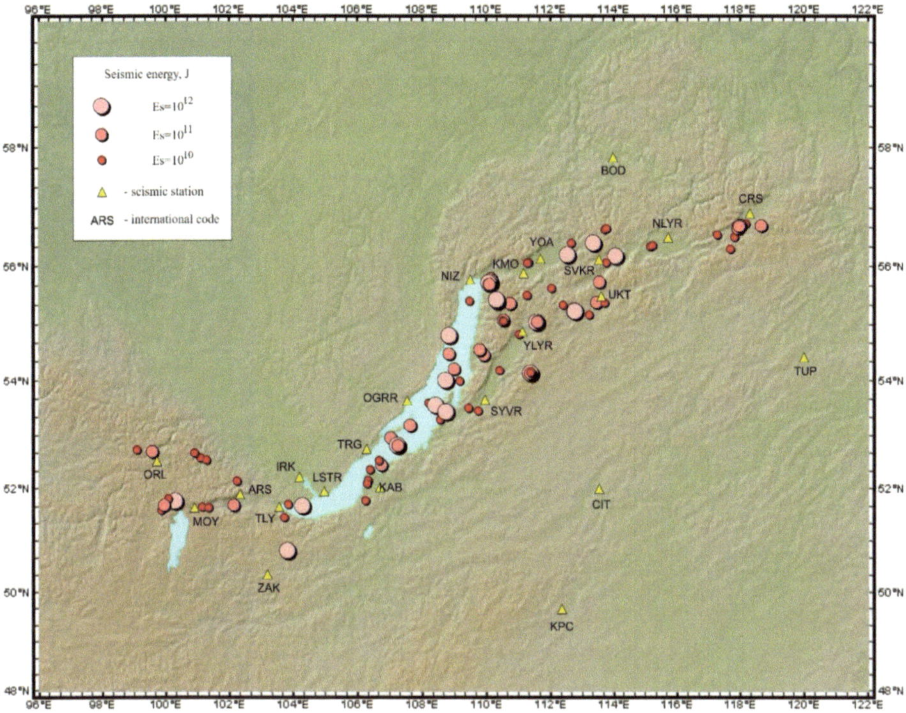

Figure 11.6 The location of the seismic stations and the selected earthquakes.

consists of 23 stations, of which 19 are located directly within the rift system (Figure 11.6 and Table 11.1). The stations are equipped with the digital seismic instrumentation of the “Baikal-10”, “Baikal-11” types, developed in the geophysical survey. The equipment set has three short-period channels of the high sensitivity, which record the vibration rates from 0.01–0.1 μm/s up to 100–1000 μm/s; and three rough channels for recording of the acceleration from 50–500 μm/s^2 up to 100–250 cm/s^2.

For the estimation of the source parameters from the regional earthquake’s catalog, 150 events with the local magnitude M_L = 3.1–4.7 (Table 11.2) occurring within the BRS during 2003–2006 were selected (Figure 11.6). The selection of data for the spectral analysis was carried out according to the following conditions:

1. The signal-to-noise ratio is no less than 5:1.
2. The clear S wave arrival.

Table 11.2 The parameters of the selected earthquakes

No	*Date, dd.mm.yy*	*Time, hh:mm*	*Latitude, °N*	*Longitude3, °E*	M_L
	1	*2*	*3*	*4*	
1	01.01.04	12:26	55.06	110.58	3.3
2	02.01.04	17:11	55.06	110.59	3.3
3	02.01.04	23:38	55.06	110.58	3.2
4	03.01.04	07:23	51.69	102.25	3.2
5	04.01.04	05:32	55.06	110.6	3.7
6	04.01.04	19:21	56.06	111.37	3.3
7	05.01.04	03:30	55.06	110.58	3.2
8	08.01.04	13:38	55.05	110.59	3.4
9	24.01.04	14:18	52.45	106.83	3.7
10	29.01.04	07:33	54.47	108.9	4.0
11	12.02.04	18:15	55.05	110.6	3.2
12	23.02.04	00:34	56.39	112.7	3.3
13	08.03.04	09:27	56.16	114.11	4.7
14	09.03.04	23:46	56.29	117.68	3.3
15	11.03.04	09:39	52.14	100.74	4.6
16	19.03.04	16:49	56.66	118.67	3.1
17	22.03.04	03:49	56.68	118.64	3.7
18	04.04.04	15:48	55.37	113.5	3.8
19	07.04.04	02:24	55.72	113.59	3.7
20	21.04.04	04:19	56.53	117.27	3.3
21	05.05.04	17:46	55.78	110.21	3.9
22	16.05.04	20:00	55.16	113.26	3.4
23	24.05.04	07:42	55.5	111.32	3.4
24	26.05.04	23:56	54.12	111.44	4.7
25	27.05.04	00:36	54.12	111.43	4.1
26	30.05.04	04:13	54.18	110.47	3.3
27	30.05.04	14:22	54.13	111.44	3.4
28	09.06.04	02:58	54.14	111.43	3.6
29	17.06.04	20:48	56.72	118.16	3.6
30	17.06.04	22:01	54.45	109.99	4.1
31	18.06.04	08:52	54.82	111.08	3.2
32	26.06.04	15:36	55.22	112.8	4.3
33	28.06.04	17:25	56.7	117.94	3.3
34	29.06.04	11:27	56.69	118	3.7
35	03.07.04	00:45	56.49	117.81	3.3
36	06.07.04	14:37	56.38	113.41	4.3
37	10.07.04	02:03	56.67	117.99	3.2
38	26.07.04	00:25	56.69	117.86	3.2
39	30.07.04	02:40	53.29	108.62	3.3
40	31.07.04	21:49	56.06	113.8	3.1

(Continued)

Table 11.2 The parameters of the selected earthquakes (Continued)

No	Date, dd.mm.yy	Time, hh:mm	Latitude, °N	Longitude3, °E	M_L
	1	2	3	4	
41	06.08.04	10:13	54.78	108.87	4.0
42	06.08.04	16:10	54.79	108.91	4.2
43	17.08.04	08:22	51.61	99.97	3.3
44	13.09.04	07:01	50.82	103.93	4.3
45	24.09.04	01:39	52.37	106.47	3.1
46	15.10.04	12:35	56.62	113.76	3.4
47	21.10.04	01:00	51.7	102.25	4.1
48	24.10.04	04:31	56.63	117.93	3.8
49	25.10.04	23:13	55.79	110.28	3.3
50	01.11.04	17:15	56.68	117.98	3.6
51	02.11.04	19:06	56.67	117.98	3.3
52	14.11.04	08:15	54.54	109.86	3.9
53	19.11.04	14:47	56.05	111.33	3.6
54	21.11.04	05:57	55.41	110.46	3.2
55	12.12.04	08:59	51.77	100.42	4.3
56	15.12.04	16:57	55.37	113.75	3.6
57	22.12.04	09:22	51.83	100.19	3.6
58	07.01.05	15:45	56.68	118	3.9
59	08.01.05	07:18	53.59	108.26	3.2
60	09.01.05	22:27	56.21	112.62	3.3
61	11.01.05	06:19	52.16	102.34	3.1
62	22.01.05	04:16	56.67	117.97	4.2
63	22.01.05	09:21	56.33	115.17	3.4
64	25.01.05	10:57	56.68	117.99	3.7
65	13.02.05	21:24	55.4	109.53	3.6
66	14.02.05	01:28	53.62	108.8	3.5
67	24.02.05	11:14	52.54	106.75	3.4
68	07.03.05	07:44	51.67	101.25	3.2
69	11.03.05	14:28	53.99	108.8	4.5
70	18.03.05	11:17	56.35	115.24	3.4
71	21.03.05	18:04	51.68	104.39	4.6
72	14.04.05	15:40	55.8	110.12	3.5
73	06.05.05	03:09	55.77	110.12	3.5
74	11.05.05	02:01	51.7	100.07	3.9
75	11.05.05	23:06	52.17	106.41	3.2
76	18.05.05	03:09	55.68	110.53	3.3
77	10.06.05	17:09	55.62	112.08	3.6
78	11.06.05	17:02	51.72	103.94	3.4
79	14.06.05	01:13	51.53	100.4	3.8
80	26.06.05	04:02	55.33	112.44	3.2

(Continued)

Table 11.2 The parameters of the selected earthquakes *(Continued)*

No	*Date, dd.mm.yy*	*Time, hh:mm*	*Latitude, °N*	*Longitude3, °E*	M_L
	1	*2*	*3*	*4*	
81	11.07.05	23:57	53.45	109.8	3.4
82	31.07.05	16:04	53.5	109.5	3.4
83	21.08.05	22:31	55.21	112.83	4.6
84	29.08.05	15:14	51.79	106.33	3.2
85	30.08.05	16:40	54.46	110.68	3.6
86	16.09.05	00:17	51.66	101.45	3.2
87	19.09.05	01:18	54	108.78	3.5
88	26.09.05	13:13	56.67	118.69	3.7
89	04.11.05	22:39	56.18	112.59	4.4
90	09.11.05	06:16	56.07	114.52	3.9
91	16.11.05	22:47	52.7	99.69	3.8
92	28.11.05	08:16	55.74	113	3.6
93	01.12.05	18:33	56.11	117.98	4.1
94	13.12.05	20:08	55.01	111.63	3.9
95	14.12.05	00:06	55.02	111.63	3.7
96	26.12.05	11:14	52.95	107.12	4.0
97	20.01.06	08:07	55.35	110.81	3.8
98	20.01.06	15:01	55.03	111.64	3.7
99	28.01.06	20:07	53.54	108.49	4.4
100	28.01.06	01:34	55.26	110.36	3.6
101	03.02.06	01:24	53.42	108.81	4.7
102	05.02.06	02:24	55.02	111.63	4.3
103	11.02.06	14:39	55.26	110.34	3.9
104	06.03.06	17:15	51.47	103.82	3.2
105	28.03.06	21:47	53.56	108.48	4.3
106	07.04.06	08:40	52.59	101.19	3.1
107	14.05.06	14:26	54.2	109.06	4.0
108	18.05.06	22:45	52.68	101.01	3.2
109	19.05.06	20:31	56.66	118.72	3.9
110	26.05.06	15:11	55.04	111.65	3.8
111	04.06.06	15:59	55.08	111.3	3.3
112	11.06.06	21:20	56.99	118.35	4.3
113	05.07.06	05:15	56.33	118.12	4.3
114	08.07.06	19:19	56.33	118.09	4.2
115	12.07.06	11:22	56.33	118.11	3.4
116	16.07.06	17:42	53.18	107.71	3.8
117	16.07.06	05:43	56.35	118.11	3.9
118	18.07.06	12:42	53.39	108.46	3.7
119	30.07.06	16:06	55.04	111.68	3.8
120	01.08.06	19:16	53.12	107.48	3.8

(Continued)

Table 11.2 The parameters of the selected earthquakes *(Continued)*

No	*Date, dd.mm.yy*	*Time, hh:mm*	*Latitude, °N*	*Longitude3, °E*	M_L
	1	*2*	*3*	*4*	
121	03.08.06	11:56	55.4	110.4	4.6
122	03.08.06	16:55	55.39	110.41	4.2
123	03.08.06	16:57	55.41	110.39	4.6
124	11.08.06	02:42	52.74	99.2	3.6
125	04.09.06	16:59	54.83	111.15	3.9
126	08.09.06	09:37	53.99	109.22	3.3
127	08.09.06	13:39	55.04	111.66	4.0
128	16.09.06	01:03	56.68	118.1	3.6
129	17.09.06	04:48	55.03	111.65	4.1
130	29.09.06	17:28	52.11	106.38	3.3
131	09.10.06	23:49	52.99	107.44	3.4
132	10.11.06	18:14	55.02	111.66	3.9
133	04.12.06	09:31	55.7	110.19	3.7
134	04.12.06	10:38	55.7	110.18	3.7
135	04.12.06	09:02	55.66	110.16	4.0
136	04.12.06	13:40	55.71	110.18	3.7
137	04.12.06	18:36	55.67	110.12	3.6
138	06.12.06	08:14	55.68	110.13	3.9
139	09.12.06	00:00	52.55	101.38	3.3
140	11.12.06	11:55	55.71	110.2	4.3
141	11.12.06	14:43	55.72	110.18	3.8
142	13.12.06	17:44	52.81	107.33	4.3
143	14.12.06	02:58	52.81	107.34	3.9
144	15.12.06	22:35	55.7	110.15	4.7
145	16.12.06	01:54	55.68	110.15	4.2
146	19.12.06	02:09	55.69	110.13	4.3
147	19.12.06	03:11	55.66	110.12	3.7
148	20.12.06	10:31	55.68	110.14	3.3
149	24.12.06	07:36	55.73	110.23	4.2
150	27.12.06	10:20	55.68	110.17	3.8

3. The number of stations, used for determining the source parameters of the earthquake, must be no less than 4.
4. The records, obtaining on the seismic station Kabansk (KAB), wasn't processed because of the unstable local site effect.

Since for the epicentral distances $\Delta \leq 1000$ km the amplitude of the S wave, as a rule, exceeds appreciably the amplitudes of the P waves (Kochetkov et al., 1985), the transverse waves were used for calculating the spectra of the displacements. There are several typical approaches for choosing the window length for calculating the source spectrum: (1) from the beginning of the S wave to the moment when the amplitude falls to 1/3 of its maximum value; (2) from the beginning of the S wave to the end of the wave packet including the maximum amplitude; (3) the fixed window equal $2(t_S - t_P)$, where t_S, t_P are the arrival times of the direct S- and P phases, respectively. In our investigation, we chose approach 3, since in this case the selected window includes all source shear waves and the possibility of operator's error is excluded. Because of a large part of the record, it inevitably contains different secondary (reflected and others) waves. The question about the influence of the secondary waves on the earthquake spectrum was considered in detail by N.T. Tarasov (Tarasov, 1982). He concluded that the amplitude and the form of the seismic spectrum are defined by the direct radiation from the source. The errors of the spectrum amplitude determination due to an influence of the secondary waves are small and in the existing accuracy of the seismic moment estimation, we can neglect them.

11.5 SOURCE SPECTRUM ESTIMATION

The spectra of earthquakes were estimated by the fast Fourier transform of the horizontal (N-S and E-W) components, the displacement spectra were obtained by dividing of the velocity or acceleration spectra on $2 \cdot \pi \cdot f$ or $(2 \cdot \pi \cdot f)^2$, respectively (here f is the frequency in Hz). For the source spectrum calculation, the spectrum obtained in the seismic station was corrected by the receiver response, the local site effect (calculated by the H/V ratio method), and the geometrical spreading and regional Q_S values (Dobrynina et al., 2016). For the seismic stations of the second group (with daily and/or season variations in the site effect), the site corrections were done by taking into account the season. The spectral parameters – the level in the low-frequency area (Ω_0) and the corner frequency (f_C) – were obtained by the comparison of the theoretical and observed spectra (Figure 11.7).

It's known that the focal mechanism exerts considerable influence on the amplitudes of the seismic waves, recorded at a station, and it can lead to a change of the seismic moment by several times (Vidale, 1989). In our case,

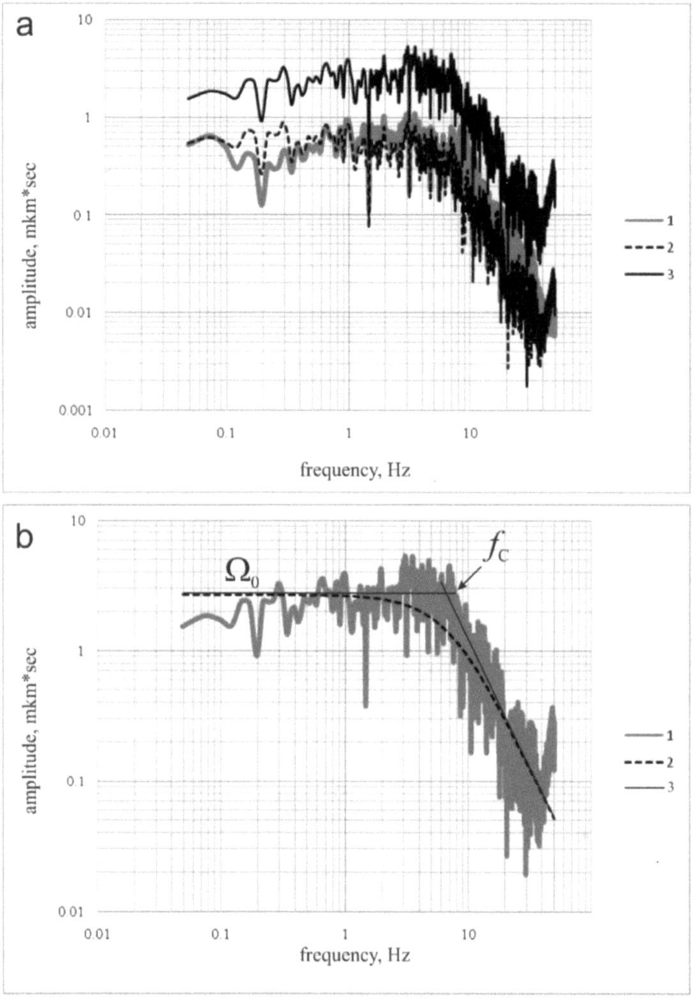

Figure 11.7 The example of the calculation of the source spectrum (a) and the spectral parameters determination (b). a: 1 – the spectrum recorded by the seismic station, 2 – the spectrum with the response and H/V-curve corrections, 3 – the spectrum with the attenuation correction; b: 1 – the estimated source spectrum (see explanation in the text), 2 – the theoretical spectrum, 3 – the approximating straight lines.

in order to minimize the error in the determination of the seismic moment, for the events with a known individual solution of the focal mechanism, their coefficients of directivity at each station were calculated (Aki and Richards, 1980). In the absence of the determination of the focal mechanism under the assumption of the self-similarity of the stress-strain state, the average coefficient of the directivity was calculated taking into account the focal mechanism of the earthquakes, localized nearby. In the absence of such events, the average coefficient of the directivity of the nucleation for the S waves was taken $R\theta\phi = 0.62$ (Fletcher, 1982).

11.6 RESULTS AND DISCUSSION

The results of the source parameters determination are presented in Table 11.3. The seismic moment of investigated 150 earthquakes changes from $2.64{\cdot}10^{13}$ N·m to $1.05{\cdot}10^{16}$ N·m. The values of the standard deviation of the logarithms of the seismic moment and the corner frequency

Table 11.3 The source parameters of the selected earthquakes

No	$M_0{\cdot}10^{13}$, N·m	σ_{M0}, log.un.	R, km	σ_R, log.un.	σ_R, %	Mw	D, sm	σ_D, %	$\Delta\sigma{\cdot}10^5$, Pa	$\sigma\Delta\sigma$, %	N
1	2	3	4	5	6	7	8	9	10	11	12
1	19.66	0.40	0.51	0.06	13	3.5	0.76	40	6.64	60	8
2	22.67	0.38	0.52	0.04	10	3.5	0.85	31	7.25	46	6
3	22.24	0.39	0.42	0.10	21	3.5	1.23	54	12.71	81	11
4	4.20	0.32	0.35	0.11	22	3.0	0.34	47	4.22	71	5
5	75.29	0.36	0.58	0.09	18	3.9	2.25	72	17.13	108	7
6	22.78	0.23	0.35	0.11	23	3.5	1.84	49	22.96	74	7
7	22.26	0.41	0.51	0.09	19	3.5	0.86	63	7.44	94	9
8	27.92	0.16	0.64	0.07	14	3.6	0.67	68	4.57	103	8
9	45.07	0.30	0.39	0.08	16	3.7	2.94	37	33.15	56	13
10	122.63	0.31	0.47	0.06	14	4.0	5.62	38	52.95	58	11
11	13.58	0.32	0.45	0.10	20	3.4	0.68	56	6.72	84	10
12	16.03	0.17	0.45	0.11	22	3.4	0.79	61	7.79	91	5
13	650.22	0.27	0.45	0.06	13	4.5	31.29	36	302.41	53	7
14	16.68	0.20	0.39	0.10	21	3.4	1.08	50	12.18	75	9
15	1045.61	0.27	0.41	0.05	11	4.6	62.20	27	668.47	41	9
16	11.37	0.18	0.55	0.06	13	3.3	0.38	45	3.02	67	4
17	42.74	0.17	0.72	0.06	12	3.7	0.83	77	5.07	116	7
18	85.37	0.27	0.45	0.08	17	3.9	4.26	46	41.92	68	7
19	46.94	0.31	0.38	0.08	16	3.7	3.22	37	37.18	56	8
20	24.03	0.23	0.40	0.06	12	3.6	1.49	28	16.27	43	5
21	65.13	0.27	0.37	0.11	22	3.8	4.84	49	58.15	73	7
22	19.06	0.32	0.36	0.09	19	3.5	1.46	40	17.79	60	9
23	20.64	0.49	0.28	0.11	22	3.5	2.55	41	39.53	61	4
24	557.90	0.26	0.46	0.07	14	4.5	26.26	40	251.08	59	6
25	145.01	0.47	0.38	0.12	24	4.1	9.74	57	111.23	85	10
26	19.87	0.39	0.29	0.11	23	3.5	2.38	41	36.31	62	9

(Continued)

Table 11.3 The source parameters of the selected earthquakes *(Continued)*

No	$M_0 \cdot 10^{13}$, *N·m*	σ_{M0}, *log.un.*	*R, km*	σ_R, *log. un.*	σ_R, %	*Mw*	*D, sm*	σ_D, %	$\Delta\sigma \cdot 10^5$, *Pa*	$\sigma\Delta\sigma$, %	*N*
1	*2*	*3*	*4*	*5*	*6*	*7*	*8*	*9*	*10*	*11*	*12*
27	22.01	0.49	0.39	0.12	23	3.5	1.46	56	16.50	84	12
28	32.26	0.47	0.34	0.12	24	3.6	2.84	51	37.18	77	10
29	55.42	0.33	0.39	0.11	23	3.8	3.55	57	39.65	85	11
30	238.71	0.42	0.46	0.08	17	4.2	11.21	47	107.07	70	8
31	16.87	0.41	0.40	0.11	23	3.5	1.07	55	11.94	83	8
32	264.72	0.49	0.48	0.08	16	4.2	11.58	48	106.74	72	4
33	34.93	0.39	0.41	0.08	16	3.7	2.06	40	22.01	60	6
34	49.84	0.51	0.56	0.07	15	3.8	1.58	58	12.40	86	5
35	19.72	0.38	0.45	0.09	18	3.5	0.97	50	9.43	74	6
36	308.01	0.16	0.45	0.10	21	4.3	15.32	58	150.60	86	6
37	16.14	0.26	0.38	0.07	16	3.4	1.10	35	12.67	53	4
38	4.71	0.16	0.31	0.07	15	3.1	0.48	29	6.84	43	3
39	9.35	0.47	0.36	0.10	21	3.3	0.71	46	8.66	69	7
40	14.99	0.22	0.39	0.09	18	3.4	0.96	43	10.71	64	7
41	185.85	0.40	0.49	0.06	12	4.1	7.80	35	70.48	53	4
42	193.80	0.38	0.56	0.07	14	4.2	6.18	52	48.63	78	6
43	7.07	0.32	0.55	0.05	11	3.2	0.23	41	1.86	61	4
44	159.30	0.19	0.43	0.09	18	4.1	8.51	47	86.73	70	5
45	7.94	0.48	0.53	0.08	16	3.2	0.28	55	2.36	83	3
46	21.71	0.60	0.80	0.05	11	3.5	0.34	104	1.84	156	4
47	52.79	0.17	0.48	0.07	14	3.8	2.30	42	21.10	63	4
48	91.90	0.33	0.42	0.10	21	3.9	5.16	54	53.87	80	7
49	8.82	0.45	0.61	0.08	17	3.3	0.24	76	1.70	115	3
50	64.06	0.36	0.47	0.09	19	3.8	2.85	57	26.44	85	10
51	32.23	0.29	0.40	0.04	9	3.6	1.99	22	21.85	32	7
52	96.37	0.43	0.35	0.13	26	4.0	7.75	57	96.94	85	11
53	24.46	0.44	0.32	0.14	27	3.6	2.31	57	31.21	86	9
54	7.35	0.25	0.29	0.14	27	3.2	0.90	50	13.89	75	6
55	198.43	0.31	0.64	0.05	11	4.2	4.77	55	32.61	82	5
56	66.24	0.38	0.43	0.09	20	3.8	3.53	52	35.93	78	10
57	21.76	0.45	0.42	0.12	25	3.5	1.20	66	12.42	100	6
58	140.82	0.35	0.41	0.08	17	4.1	8.30	43	88.82	65	11
59	13.25	0.37	0.40	0.09	19	3.4	0.84	46	9.26	69	12
60	15.24	0.33	0.38	0.08	17	3.4	1.08	39	12.64	58	9
61	3.89	0.34	0.24	0.14	28	3.0	0.69	45	12.79	67	6
62	135.37	0.44	0.41	0.04	9	4.1	8.08	21	87.04	32	12
63	30.02	0.37	0.40	0.07	16	3.6	1.89	37	20.95	55	10
64	167.62	0.28	0.40	0.07	15	4.1	10.49	36	115.58	54	9
65	30.49	0.42	0.36	0.12	25	3.6	2.40	56	29.75	83	15
66	30.35	0.38	0.47	0.04	9	3.6	1.35	26	12.54	39	6

(Continued)

Table 11.3 The source parameters of the selected earthquakes *(Continued)*

No	$M_0 \cdot 10^{13}$, *N·m*	σ_{M0}, *log.un.*	*R, km*	σ_R, *log. un.*	σ_R, *%*	*Mw*	*D, sm*	σ_D, *%*	$\Delta\sigma \cdot 10^5$, *Pa*	$\sigma\Delta\sigma$, *%*	*N*
1	*2*	*3*	*4*	*5*	*6*	*7*	*8*	*9*	*10*	*11*	*12*
67	17.88	0.46	0.33	0.11	22	3.5	1.64	45	21.82	67	13
68	7.20	0.28	0.37	0.12	24	3.2	0.53	55	6.29	83	7
69	263.29	0.32	0.45	0.11	22	4.2	12.83	61	124.77	92	11
70	18.72	0.23	0.29	0.12	24	3.5	2.19	45	32.97	68	11
71	503.18	0.41	0.49	0.10	21	4.4	21.13	64	190.83	96	10
72	30.47	0.44	0.39	0.10	21	3.6	2.02	50	22.95	76	4
73	33.28	0.47	0.57	0.09	18	3.6	1.03	69	8.01	103	5
74	72.18	0.45	0.59	0.06	14	3.9	2.05	57	15.26	85	6
75	12.65	0.52	0.35	0.14	27	3.4	1.00	62	12.43	92	11
76	17.91	0.54	0.29	0.11	22	3.5	2.16	40	33.12	60	5
77	40.30	0.32	0.49	0.07	15	3.7	1.65	46	14.71	69	7
78	10.58	0.30	0.35	0.13	26	3.3	0.86	57	10.71	86	8
79	46.90	0.21	0.39	0.05	11	3.7	3.12	25	35.38	38	4
80	19.83	0.33	0.42	0.10	21	3.5	1.13	55	11.85	83	7
81	16.93	0.55	0.32	0.15	30	3.5	1.61	62	21.95	94	5
82	29.77	0.26	0.42	0.11	23	3.6	1.66	61	17.34	91	12
83	777.07	0.35	0.49	0.05	12	4.6	31.98	35	285.85	52	4
84	8.84	0.41	0.29	0.16	31	3.3	1.08	60	16.59	91	7
85	52.98	0.10	0.56	0.05	11	3.8	1.69	42	13.25	63	12
86	5.10	0.20	0.48	0.10	20	3.1	0.22	59	2.07	89	4
87	30.62	0.22	0.33	0.07	16	3.6	2.81	31	37.44	46	9
88	69.78	0.03	0.64	0.04	10	3.9	1.70	46	11.67	69	5
89	289.72	0.25	0.64	0.08	16	4.3	7.01	78	48.01	118	8
90	179.07	0.29	0.41	0.10	20	4.1	10.55	51	112.88	77	8
91	50.08	0.34	0.48	0.07	14	3.8	2.20	42	20.36	63	7
92	55.34	0.32	0.34	0.13	26	3.8	4.74	55	61.07	83	12
93	251.78	0.31	0.39	0.09	20	4.2	16.45	46	185.27	69	11
94	88.93	0.42	0.45	0.08	17	3.9	4.34	48	42.28	72	11
95	51.04	0.48	0.50	0.09	19	3.8	2.03	60	17.83	89	4
96	106.50	0.41	0.56	0.07	15	4.0	3.36	58	26.32	87	14
97	52.62	0.27	0.44	0.10	20	3.8	2.74	55	27.51	82	9
98	60.86	0.34	0.44	0.10	20	3.8	3.10	54	30.86	81	9
99	374.58	0.37	0.53	0.09	19	4.3	13.06	69	107.49	103	11
100	43.83	0.47	0.42	0.11	23	3.7	2.42	60	25.13	91	9
101	816.71	0.41	0.59	0.09	18	4.6	23.60	75	176.81	113	12
102	235.43	0.26	0.78	0.06	12	4.2	3.84	105	21.61	158	7
103	133.39	0.33	0.43	0.11	23	4.1	7.19	63	73.59	94	13
104	9.72	0.47	0.52	0.08	18	3.3	0.36	60	3.01	90	6
105	352.62	0.33	0.53	0.06	12	4.3	12.57	42	104.57	63	10
106	2.64	0.23	0.34	0.13	25	2.9	0.23	53	3.07	80	6

(Continued)

Table 11.3 The source parameters of the selected earthquakes *(Continued)*

No	$M_0 \cdot 10^{13}$, N·m	σ_{M0}, log.un.	R, km	σ_R, log. un.	σ_R, %	Mw	D, sm	σ_D, %	$\Delta\sigma \cdot 10^5$, Pa	$\sigma\Delta\sigma$, %	N
1	2	3	4	5	6	7	8	9	10	11	12
107	142.19	0.28	0.50	0.10	21	4.1	5.57	68	48.60	102	13
108	5.55	0.25	0.54	0.08	17	3.1	0.19	61	1.57	91	5
109	182.69	0.29	0.59	0.07	14	4.1	5.14	60	38.04	90	9
110	93.13	0.27	0.45	0.10	20	3.9	4.52	56	43.86	83	7
111	22.92	0.23	0.33	0.05	11	3.5	2.07	21	27.34	31	11
112	207.13	0.09	0.64	0.02	5	4.2	5.10	23	35.23	35	8
113	302.73	0.14	0.75	0.02	4	4.3	5.32	25	31.10	38	6
114	197.65	0.04	0.53	0.06	12	4.2	7.05	40	58.67	61	10
115	21.77	0.16	0.40	0.15	29	3.5	1.37	74	15.10	110	4
116	62.65	0.47	0.36	0.11	22	3.8	4.83	49	59.06	74	11
117	68.19	0.17	0.58	0.05	12	3.9	2.02	46	15.37	69	4
118	146.56	0.19	0.46	0.06	13	4.1	6.99	37	67.31	55	4
119	30.39	0.14	0.36	0.10	21	3.6	2.31	46	28.04	69	7
120	147.06	0.18	0.57	0.03	8	4.1	4.58	28	35.63	42	9
121	311.47	0.44	0.59	0.08	16	4.3	8.87	66	65.95	100	7
122	97.59	0.33	0.53	0.05	12	4.0	3.49	38	29.12	57	4
123	435.96	0.31	0.61	0.08	17	4.4	11.54	76	82.70	114	7
124	19.03	0.34	0.46	0.02	5	3.5	0.91	14	8.84	21	4
125	37.11	0.07	0.55	0.06	12	3.7	1.20	45	9.51	68	11
126	13.15	0.39	0.38	0.12	25	3.4	0.89	60	10.15	90	6
127	152.30	0.46	0.44	0.11	23	4.1	7.75	64	77.09	96	9
128	64.79	0.16	0.48	0.02	4	3.8	2.84	11	26.21	16	6
129	104.47	0.58	0.46	0.09	19	4.0	5.02	52	48.51	78	5
130	11.75	0.47	0.54	0.07	16	3.3	0.40	55	3.31	82	8
131	46.03	0.35	0.52	0.05	11	3.7	1.69	37	14.30	56	5
132	69.26	0.41	0.35	0.13	26	3.9	5.74	58	72.80	87	9
133	24.72	0.25	0.23	0.06	13	3.6	4.68	20	89.78	29	4
134	43.86	0.41	0.51	0.09	19	3.7	1.71	63	14.86	95	7
135	46.59	0.45	0.66	0.09	19	3.7	1.08	101	7.24	152	4
136	52.44	0.43	0.56	0.08	17	3.8	1.65	63	12.93	94	5
137	41.16	0.41	0.47	0.08	16	3.7	1.82	46	16.88	70	9
138	161.85	0.42	0.64	0.04	8	4.1	3.93	38	27.02	56	4
139	6.15	0.24	0.33	0.12	24	3.2	0.55	50	7.26	75	7
140	463.22	0.38	0.57	0.08	18	4.4	14.12	69	108.69	104	5
141	88.17	0.32	0.44	0.11	22	3.9	4.59	60	46.14	90	8
142	452.82	0.33	0.60	0.07	15	4.4	12.63	65	92.95	97	7
143	105.97	0.29	0.60	0.08	18	4.0	2.96	75	21.81	113	10
144	700.50	0.46	0.72	0.06	14	4.5	13.42	89	81.82	134	8
145	267.73	0.37	0.54	0.09	18	4.3	9.06	66	73.50	99	4
146	521.40	0.20	0.58	0.09	19	4.4	15.42	77	116.86	115	5

(Continued)

Table 11.3 The source parameters of the selected earthquakes *(Continued)*

No	$M_0 \cdot 10^{13}$, N·m	σ_{M0}, log.un.	R, km	σ_R, log. un.	σ_R, %	Mw	D, sm	σ_D, %	$\Delta\sigma \cdot 10^5$, Pa	$\sigma\Delta\sigma$, %	N
1	2	3	4	5	6	7	8	9	10	11	12
147	102.13	0.49	0.58	0.04	9	4.0	2.98	37	22.45	55	6
148	23.44	0.40	0.34	0.10	20	3.5	1.97	42	25.20	63	7
149	227.20	0.37	0.55	0.10	20	4.2	7.48	73	59.83	110	8
150	96.19	0.56	0.61	0.08	16	4.0	2.59	70	18.69	105	7

Note: 1 – the serial numbers of the earthquake according to the Table 11.2; 2 – the seismic moment; 3 – the standard deviation of the seismic moment according to (Archuleta et al., 1982); 4 – the source radius according to (Brune, 1970); 5,6 – the standard deviation of the corner frequency according to (Archuleta et al., 1982) in the logarithmic units and the percentage; 7 – the moment magnitude; 8 – the amplitude of the displacement in the earthquake source according to (Aki, 1966); 9 – the standard deviation of the amplitude of the displacement (%) according to (Fletcher et al., 1984); 10 – the stress drop according to (Brune, 1970; 1971); 11 – the standard deviation of the stress drop (%) according to (Fletcher et al., 1984); 12 – the quantity of the seismic stations used for the source parameter determinations.

derived following (6), on average, amount to 0.33 and 0.09 logarithmic units, respectively. The maximum error of the corner frequency determination is equal to 32%. Such significant error may be explained by the strong dependence of the corner frequency on the directivity of the rupture propagation and the distance "source-receiver".

For 34 events, the source parameters determination was carried out earlier without taking the regional attenuation model and the local site effect into account (Dobrynina, 2009). In some cases, the use of the corrections on the Q factor and the site effect results in changing of the seismic moment value twice. On average, the standard deviation between the seismic moment values with corrections and ones without corrections is equal to 0.96.

11.6.1 The seismic moment versus the local magnitude for the Baikal rift system

The seismic moment as the function of the local magnitude is shown in Figure 11.8a. The linear trend is evident; the linear least squares fit to the data gives the relation:

$$\log(M_0) \pm 0.25 = (1.23 \pm 0.07) \cdot M_L + (10.14 \pm 0.03). \qquad (11.10)$$

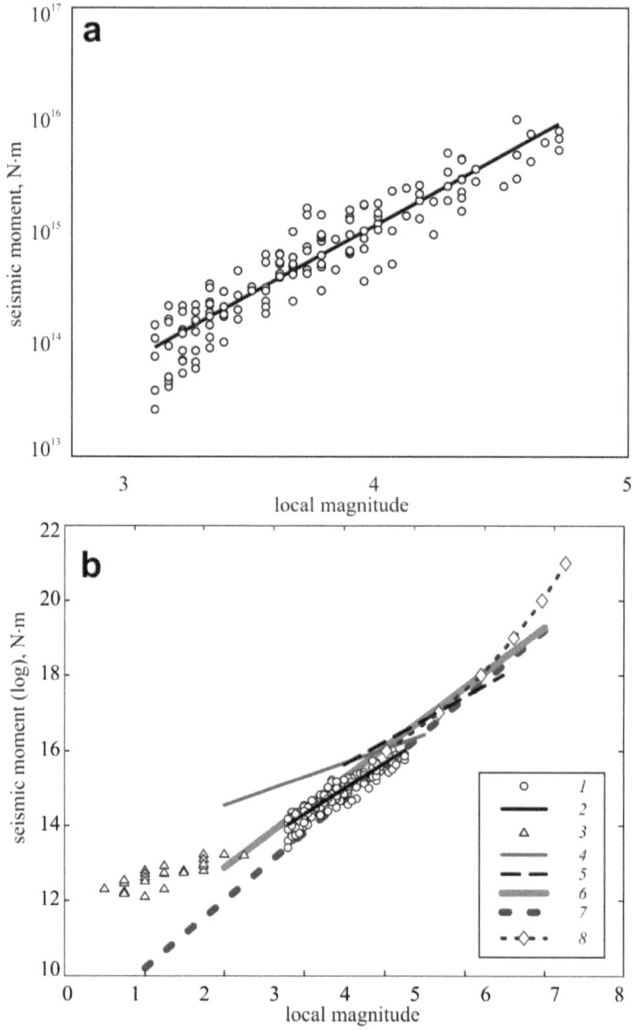

Figure 11.8 The seismic moment versus the local magnitude (a) this study, (b): 1, 2 – the seismic moments and the regression line estimated in this study, 3 – the seismic moments according to (Danzig, 1981), 4–7 –the relations for: 4 – the North Muya (Kochetkov et al., 1985), 5 – for BRS obtained by the express method (Klyuchevskii and Demjanovich, 2002), 6 – for the crustal earthquakes (Rautian et al., 1981), 7 – (Riznichenko, 1985), 8 – (Gusev and Melnikova, 1990).

The correlation coefficient R is equal to 0.94; the data obtained are well described by the empirical regression line (relation 11.10) – the variations in the logarithm of the seismic moment ±0.25 include 94% of the obtained values.

The obtained values of the seismic moment were compared with ones obtained for the weak earthquakes located in the Barguzin depression ($M_L = 0.6$–2.5) using the S-waves spectra (Danzig, 1981); with the regression ratio M_0–M_L for the North Muya region ($M_L = 2.2$–5.0): $\log(M_0) \pm 0.40 = 0.68 \cdot M_L + 12.07$ (Kochetkov et al., 1985) and with the ratio M_0–M_L for all Baikal rift estimated using the express method by A. Klyuchevskii: $\log(M_0) \pm 0.58 = 1.08 \cdot M_L + 11.43$ (Kluchevsky, 2002) (Figure 11.8b).

Figure 11.8b also shows several regression ratios of the seismic moment (N·m), the seismic energy (J), and the local magnitude, determined by different researchers based on the world data:

1. the ratio of T.G. Rautian (Rautian et al., 1981), obtained for the crustal local ($\Delta < 200$ km) and teleseismic (200 km $< \Delta <$ 1000 km) earthquakes ($M_L = 2.0$–7.0), the seismic moment in this case was determined by the seismic coda: $\log(M_0) \pm 0.60 = 1.44 \cdot M_L + 9.66$;
2. the ratio of Yu.V. Riznichenko (1985), obtained on the basis of the field and instrumental data on the seismic moments of weak, moderate, and strong earthquakes, as well as microearthquakes (the range of magnitudes M = 0–8.9): $\log(M_0) \pm 0.60 = 1.6 \cdot M + 8.4$;
3. and the world average nonlinear dependence of the local magnitude on the seismic moment ($M_L = 4.6$–7.2), according to (Gusev and Melnikova, 1990).

The results of E.A. Steiman (Kochetkov et al., 1985) in the area of the low seismic energy ($E_S = 10^{10}$–10^{11} J) are more than the order of the magnitude higher than the order obtained by other researchers (Figure 11.7b). This discrepancy can be explained by the fact that in the work of E.A. Steiman (1), the spectra of SV waves were used and (2) the effect of the low-velocity near-surface layer was taken into account. In addition, E.A. Steiman investigated the earthquakes in the local Severomuisk region, while the definitions presented in this work were performed for the events localized throughout the rift system. Within the limits of the definition errors, there is a fairly good agreement of the "seismic moment – seismic energy" ratios obtained for the entire BRS territory in this work and determined by A.V. Klyuchevskii (Figure 11.8b).

There is also a good agreement in the values of M_0 obtained in the present study and reconstructed from the world average dependences according to the data (Rautian et al., 1981; Riznichenko, 1985; Gusev and Melnikova, 1990) (Figure 11.8b).

The comparison data obtained in the present study and the ones estimated for the Barguzin depression is difficult because their energy ranges aren't overlapping. At the time the nonlinear relationship of the seismic moment and the local magnitude is observed for these events. And the critical point (or the inflection) lies in the interval $M_L = 2.5–3.1$. The similar behavior of M_0 with increasing the magnitude was observed for the intraplate earthquakes (the critical point $m_b = 4.5$, here m_b – the body-wave magnitude) (Hasegawa, 1983; Nuttli, 1983; Street and Turcotte, 1977). Also, the similar nonlinear character of the relationship "seismic moment – local magnitude" was shown for the California earthquakes; and two critical points are observed: $M_L = 3.6$ and $M_L = 5.0$ (Bormann, 2002) (Figure 11.9). In any case, the question about the form of the relationship between the seismic moment and the local magnitude for the whole energy range of the earthquakes that occurred in the BRS is open and requires a single detailed study using numerous original data.

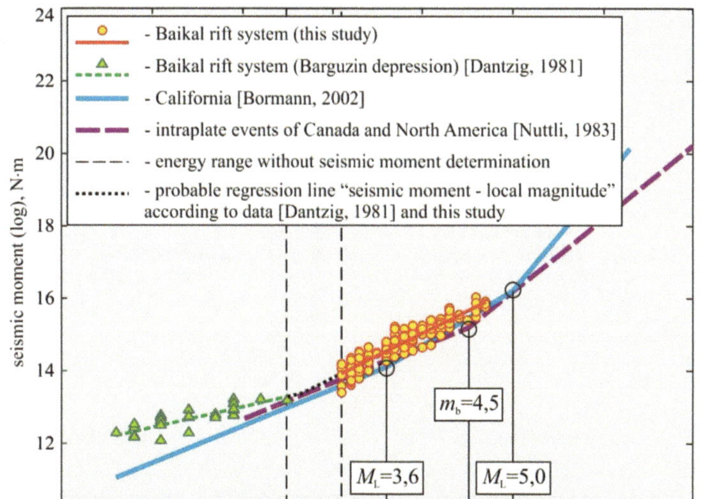

Figure 11.9 The relations "seismic moment – local magnitude".

Within our study, we can say that the resulting relationship between M_0 and the local magnitude is linear for the energy range $M_L = 3.1–4.7$.

11.6.2 Source parameter's scaling relations

The radii of earthquake foci obtained, according to J. Brune's formula (Brune, 1970, 1971), vary from 0.23 km to 0.8 km for the earthquakes with $M_L = 3.1–4.7$ (Table 11.3), respectively. The relationship between the seismic moment M_0 (N·m) and the source radius r (km) for the investigated earthquakes is $\log(M_0) \pm 0.70 = (2.58 \pm 0.78) \cdot \log(r) + (15.61 \pm 0.08)$; the correlation coefficient $R = 0.47$. The variations in $\log(M_0) \pm 0.70$ include 81% of the values obtained (Figure 11.10a). The relationship

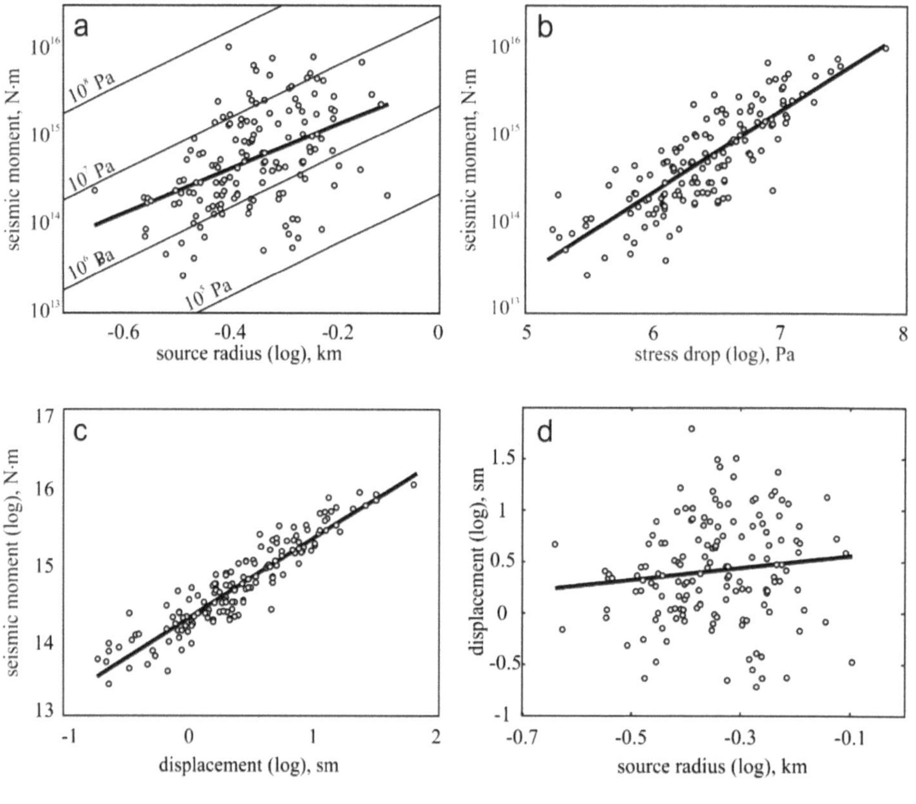

Figure 11.10 The source parameter's scaling relations: (a) – the seismic moment and the source radius, the thin lines – the lines of the constant stress drop, according to J. Brune's model (Brune, 1970, 1971). (b) – the seismic moment and the stress drop, (c) – the seismic moment and the amplitude of the displacement in the earthquake source, (d) – the amplitude of the displacement in the earthquake source and the source radius.

between the source radius r (km) and the local magnitude M_L is $\log(r) \pm 0.15 = (0.11 \pm 0.04) \cdot M_L - (0.76 \pm 0.01)$; $R = 0.46$. The variations in $\log(r) \pm 0.15$ include 90% of the obtained values (Figure 11.10a).

The obtained values of the stress drop $\Delta\sigma$, according to the model of J. Brune (Brune, 1970, 1971), vary from $2 \cdot 10^5$ to $6.7 \cdot 10^7$ Pa for the events with $M_L = 3.1$–4.7 (Table 11.3), respectively, which is consistent with data obtained for the number of extension zones (Shapira and Hofstetter, 1993; Zobin and Havskov, 1995; Margaris and Hatzidimitriou, 2002). Errors in determining the values of the stress drop are determined by the formula (Fletcher et al., 1984):

$$\sigma_{\Delta\sigma} = \Delta\sigma \cdot \sqrt{\left(\frac{\sigma_{M0}}{M_0}\right)^2 + 9 \cdot \left(\frac{\sigma_R}{R}\right)^2}, \tag{11.11}$$

where $\sigma\Delta\sigma$ – the error in determining the value of the stress drop. The obtained error in the calculation of $\Delta\sigma$ varies from 16% to 158%, depending mainly on the error in determining the average corner frequency of the earthquake. The ratio between the stress drop $\Delta\sigma$ (Pa) and the seismic moment M_0 (N·m): $\log(M_0) \pm 0.40 = (0.95 \pm 0.10) \cdot \log(\Delta\sigma) + (8.67 \pm 0.05)$, $R = 0.83$ (Figure 11.10b). The variations ±0.40 include 76% of the values obtained. The dependence of the value of the stress drop $\Delta\sigma$ (Pa) on the local magnitude: $\log(\Delta\sigma) \pm 0.40 = (0.90 \pm 0.12) \cdot M_L + (3.05 \pm 0.05)$, $R = 0.78$. The variations $\log(\Delta\sigma) \pm 0.40$ include 75% of the obtained values.

The maximum displacement amplitude in the source (D) – 62 cm was obtained for one of the strongest of the considered events with $M_L = 4.6$ (number 15 in Tables 11.2 and 11.3). The error in calculating D is calculated similarly to the error $\Delta\sigma$ (formulae [11]) and varies within 11–105%.

The amplitude of the displacement along the fracture D (cm) increases linearly with an increase in the seismic moment M_0 (N·m): $\log(M_0) \pm 0.30 = (1.05 \pm 0.07) \cdot \log(D) + (14.29 \pm 0.03)$, $R = 0.93$. The variations in $\log(M_0) \pm 0.30$ include 90% of the values obtained (Figure 11.10c). The relationship between the amplitude of the displacement in the source D (cm) and the local magnitude M_L: $\log(D) \pm 0.30 = (1.01 \pm 0.09) \cdot M_L - (3.35 \pm 0.04)$, $R = 0.87$, the variations in $\log(D) \pm 0.30$ include 75% of the obtained values.

11.6.3 Scaling law

The concept of the seismic spectrum similarity (the scaling law) was first proposed by K. Aki (Aki, 1967). It is based on the assumption that all earthquakes have the constant stress drop regardless of the seismic moment, in other words, the shape of the seismic spectrum changes with the magnitude of the earthquake according to the law:

$$M_0 \propto f_C^{-\gamma}, \tag{11.12}$$

where $\gamma = 3$. The comparison of the seismic moments, the corner frequencies, and the values of the stress drop for a wide range of magnitudes showed that the similarity law is not always fulfilled, and the $M_0 \propto f_C^{-\gamma}$ ratio depends on the tectonic conditions, the magnitude of the earthquake, the nature of the seismic energy release (the single earthquake, swarm, aftershock sequence), and so on.

According to the results of studies by several authors based on the $M_0 \propto f_C^{-\gamma}$ ratio, three main earthquake models can be distinguished (Dobrynina, 2009):

1. K. Aki's model (Aki, 1967) with the constant stress drop (Hanks, 1977; Abercrombie and Leary, 1993 and others);
2. the model for which the stress drop is an option, $\gamma \neq 3$ (Rautian et al., 1981; Archuleta et al., 1982; Shapira and Hofstetter, 1993; Zobin and Havskov, 1995 and others);
3. the transient model, in which for the different intervals of the values of the seismic moment and the geometrical dimensions of the source there are their relations $M_0 \propto f_C^{-\gamma}$ (Jin et al., 2000).

Figure 11.10a shows the ratio of the seismic moments M_0 (N·m) and the source radii r (km), inversely proportional to the corner frequency f_C (Hz), for the earthquakes under study. It can be seen that the ratio of the seismic moments and the corner frequencies for the earthquakes under consideration does not satisfy the law of similarity of the seismic spectrum of K. Aki, and the exponent $\gamma = 2.58$.

On the other hand, there is also an increase in the value of the stress drop with an increase in the seismic moment (see Figure 11.10b). The stress drop characterizes the ratio of the amplitude of the displacement in the focus to its size:

$$\Delta\sigma \propto \frac{D}{r}. \tag{11.13}$$

In K. Aki's model, the stress drop does not depend on the value of the seismic moment ($\Delta\sigma = \text{const}$), that is, the size of the source and the amplitude of the displacement should be linearly related (Aki, 1967).

The ratio of the amplitude of the dislocation and the radius of the source for the events under study has the form: $\log(D) \pm 0.60 = (0.58 \pm 0.78)\cdot\log(r) + (0.61 \pm 0.08)$, here, the amplitude of the displacement along the fault is D in m, the radius of the source r in km (the correlation coefficient $R = 0.11$). A very large scatter of the values is observed for the events with the same geometric dimensions of the source, the amplitude of the displacement in the source can vary within more than two orders of magnitude (Figure 11.10d). This scatter can be explained, first, by the different types of the movements in the earthquake sources (the strike-slip, normal and reverse fault), and, second, by the existing shortcomings in determining the geometric dimensions of the source (the poor azimuthal distribution of the seismic stations, averaging of the corner frequency, using of the circular dislocation model).

Thus, it seems likely that there is a nonlinear (power) dependence between the displacement amplitude and the source radius for the earthquakes under consideration (Dobrynina, 2009). The nonlinear relationship between the displacement amplitude and the geometric dimensions of the earthquake source (the length and width) was also established by V.V. Ruzhich and S.I. Sherman in the statistical processing of the results of the field observations of the faults in the BRS, the Eastern Siberia, Mongolia, and other regions (Ruzhich and Sherman, 1978). This character of the dependence can be explained by the relative change in the elastic properties of the medium for the earthquakes of different magnitudes. That is, with an increase in the magnitude of the earthquake, the medium behaves no longer as fragile, but as an elastic-viscous (elastic-plastic) body, for which the ratio of the displacement amplitude to the length of the rupture changes nonlinearly.

Thus, the difference between the exponent in equation (11.12) from 3 ($\gamma = 2.58$), the significant variations in the the M_0–r ratio (Figure 11.10a), an increase in the values of the stress drop with an increase in the seismic moment (Figure 11.10b) and the nonlinear nature of the relationship between the amplitude of the displacement in the source and the size of the source allow us to conclude that the theoretical model of K. Aki for the homogeneous source with the constant stress drop is not suitable for describing the considered earthquakes of the BRS (Dobrynina, 2009).

11.6.4 Comparison of the seismic moments for different tectonic regions

The comparison of the results of determining the seismic moment for BRS and other tectonically active regions of the world. The results of determining the seismic moments obtained in this work for the BRS were compared with the data published for other zones of lithosphere extension – for the rifts of the North Sea, Greece, and the Basin and Range Province (Zobin and Havskov, 1995; Margaris and Boore, 1998; Benetatos et al., 2002; Margaris and Hatzidimitriou, 2002; Roumelioti et al., 2002; Chouliaras and Stavrakakis, 2007; Pechmann et al., 2007).

In the work of V.M. Zobin and J. Havskov, the seismic moments of 22 local earthquakes ($M_L = 2.5$–3.5) were calculated from the P-wave spectra (Zobin and Havskov, 1995); for 47 earthquakes in the Basin and Range Province with magnitude $M_L = 3.0$–6.5, the values of the seismic moments were obtained by the inversion method (Pechmann et al., 2007). The seismic moments of 197 earthquakes in Greece with magnitudes $M_L = 2.8$–6.4 were determined from the spectra of the direct P- and S waves (Margaris and Boore, 1998; Benetatos et al., 2002; Margaris and Hatzidimitriou, 2002; Roumelioti et al., 2002; Chouliaras and Stavrakakis, 2007).

For the convenience of comparing the data for different regions, the local magnitude M_L was used. For the BRS earthquakes, the local magnitude was recalculated from the seismic energy using the formula of T.G.

Rautian (Rautian, 1960). In other cases, the values of the M_L magnitude given in the sources were used.

Figure 11.11a shows the relationship "seismic moment – local magnitude" for the specified regions. In general, there is the good agreement in the values of the seismic moment for the different zones of the lithosphere extension. The M_0–M_L relation obtained for the rift zones is in the good agreement with the relationship between the seismic moment and the local magnitude for BRS (see Section 6.1) and is well described by the log-linear dependence in the range of $M_L = 2.5$–6.4 magnitudes (Dobrynina, 2009): $\log(M_0) \pm 0.50 = (1.58 \pm 0.07) \cdot M_L + (8.98 \pm 0.06)$, $R = 0.96$. The obtained empirical dependence well describes the entire volume of the observed data – the correlation coefficient is 0.93. The variations in $\log(M_0) \pm 0.50$ include about 88% of all values (Figure 11.11a). On this basis, it can be assumed that the earthquake sources in extension zones are characterized by a certain commonality, reflecting the processes occurring in the lithosphere.

For the comparison, Figure 11.11b shows the averaged regression relationships "seismic moment – local magnitude" for the zones with the different tectonic regimes:

1. the zones of extension of the lithosphere (the Baikal rift system, North Sea rift, Greece, the Basin and Range Province);
2. the compression zones (Canada, Italy): $\log(M_0) = (1.29 \pm 0.04) \cdot M_L + (9.70 \pm 0.07)$;
3. the shear zones – the large transform faults (the San Andreas [California], the Alpine fault [New Zealand], the North Anatolian fault [Turkey]): $\log(M_0) = (1.34 \pm 0.04) \cdot M_L + (9.48 \pm 0.04)$;
4. the transform fault complicated by the pull-apart system (the system Dead Sea – the Jordan transform fault [Israel]): $\log(M_0) = (1.25 \pm 0.08) \cdot M_L + (10.29 \pm 0.06)$;
5. and the relict subduction zone (the Vrancea zone): $\log(M_0) = 1.00 \cdot M_L + 10.39$.

The relations in pp. 1–4 were obtained from the data published in (Shapira and Hofstetter, 1993; Hofstetter et al., 1996; Bindi et al., 2005; Ristau et al.,

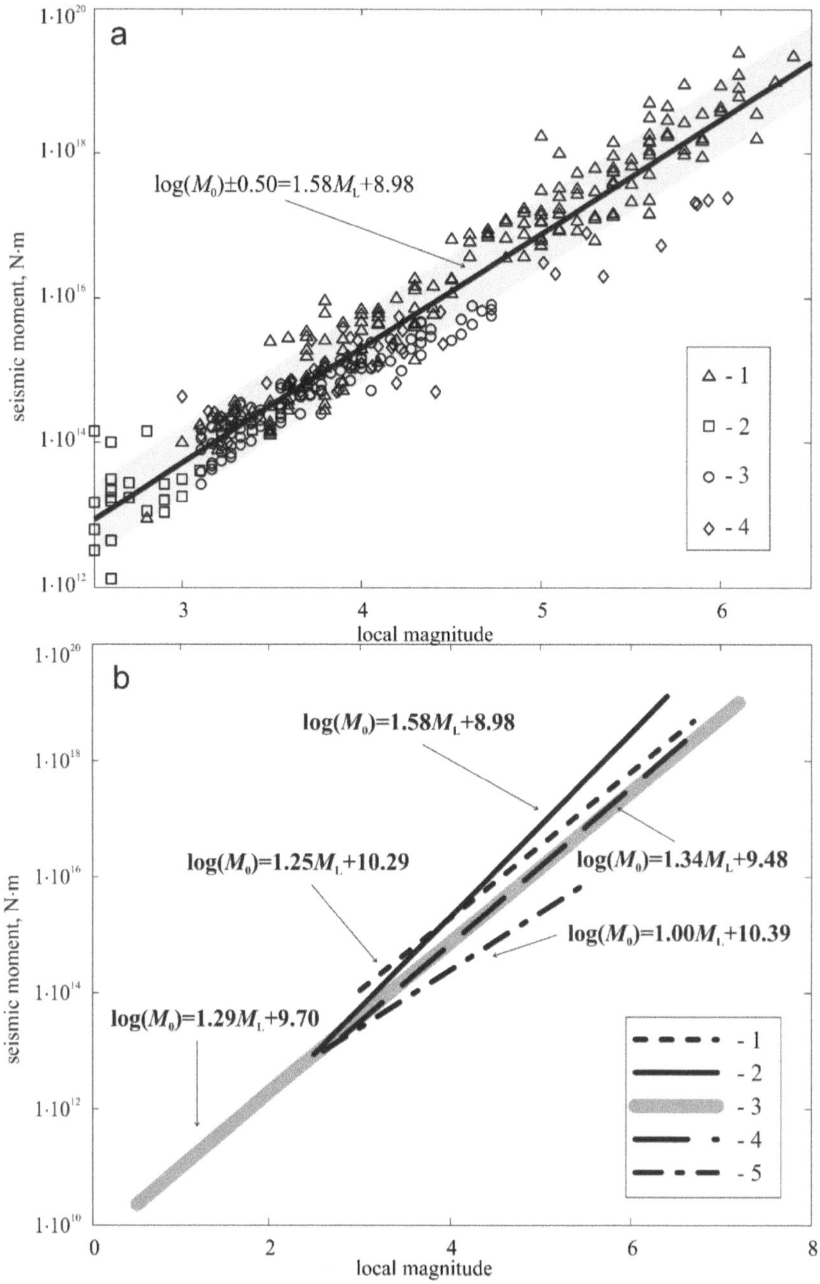

Figure 11.11 The comparison of the seismic moments for the different tectonic regions – the extension zones (a): 1 – Greece, 2 – the North Sea rift, 3 – the Baikal rift system, 4 – Basin and Range Province; (b): 1 – the extension zones, 2 – the compression zones, 3 – the shear zones and the large transform faults, 4 – the transform fault complicated by the pull-apart system, 5 – the relict subduction zone (the Vrancea zone).

2005; Parolai et al., 2007; Ristau, 2009; Hutton et al., 2010), and according to the data obtained in this work. Relation in p. 5 for the Vrancea zone is borrowed from (Radulian and Popa, 1996; Trifu and Radulian, 1991).

The comparison of the M_0–M_L dependences for different regions (Figure 11.11b) showed that there is a tendency of the incremental slope in the log-linear graph when passing from the Vrancea zone to the extension zones. In general, one can see the difference in the regression lines related to the zones with the different tectonic regimes. This allows us to say that the focal parameters of the earthquakes in the rift systems are characterized by the certain commonality, reflecting the general patterns of the destruction of the lithosphere and the seismicity in the zones of the extension of the lithosphere (Dobrynina, 2009).

11.7 SUMMARY

The results of the source parameter studies related to the earthquakes of the BRS and their comparison with the data for the extension zones and the other seismically active regions led us to make the following main conclusions:

1. The good agreement of the source parameters of the earthquakes for the different rift zones makes it possible to speak about the existence of the general regular patterns of the destruction of the lithosphere and the seismicity in the extension zones of the lithosphere. The tendency to change the slope of the linear relation "seismic moment – local magnitude" from the compression zones to the extension zones is observed.
2. The attained relationship between the seismic moment and the corner frequency indicates that the scaling law of the seismic spectrum of K. Aki has not been fulfilled for the studied earthquakes of the BRS.

DATA AND RESOURCES

The data used in the work were obtained with large-scale research facilities "Seismic infrasound array for monitoring Arctic cryolitozone and continuous seismic monitoring of the Russian Federation, neighbouring territories and the world".

ACKNOWLEDGEMENTS

This work involved the South Baikal instrumental complex for monitoring hazardous geodynamic processes in frame of the Centre of Geodynamics and Geochronology at the Institute of the Earth's Crust, Siberian Branch of the Russian Academy of Sciences (grant №075-15-2021-682). The reported study was funded by RFBR and NSFC, project number 21-55-53019.

REFERENCES

Abercrombie, R.E. and P.C. Leary. (1993). Source parameters of small earthquakes recorded at 2.5 km depth, Cajon Pass, Southern California: Implications for earthquake scaling. *Geophys. Res. Lett.*, 20, 1511–1514.

Aki, K. and P. Richards. (1980). *Quantitative Seismology. Theory and Methods*, 1, 1–520.

Aki, K. (1966). Generation and propagation of G-waves from the Niigata earthquake of June 16, 1964: Part 2. Estimation of earthquake moment, released energy and stress drop from the G-wave spectra. *Bull. Earthq. Res. Inst, Univ. Tokyo*, 44, 73–88.

Aki, K. (1967). Scaling law of seismic spectrum. *J. Geophys. Res.*, 72 (4), 1217–1231.

Aptekman, Z.Y., Y.F. Belavina and A.I. Zakharova et al. (1989). The spectra of P-waves in the problem of determination of the dynamic parameters of earthquake sources. The transition from the stationary spectrum to the source spectrum and calculation of the source dynamic parameters. *Vulkanologiya i Seismologiya*, (2), 66–79.

Archuleta, R.J., E. Cranswinck, C. Mueller and P. Spudich. (1982). Source parameters of the 1980 Mammoth Lakes, California Earthquake sequence. *J. Geophys. Res.*, 87, 4595–4607.

Aref'ev, S.S. (1985). The step-wise structure of the spectra of close earthquakes. Macroseismic and instrument studies of strong earthquakes. In *Questions of Engineering Seismology*, edited by N.N. Shebalina, pp. 134–141. Moscow: Nauka.

Benetatos, C., Z. Roumelioti, A. Kiratzi and N.S. Melis. (2002). Source parameters of the M = 6.5 Skyros Island (North Aegean Sea) earthquake of July 26, 2001. *Annals of Geophysics*, 45, 513–526.

Bindi, D., D. Spallarossa and C. Eva et al. (2005). Local and duration magnitudes in Northwestern Italy, and seismic moment versus magnitude relationships. *Bull. Seismol. Soc. Am.*, 95, 592–604.

Bonilla, L.F., J.H. Steidl and G.T. Lindley et al. (1997). Site amplification in the San Fernando Valley, California: variability of site-effect estimation using the S–wave, coda, and H/V methods. *Bull. Seismol. Soc. Am.*, 87, 710–730.

Bormann, P., M. Baumbach, G. Bock, H. Grosser, G.L. Choy and J. Boatwright. (2002). Seismic Sources and Source Parameters in "New Manual of Seismological Observatory Practice – NMSOP". *Geo Forschungs Zentrum Potsdam*, 3–259. doi: 10.2312/gfz.nmsop-2_ch3.

Brune, J.N. (1971). Correction: Tectonic stress and the spectra of seismic shear waves from earthquakes. *J. Geophys. Res.*, 76 (20), 5002.

Brune, J.N. (1970). Tectonic stress and the spectra of seismic shear waves from earthquakes. *J. Geophys. Res.*, 75 (26), 4997–5009.

Chouliaras, G. and G.N. Stavrakakis. (2007). Seismic source parameters from a new dial-up seismological network in Greece. *Pure Appl. Geophys.*, 150, 91–111.

Danzig, L.G. (1981). The displacement spectra of the body waves and the source parameters of the weak earthquakes of the Barguzin region of the Baikal rift zone. In *Seismic Studies of East Siberia*, edited by V.A. Rogozhina, pp. 43–48. Moscow: Nauka.

Delouis, B., J. Deverchere and V. Melnikova et al. (2002). A reappraisal of the 1950 (Mw 6. 9) Mondy earthquake, Siberia, and its relationship to the strain pattern at the south-western end of the Baikal rift zone. *Terra Nova*, 6, 491–500.

Deverchere, J., C. Petit and N. Gileva et al. (2001). Depth distribution of earthquakes in the Baikal rift system and its implications for the rheology of the lithosphere. *Geophys. J. Int.*, 146, 714–730.

Deverchere, J., F. Houdry and M. Diament. (1991). Evidence for seismogenic upper mantle and lower crust in the Baikal rift. *Geophys. Res. Lett.*, 18, 1099–1102.

Dobrynina, A.A. (2011). Coda-wave attenuation in the Baikal rift system lithosphere. *Phys. Earth Planet. Inter.*, 188, 121–126. doi:10.1016/j.pepi.2011.05.008.

Dobrynina, A.A. (2009). Source parameters of the earthquakes of the Baikal rift system. *Izv., Phys. Solid Earth*, 45 (12), 1093–1109.

Dobrynina, A.A., A.V. Sankov and A.N. Shagun. (2018). Seasonal variations in the amplitude-frequency characteristics of microseismic oscillations according to the data of the seismic stations of the Baikal network. *Izvestia of the Siberian Branch of the Earth Sciences Section of the Russian Academy of Natural Sciences. Geology, Exploration and Development of Mineral Deposits*, 41 (3), 22–34. doi: 10.21285/2541-9455-2018-41-3-22-34

Dobrynina, A.A., V.A. Sankov, V.V. Chechelnitsky and J. Déverchère. (2016). Spatial changes of seismic attenuation and multiscale geological heterogeneity in the Baikal Rift and surroundings from analysis of coda waves. *Tectonophysics* 675, 50–68. doi: 10.1016/j.tecto.2016.03.010

Doser, D.I. (1991). Faulting within the western Baikal rift as characterized by earthquake studies. *Tectonophysics*, 196, 87–107.

Douglas, B.M. and A. Ryall. (1972). Spectral characteristics and stress droop for microearthquakes near Fairview Peak, Nevada. *J. Geophys. Res.*, 77 (2), 351–359.

Field, E.H. and K.H. Jacob. (1995). A comparison and test of various site response estimation techniques, including three that are not reference site dependent. *Bull. Seismol. Soc. Am.*, 85, 1127–1143.

Field, E.H. and K.H. Jacob. (1993). The theoretical response of sedimentary layers to ambient seismic noise. *Geophys. Res. Lett.*, 20, 2925–2928.

Fletcher, J.B. (1982). A comparison between the tectonic stress measured in situ and stress parameters from induced seismicity at Monticello Reservoir, South Carolina. *Geophys. Res.*, 87, 6931–6944.

Fletcher, J.B., J. Boatwright and L. Hart et al. (1984). Source parameters for aftershocks of the Oroville, California, earthquake. *Bull. Seismol. Soc. Am.*, 74, 1101–1123.

Golenetskii, S.I. and F.V. Novomeiskaya. (1975). About the thickness of the earth's crust according to observations at the seismic stations of the Baikal region. In *Baikal Rift. Vyp. 2*, edited by N.A. Florensov, pp. 34–43. Novosibirsk: Nauka.

Golenetsky, S.I. (1990). Problems of seismicity of the Baikal rift zone. *J. Geodyn.*, 11, 293–307.

Gusev, A.A., and V.N. Melnikova. (1990). The connections between the magnitudes are world average and for Kamchatka. *Volcanology and Seismology*, (6), 55–63.

Hanks, T. and M. Wyss. (1972). The use of body-wave spectra in the determination of seismic-source parameters. *Bull. Seismol. Soc. Am.*, 62, 561–589.

Hanks, T. (1977). Earthquake stress-drops, ambient tectonic stress and stress that drive plates. *Pure Appl. Geophys.*, 115, 441–458.

Hasegawa, H.S. (1983). Lg spectra of local earthquakes recorded by the Eastern Canada Telemetered Network and spectral scaling. *Bull. Seismol. Soc. Am.* 73, 1041–1061.

Haskell, N. (1964). Total energy and energy spectral density of elastic wave radiation from propagating faults. *Bull. Seismol. Soc. Am.*, 54, 1811–1841.

Hofstetter, A., T. van Eck and A. Shapira. (1996). Seismic activity along fault branches of the Dead Sea – Jordan Transform System: The Carmel – Tirtza fault system. *Tectonophysics*, 267, 317–330.

Honda, H. (1962). Earthquake mechanism and seismic waves. *J. Phys. Earth*, 10 (2), 1–97.

Hutton, K., J. Woessner and E. Hauksson. (2010). Earthquake monitoring in Southern California for seventy–seven years (1932–2008). *Bull. Seismol. Soc. Am.*, 100, 423–446.

Jin A., C.A. Moya and M. Ando. (2000). Simultaneous determination of site responses and source parameters of small earthquake along the Atotsugawa fault zone, Central Japan. *Bull. Seismol. Soc. Am.*, 90, 1430–1445.

Klyuchevskii, A.V. (1986). Comparative study of seismometric channels with magnetic and galvanometric registration: Abstract of the thesis of Ph.D. tech. sciences – M: IFZ AN SSSR, 21.

Klyuchevskii, A.V. and V.M. Demjanovich. (2002). Source amplitude parameters of strong earthquakes in the Baikal seismic zone. *Izv., Phys. Solid Earth*, 38 (2), 139–148.

Kochetkov, V.M., N.S. Borovik and A.V. Solonenko et al. (1985). The detailed seismological studies in the North Muya region. In *Geology and Seismicity of the Baikal-Amur Mainline Railway Zone*, edited by S.L. Solov'ev, pp. 123–180. Novosibirsk: Nauka.

Konno, K.A. and T. Ohmachi. (1995). Smoothing function suitable for estimation of amplification factor of the surface ground from microtremor and its application. *J. JSCE.*, 525, 247–259 (in Japanese with English abstract).

Lermo, J. and F.J. Chavez-Garcia. (1994). Are microtremors useful in site response evaluation? *Bull. Seismol. Soc. Am.*, 84, 1350–1364.

Margaris, B.N. and P.M. Hatzidimitriou. (2002). Source spectral scaling and stress release estimates using strong-motion records in Greece. *Bull. Seismol. Soc. Am.*, 92 (3), 1040–1059.

Margaris, B.N. and D.M. Boore. (1998). Determination of Δσ and k0 from response spectra of large earthquakes in Greece. *Bull. Seismol. Soc. Am.*, 88, 170–182.

Masalskii, O.K., V.V. Chechelnitskii and N.A. Gileva. (2007). The contemporary state of seismic observations in the Baikal region. In *Contemporary Methods of Processing and Interpretation of Seismological Data. Proceedings of the Second International Seismological School*, edited by V.I. Utkin, pp. 140–144. Obninsk: Geophysical Service, Russian Academy of Sciences.

Nakamura, Y. (1989). A method for dynamic characteristics estimation of subsurface using microtremor on the ground surface. *QR Railw. Tech. Res. Inst.*, 30, 25–33.

Nogoshi, M. and T. Igarashi. (1971). On the amplitude characteristics of microtremor (part 2). *J. Seism. Soc. Japan*, 24, 26–40.

Nuttli, O.W. (1983). Average seismic source-parameter relations for midplate earthquakes. *Bull. Seismol. Soc. Am.*, 73, 519–535.

Ohmachi, T., K. Konno and T. Endoh et al. (1994). Refinement and application of an estimation procedure for site natural periods using microtremor. *J. JSCE.*, 489, 251–261 (in Japanese with English abstract).

Parolai, S., D. Bindi and M. Baumbach et al. (2004). Comparison of different site response estimation techniques using aftershocks of the 1999 Izmit earthquake. *Bull. Seismol. Soc. Am.*, 94, 1096–1108.

Parolai, S., D. Bindi and E. Durukal et al. (2007). Source parameters and seismic moment – Magnitude scaling for northwestern Turkey. *Bull. Seismol. Soc. Am.*, 97, 655–660.

Pechmann, J.C., S.J. Nava and F.M. Terra et al. (2007). Local magnitude determinations for intermountain seismic belt earthquakes from broadband digital data. *Bull. Seismol. Soc. Am.*, 97, 557–574.

Petit, C., J. Deverchere and F. Houdry et al. (1996). Present-day stress field changes along the Baikal rift and tectonic implications. *Tectonics*, 15, 1171–1191.

Radulian, M. and M. Popa. (1996). Scaling of source parameters for Vrancea (Romania) intermediate depth earthquakes. *Tectonophysics*, 261, 67–81.

Radziminovich, N.A. (2010). Focal depths of earthquakes in the Baikal region: A review. *Izv., Phys. Solid Earth*, 46 (3), 216–229.

Radziminovich, N.A. and M.G. Ochkovskaya. (2013). Identification of earthquake aftershock and swarm sequences in the Baikal rift zone. *Geodynamics & Tectonophysics*, 4(2), 169–186. (In Russ.) https://doi.org/10.5800/GT-2013-4-2-0096

Rautian, T.G. (1960). Attenuation of Seismic Waves and Energy of Earthquakes. *Trudy TISS Academy of Sciences of TadzhikSSR*, (7), 41–96.

Rautian, T.G., V.I. Khalturin, M.S. Zakirov, A.G. Zemchova, A.P. Proskuriin, B.G. Pustovitenko, A.N. Pustovitenko, L.G. Sinelnikova, A.G. Filinova and I.S. Tchengelia. (1981). Experimental studies of the seismic coda. Moscow: Nauka. 144 p. (in Russian)

Ristau, J. (2009). Comparison of magnitude estimates for New Zealand earthquakes: Moment magnitude, local magnitude, and teleseismic body – Wave magnitude. *Bull. Seismol. Soc. Am.*, 99, 1841–1852.

Ristau, J., G.C. Rogers and J.F. Cassidy. (2005). Moment magnitude – Local magnitude calibration for earthquakes in Western Canada. *Bull. Seismol. Soc. Am.*, 95, 1994–2000.

Riznichenko, Y.V. (1985). *Problems of Seismology*. Moscow: Nauka. 405 p. (in Russian)

Roumelioti, Z., A. Kiratzi and N. Theodoulidis et al. (2002). S-wave spectral analysis of the 1995 Kozani–Grevena (NW Greece) aftershock sequence. *J. Seismol.*, 6, 219–236.

Ruzhich, V.V. and S.I. Sherman. (1978). Estimation of the relation between the length and the amplitude of fracture dislocations. In *Dynamics of the Earth's Crust of East Siberia*, edited by S.A. Kashika, pp. 52–57. Novosibirsk: Nauka.

Sankov, V.A., A.V. Lukhnev, A.I. Miroshnichenko, S.V. Ashurkov, L.M. Byzov, E. Calais and J. Déverchère. (2009). Extension in the Baikal rift: Present-day kinematics of passive rifting. *Doklady Earth Sci.*, 425, 205–209.

Sankov, V.A., A.V. Lukhnev and A.A. Dobrynina. (2012). Modern geodynamic, seismoelectromagnetic and seismoionospheric processes in the Baikal rift zone. In *Seismoionospheric and electromagnetic processes in the Baikal rift zone*, edited by G.A. Zherebtsov, p. 304. Novosibirsk: SB RAS Publishing.

Sankov, V.A. and A.A Dobrynina. (2018). Active faulting in the earth's crust of the Baikal Rift system based on the earthquake focal mechanisms. In *Moment Tensor Solutions – A Useful Tool for Seismotectonics*, edited by S. D'Amico, pp. 599–618. Cham: Springer. doi: https://doi.org/10.1007/978-3-319-77359-9_27

Shapira, A. and A. Hofstetter. (1993). Source parameters and scaling relationships of earthquakes in Israel. *Tectonophysics*, 217, 217–226.

Street, R.L. and F.T. Turcotte. (1977). A study of northeastern North American spectral moments, magnitudes and intensities. *Bull. Seismol. Soc. Am.*, 67, 599–614.

Tarasov, N.T. (1982). Detailed study of the structure of direct body waves of local earthquakes and methods of their identification on seismograms: Abstract of the thesis of Ph.D. geol.-mineral. sciences. *Cand. dis.*, IFZ AN SSSSR. 21 p.

Thatcher, W. and T. Hanks. (1973). Source Parameters of Southern California Earthquakes. *J. Geophys. Res.*, 78, 8547–8576.

Trifu, C.I. and M. Radulian. (1991). Frequency magnitude distribution of earthquakes in Vrancea: Relevance for a discrete model. *J. Geophys. Res.*, 96, 4301–4311.

Trifunac, M.D. (1972). Tectonic stress and the source mechanics of the Imperial Valley, California, earthquake of 1940. *Bull. Seismol. Soc. Am.*, 62, 1283–1302.

Vidale, J.E. (1989). Influence of focal mechanism on peak ground accelerations of strong motions of the Whittier Narrows, California, earthquake and an aftershock. *J. Geophys. Res.*, 94, 9607–9613.

Zobin, V.M. and J. Havskov. (1995). Source spectral properties of small earthquakes in the northern North Sea. *Tectonophysics*, 248, 207–218.

Chapter 12

Seismic anisotropy for understanding the dynamics of crust and upper mantle

A case study in the northwest Himalaya and eastern Ladakh-Karakoram zone

Arpita Paul and Devajit Hazarika

12.1 INTRODUCTION

Seismic anisotropy is a unique property of underlying rocks/foliated formations which affect the passage of seismic waves leading to directional dependence of seismic wave velocity (Babuška and Cara, 1991; Savage, 1999). There are several ways through which the anisotropic characteristic of the propagating medium gets manifested in seismic wave propagation. These include the birefringence or splitting of shear waves, the directional dependence of *P*-wave travel times, the characteristic conversion of *P*-to-*S* waves, the splitting of normal modes, and the difference between the propagation of Love and Rayleigh surface waves, and so on (Wei et al., 2016; Alvizuri and Tanimoto, 2011). However, the most popular tool to study the physical phenomenon of anisotropy is shear-wave splitting which has emerged as a standard seismological tool over the past several decades (Mainprice and Silver, 1993; Savage, 1999; Park and Levin. 2002; Long and Becker, 2010; Hazarika et al., 2013; Long, 2013; Wu et al., 2015; Zhao et al., 2016; Gao et al, 2020; Liu et al., 2020; Paul et al., 2017, 2021; Sivaram et al., 2022). The principle of the shear wave splitting technique is that, when a shear wave propagates through an anisotropic medium, it splits into two components (quasi-*S* waves) that have different polarizations and

DOI: 10.1201/9781003177692-12

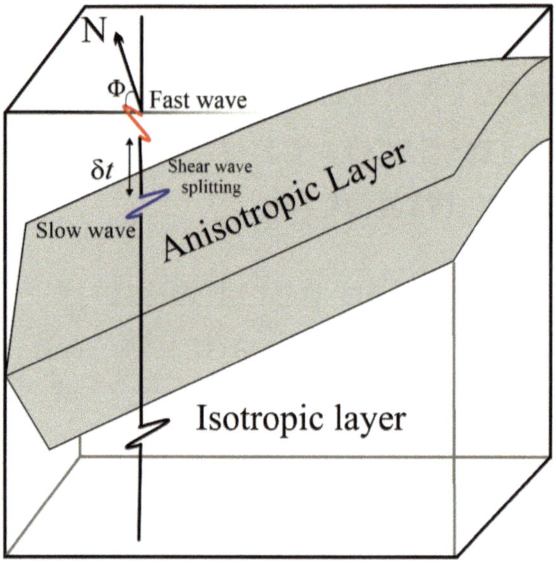

Figure 12.1 Illustrations of shear-wave splitting of S wave into two orthogonally polarized fast and slow shear-waves propagating in an anisotropic medium.

travel at different speeds (Figure 12.1). This phenomenon is referred to as shearwave splitting (Babuška and Cara, 1991; Savage, 1999). The shear-wave splitting mechanism is the elastic analog of the birefringence phenomenon observed in the polarized microscopy technique (Savage, 1999; Park and Levin 2002). The polarization direction of the faster wave is termed as 'fast polarization direction' (FPD or Φ) and the time difference between the fast and slow split waves is called delay time (δt). Both these parameters' orientation of anisotropy and delay time provide an estimate of the strength of anisotropy. Seismic anisotropy has been observed in different depth ranges of the Earth's interior, including the crust, the upper mantle, the transition zone, the D″ layer, and the inner core. Most of the scientific interest in delineating and interpreting seismic anisotropy is driven by the link between deformational processes and anisotropic structure (Crampin and Lovell, 1991; Mc Namara and Owens, 1993; Silver, 1996). Deformation in the Earth often leads to seismic anisotropy, either through the crystallographic or lattice-preferred orientation (CPO, LPO) of anisotropic constituent minerals or through the shape-preferred orientation (SPO) of materials. Because of this link between deformation and anisotropy, the characterization of anisotropic structure can yield some of the most direct constraints on dynamic processes in the Earth's interior.

Since anisotropy is the intrinsic property of the propagating medium. The elastic wave propagation in an anisotropic medium follows Hooke's law as described in the following section. The characteristic property of the anisotropic medium and how the seismic anisotropy varies in different layers of the Earth and its controlling factors are explained in this chapter. A case study explaining the anisotropy in the crust and upper mantle in the Northwest Himalaya and Ladakh-Karakoram zone has been discussed in this chapter to portray the importance of the shear-wave splitting study in understanding geodynamics in a tectonically active region.

12.2 PRINCIPLES OF SEISMIC ANISOTROPY

12.2.1 Hooke's law

The velocity of seismic waves traveling through a medium depends on the elastic constants of the medium. Seismic anisotropy is the intrinsic property of the propagating medium and depends on the elastic constrains. In earlier days, seismic waves are assumed to be propagating through Earth made up of purely isotropic, linearly elastic material. In such material, the stresses are linearly proportional to strains via Hooke's law (Babuška and Cara, 1991; Shearer 2009):

$$\sigma_{\mathrm{ij}} = \mathrm{C}_{\mathrm{ijkl}} \in_{kl} \tag{12.1}$$

where $i, j, k, l = 1, 2, 3, 4$ and σ_{ij} is a second-order stress tensor. C_{ijkl} is the stiffness tensor and $\in_{kl}$ is the strain tensor of a solid elastic medium. C_{ijkl} is a fourth-order tensor, which consists of 81 entries and describes the elastic properties of a medium. This tensor links the applied stress to the resulting deformation of the medium. In general, Hooke's law can lead to complicated relations, but symmetry considerations remarkably simplify the equations by reducing the number of independent parameters from 81 to 36. Elastic parameters of different crystal symmetry systems are mentioned in Table 12.1.

Table 12.1 Number of independent elastic coefficients for selected symmetry systems and typical minerals or Earth's materials (Babuška and Cara, 1991)

Serial number	*Type of symmetric system*	*Number of elastic parameters*
1	Triclinic	21
2	Monoclinic	13
3	Orthorhombic	9
4	Tetragonal	6
5	Trigonal I	7
6	Trigonal II	6
7	Hexagonal	5
8	Cubic	3
9	Isotropic Solid	2

$$C_{ijkl} = C_{jikl} = C_{ijlk} = C_{jikl} \tag{12.2}$$

$$\begin{bmatrix} C_{1111} & C_{1122} & C_{1133} & C_{1123} & C_{1113} & C_{1112} \\ C_{2211} & C_{2222} & C_{2233} & C_{2223} & C_{2213} & C_{2212} \\ C_{3311} & C_{3322} & C_{3333} & C_{3323} & C_{3313} & C_{3312} \\ C_{2311} & C_{2322} & C_{2333} & C_{2323} & C_{2313} & C_{2312} \\ C_{1311} & C_{1322} & C_{1333} & C_{1223} & C_{1313} & C_{1312} \\ C_{1211} & C_{1222} & C_{1233} & C_{1223} & C_{1213} & C_{1212} \end{bmatrix}$$

$$= \begin{bmatrix} C_{11} & C_{12} & C_{13} & C_{14} & C_{15} & C_{16} \\ C_{21} & C_{22} & C_{23} & C_{24} & C_{25} & C_{26} \\ C_{31} & C_{32} & C_{33} & C_{34} & C_{35} & C_{36} \\ C_{41} & C_{42} & C_{43} & C_{44} & C_{45} & C_{46} \\ C_{51} & C_{52} & C_{53} & C_{54} & C_{55} & C_{56} \\ C_{61} & C_{62} & C_{63} & C_{64} & C_{65} & C_{66} \end{bmatrix}$$

Moreover, general thermodynamics requires the existence of a unique strain energy potential, and therefore:

$$C_{ijkl} = C_{klij} \tag{12.3}$$

which reduces the number of independent entries in the stiffness tensor to 21 (Babuška and Cara, 1991). These 21 parameters are necessary to fully describe the stress-strain relationship of an elastic body in its most general form. These parameters may vary with direction, in which case the medium is called anisotropic. In contrast, the properties of the isotropic medium are the same in every direction.

12.2.2 Wave propagation in anisotropic media

Seismic anisotropy distorts the particle motion and no clear distinction can be made between the modes of elastic waves such as the *P* and *S* waves in the isotropic case. This implies seismic velocity variations as a function of (1) propagation direction and (2) polarization direction of the seismic waves. The solution of the wave equation in an anisotropic medium is explained in this section. A homogeneous anisotropic medium and a plane wave solution with phase velocity c, amplitude and polarization vector a, and propagation direction n, has been assumed:

$$u(r,t) = af\left(t - \frac{n \cdot r}{c}\right) \tag{12.4}$$

The problem is to find a suitable combination of c and $\vec{a}$ that satisfy the anisotropic wave equation for a wave travelling in the n can be related as:

$$c_{ijkl}\delta_j\delta_k u_l = \rho\delta_{tt}\, u_i.$$

Substitution of the plane wave solution $\vec{u}$ in the wave equation:

$$c_{ijkl}\delta_j\delta_k\left[a_l f\left(t - \frac{n \cdot r}{c}\right)\right] = \rho\delta_{tt}\left[a_i f\left(t - \frac{n \cdot r}{c}\right)\right]$$

$$c_{ijkl}a_l n_j n_k \frac{1}{c^2} = \rho a_i$$

$$m_{il}a_l = c^2 a_i$$

$$m_{il} = \frac{1}{\rho}c_{ijkl}n_j n_k.$$

The matrix M is called the Christoffel matrix. The eigenvalues and eigenvectors of the matrix, m_{il} have the phase velocities and the propagation directions of the independent wave types. This will be illustrated in the following for hexagonal media.

12.2.3 Hexagonal symmetry

For hexagonal symmetry with a vertical (x_3) symmetry axis (transverse isotropy), the stiffness matrix is:

$$C = \begin{pmatrix} A & A-2N & F & 0 & 0 & 0 \\ A-2N & A & F & 0 & 0 & 0 \\ F & F & C & 0 & 0 & 0 \\ 0 & 0 & 0 & L & 0 & 0 \\ 0 & 0 & 0 & 0 & L & 0 \\ 0 & 0 & 0 & 0 & 0 & N \end{pmatrix}$$

with the coefficients A, C, F, L, and N, according to Love (2013). If

$$\vec{n} = \begin{pmatrix} 1 \\ 0 \\ 0 \end{pmatrix}$$

then,

$$m_{il} = \frac{1}{\rho} c_{ijkl} n_j n_k = \frac{1}{\rho} c_{i11l}$$

or,

$$M = \begin{pmatrix} m_{11} & m_{12} & m_{13} \\ m_{21} & m_{22} & m_{23} \\ m_{31} & m_{32} & m_{33} \end{pmatrix}$$

$$= \frac{1}{\rho} \begin{pmatrix} c_{1111} & c_{1112} & c_{1113} \\ c_{2111} & c_{2112} & c_{2113} \\ c_{3111} & c_{3112} & c_{3113} \end{pmatrix} \frac{1}{\rho} \begin{pmatrix} C_{11} & C_{16} & C_{15} \\ C_{61} & C_{66} & C_{65} \\ C_{51} & C_{56} & C_{55} \end{pmatrix}$$

or,

$$M = \frac{1}{\rho} \begin{pmatrix} A & 0 & 0 \\ 0 & N & 0 \\ 0 & 0 & L \end{pmatrix}.$$

So, the eigenvalues are:

$$\lambda_1 = \frac{A}{\rho}$$

$$\lambda_2 = \frac{N}{\rho}$$

$$\lambda_3 = \frac{L}{\rho}$$

and find the corresponding phase velocities:

$$c_1 = \sqrt{\frac{A}{\rho}} \text{ ‘}P\text{’}$$

$$c_2 = \sqrt{\frac{N}{\rho}} \text{ ‘}S\text{’}$$

$$c_3 = \sqrt{\frac{L}{\rho}} \text{ ‘}S\text{’}$$

with the polarization vectors:

$$\vec{a}_1 = \begin{pmatrix} 1 \\ 0 \\ 0 \end{pmatrix}$$

$$\vec{a}_2 = \begin{pmatrix} 0 \\ 1 \\ 0 \end{pmatrix}$$

$$\vec{a}_3 = \begin{pmatrix} 0 \\ 0 \\ 1 \end{pmatrix}$$

This means, for propagation along the x_1 axis (in a medium with hexagonal symmetry with a vertical symmetry axis such as a layered medium), three waves with different velocity and polarization, one of P-type with phase velocity $\alpha_H = qA\rho$, and two of S-type are there. If an S wave enters the medium along the x_1 axis with components of polarization along both the x_2 and x_3 axes, the two components along these axes will propagate at different speeds. This is called shear-wave splitting.

Now if,

$$\vec{n} = \begin{pmatrix} 0 \\ 0 \\ 1 \end{pmatrix}$$

For, vertically propagating waves,

$$m_{il} = \frac{1}{\rho} c_{ijkl} n_j n_k = \frac{1}{\rho} c_{i33l} = \frac{1}{\rho} \begin{pmatrix} L & 0 & 0 \\ 0 & L & 0 \\ 0 & 0 & C \end{pmatrix}$$

and find waves with

$$\lambda_1 = \frac{L}{\rho} \quad c_1 = \sqrt{\frac{L}{\rho}} \quad \text{`}S\text{'}$$

$$\lambda_2 = \frac{L}{\rho} \quad c_2 = \sqrt{\frac{L}{\rho}} \quad \text{`}S\text{'}$$

$$\lambda_3 = \frac{C}{\rho} \quad c_3 = \sqrt{\frac{C}{\rho}} \quad \text{`}P\text{'}$$

with polarization vectors

$$\vec{a}_1 = \begin{pmatrix} 1 \\ 0 \\ 0 \end{pmatrix}$$

$$\vec{a}_2 = \begin{pmatrix} 0 \\ 1 \\ 0 \end{pmatrix}$$

$$\vec{a}_3 = \begin{pmatrix} 0 \\ 0 \\ 1 \end{pmatrix}$$

8	Cubic	3
9	Isotropic Solid	2

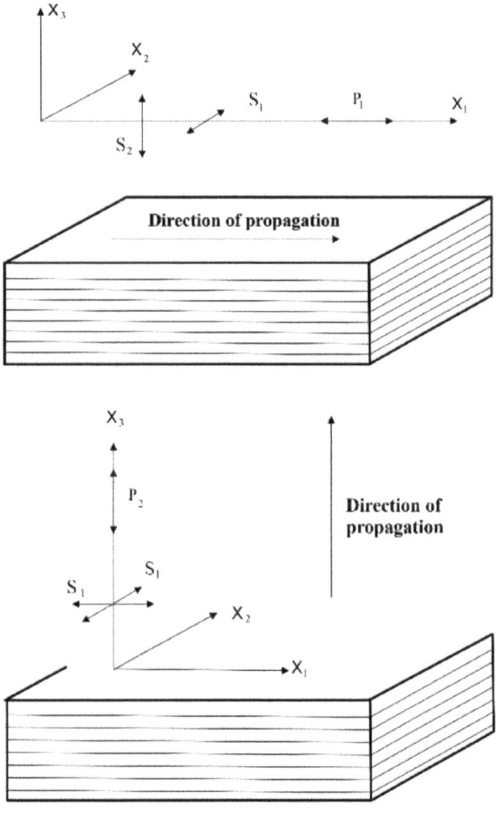

Figure 12.2 (Top) Three possible seismic plane waves propagating in the plane perpendicular to the symmetry axis (x_3 here). (Bottom) *P* and *S* waves propagating in the direction of the symmetry axis. (Savage, 1999.)

The first two solutions are of *S*-type. There is no shear-wave splitting for propagation in the x_3 direction (Figure 12.2). However, the *P*-wave in the x_1 direction travels with a different speed than the one in the x_3 direction. For a stack of isotropic layers, one can show that $A > C$ and $N > L$, so that $\alpha_H > \alpha_V$ and $\beta_H > \beta_v$. For intermediate directions, the polarization vector $\vec{a}$ is not necessarily parallel (or perpendicular) to the propagation direction $\vec{n}$.

12.3 ORIGINS OF SEISMIC ANISOTROPY

The basic mechanisms which control seismic anisotropy are discussed here.

12.3.1 Lattice-preferred orientation (LPO)

The elastic properties of any material are described by a fourth-order tensor, C_{ijkl}, with 81 independent components. However, symmetry considerations simplify the tensor and thus reduce the number of independent components. All possible crystals belong to one of nine possible symmetry systems. The more complicated the symmetry, the more independent constants are needed to completely describe the tensor. In geosciences, mostly orthorhombic (olivine; eight constants) and hexagonal (quartz; five constants) are considered. An orthorhombic symmetry has different velocities along three mutually perpendicular symmetry axes, whereas a hexagonal symmetry has different velocities parallel and perpendicular to its cylindrical symmetry axis. The complicated triclinic structure of feldspars is usually neglected. Olivine is the most abundant mineral in the upper mantle. Its seismic velocities are strongly anisotropic (25% VP and 12% VS) (Nicolas and Christensen, 1987; Crampin, 1991). While initially oriented at random, deformation leads to a preferred orientation of the minerals under certain temperature and pressure conditions (Nicolas and Christensen, 1987). Laboratory experiments with dunite (a rock comprised mainly of olivine and minor pyroxene) show that progressive simple shear will encourage the development of LPO of the olivine minerals such that the *a* axis will rotate towards the direction of shear (Figure 12.3). The main mechanism is dislocation creep within the mineral, that is, intercrystalline slip of atoms (Nicolas and Christensen, 1987).

Similarly, if the dunite is subjected to uniaxial strain (compressive or extensive), an LPO can also develop. In this case, the *a* axis is rotated towards the direction of elongation (Ismail and Mainprice, 1998). A change in stress regime (and thus deformation type and direction) erases any asthenospheric LPO relatively quickly by recrystallization (Nicolas et al., 1973; Mainprice and Silver, 1993). Therefore, any anisotropy in the lithosphere is assumed to reflect the last active deformation.

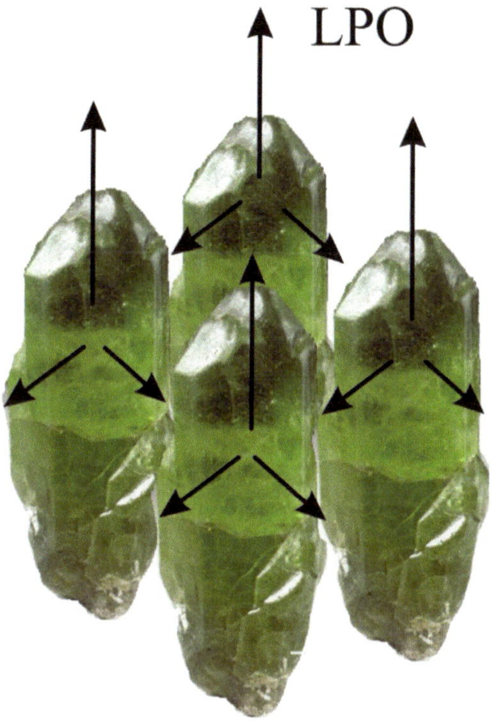

Figure 12.3 Lattice-preferred orientation of hexagonal minerals. (Nicholas and Cristensen, 1987.)

In stable tectonic regimes, this last active deformation is then "frozen" in the anisotropy of the lithospheric rocks during post-tectonic thermal relaxation (Vauchez and Nicolas, 1991; James and Assumpção, 1996; Barruol et al., 1997; Barruol et al., 1998; Heintz and Kennett, 2006).

12.3.2 Shape-Preferred orientation (SPO)

A preferred orientation of cracks, faults, fractures, or compositional lamellae makes an otherwise homogeneous medium effectively anisotropic for wavelengths much larger than the spacing interval of the structure (Backus, 1962). Such shape preferred orientations (SPO) result in an effective fast and slow orientation of seismic wave propagation, with the fast S wave, polarized along the strike of the cracks. In the mantle, mainly melt-filled cracks or lenses, or compositional lamellae are the sources of SPO anisotropy (Kendall, 1994; Mainprice, 1997; Walker et al., 2004). The SPO is thought to be a significant cause of seismic anisotropy along the active plate margins like beneath mid-ocean ridges, subduction systems, intra-continental rift zones and transform fault boundaries, and so on. The orientation of cracks and mineral lineation depend on the predominant tectonic force at these active plate boundaries (Park and Levin, 2002; Hall et al., 2004; Kern et al., 2008; Faccenda et al., 2019). There are two different types of SPOs based on the orientation of the cracks developed in an anisotropic medium. They are as follows.

12.3.2.1 Transverse isotropy/Radial anisotropy

It occurs for a stack of layered materials. Each layer is isotropic in its properties but these properties differ between layers. Thus, the elastic properties and hence seismic velocities of the stack as a whole are identical regardless of the amount of rotation about the axis of symmetry which is perpendicular to the layers. However, these aggregate properties differ in the perpendicular directions and the axis of symmetry is vertical to the layers (Figure 12.4a). That is why it is also called the vertical transverse isotropy (VTI) model (Cai et al., 2009). This phenomenon is known as transverse isotropy or radial anisotropy.

12.3.2.2 Azimuthal anisotropy

It is a type of anisotropy in which velocities vary as a function of the horizontal direction. In general, *P*-wave velocity varies with azimuth. One way to obtain this is to have transverse isotropy with the x_3 axis turned to horizontal which is analogous to standing plywood vertically. The symmetry

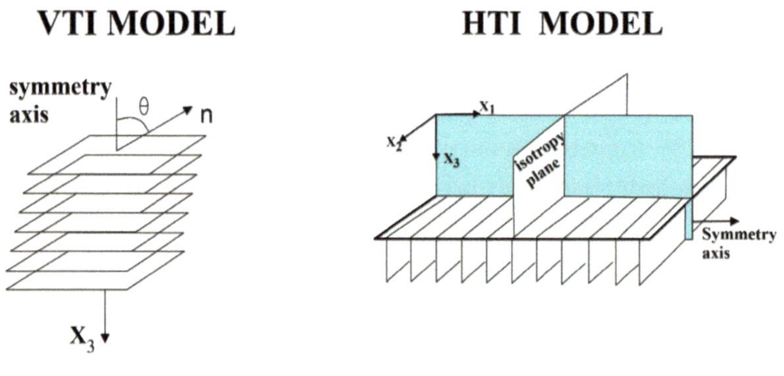

Figure 12.4 (a). Transverse isotropy or radial anisotropy is represented in this figure. (b) Azimuthal anisotropy with vertically oriented layers.

axis is horizontally oriented with respect to the isotropic planes, that is, it is also named horizontal transverse isotropy (HTI) illustrated in Figure 12.4b (Vinnik et al., 1992; Maggi et al., 2006; Zheng et al., 2021).

12.4 STRENGTH OF ANISOTROPY

The strength of anisotropy is calculated using the equations of Kern and Richter (1981).

$$\mathrm{k} = (V_{max} - V_{min}) / V_{mean} \tag{12.5}$$

where, V_{max} and V_{min} are maximum and minimum seismic velocities, respectively. The upper mantle mainly consists of olivine which is highly anisotropic in nature. For *P* waves in the olivine crystal, the minimum and maximum velocities are 9.89 and 7.72 km/s, respectively, so the strength of anisotropy is nearly 25%. For, *S* waves in olivine, the strength of anisotropy is 22% with minimum and maximum velocities of 5.53 and 4.42 km/s, respectively. While some very isotropic minerals like garnet have strengths ≤ 1%, the value of *k* may reach up to 60% for *P* waves and 116% for *S* waves for some highly anisotropic minerals like mica.

The anisotropy discussed previously is the intrinsic anisotropy of the material, defined as the difference between the minimum and maximum velocities of the medium. However, the percentage of anisotropy (k_s) in terms of the splitting parameters is different from the intrinsic anisotropy and is defined as the percentage of anisotropy.

$$k_s = 200(V_{s1} - V_{s2}) / (V_{s1} + V_{s2}) \tag{12.6}$$

This anisotropy is a little bit lower than the intrinsic anisotropy since these values vary depending upon the propagation direction.

12.5 CAUSES OF ANISOTROPY IN DIFFERENT LAYERS OF THE EARTH

According to the layered half-space model, the Earth is considered to be formed of homogeneous isotropic layers but to understand the accurate structure of the Earth, one has to consider the anisotropic characteristics of the Earth. The causative factors of anisotropy vary from layer to layer within the Earth. Various geochemical and geophysical techniques allow inferring the distribution of minerals with various intrinsic anisotropies, which dominate in different depth levels (Mainprice et al., 2000). The causes of seismic anisotropy in different layers of the Earth are hereby discussed.

12.5.1 Causes of seismic anisotropy in the crust

12.5.1.1 Extensive-Dilatancy anisotropy

It refers to the anisotropy where micro-cracks perpendicular to the maximum compressive stress in the region are closed and the cracks parallel to the maximum compressive stress remain open (Crampin, 1984) creating crack planes. Extensive-dilatancy anisotropy (EDA) is inferred when the majority of FPD measurements are parallel/subparallel to the maximum compressive stress (Figure 12.5). Crustal anisotropy resulting from such aligned cracks can be used to determine the state of stress in the crust

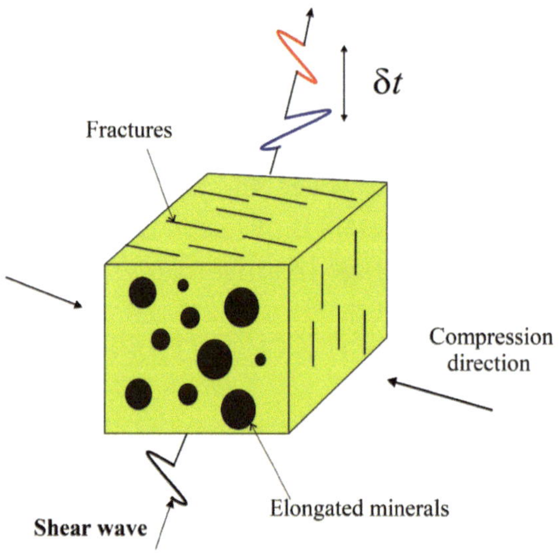

Figure 12.5 Schematic illustration of shear-wave splitting. A shear wave entering a region of aligned cracks necessarily splits into two phases with different polarizations and different velocities. The phase polarized parallel to the face of the cracks meets less impedance and travels faster and is less attenuated than the phase polarized perpendicular to the crack. (Modified after Crampin and Lovell, 1991.)

since in many cases cracks are preferentially aligned with their flat faces oriented in the direction of minimum compressive stress. In tectonically active areas, such as near faults and volcanoes, anisotropy can be used to look for changes in the preferred orientation of cracks that may indicate a rotation of the stress field.

12.5.1.2 The fracturing of rock in fault zones

Shearing within a fault zone can cause minerals to align with the direction of shear. This may lead to fault parallel FPD measurements for waves that traverse the fault zone (Collings et al., 2013; Li et al., 2020).

12.5.1.3 Mineral alignment intrinsic to the rock

Anisotropy from mineral alignment occurs when minerals align to form foliation planes and are inferred to be present when FPD aligns with known foliations. In the presence of EDA cracks, the anisotropy due to the foliations has to be higher than that of the cracks to be detected. A high percentage, mineral aligned anisotropy (10–20%) has been found for metamorphic schists and gneisses.

12.5.1.4 Horizontal layering or inter-bedding of rock

For distant earthquakes, the majority of the ray path is near horizontal and that horizontal layering/inter-bedding with a smaller thickness than the dominant wavelength can cause the SH and SV components to split, resulting in FPD measurements perpendicular to the back azimuth.

12.5.2 Anisotropy in the lithosphere

If crust and subcontinental mantle deform coherently during tectonically active episodes, it is called vertically coherent deformation (VCD) (Bormann et al., 1993; Silver, 1996). Such deformation leads to the development of LPO of olivine crystals. Thermal cooling of the lithosphere freezes the LPO to the lithosphere (Figure 12.6). Therefore, spatial variations in splitting parameters should track geologic variations and the measured Φ should, in the simplest case, follow the general tectonic trend. In the case of multiple episodes of deformation, only the last significant

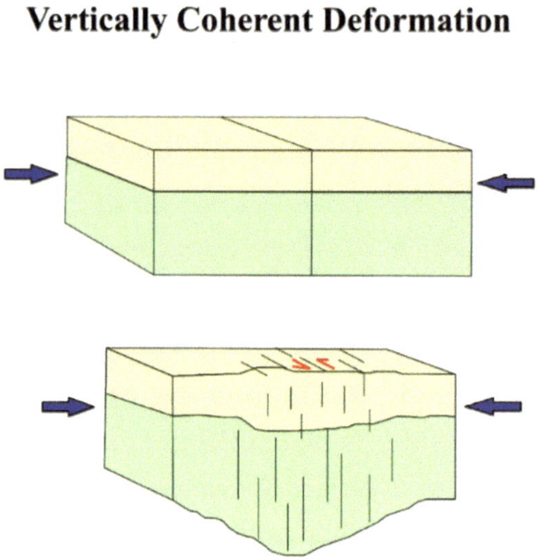

Figure 12.6 Vertically coherent deformation of crust and lithosphere where the minerals are oriented perpendicular to the direction of stress. (Silver, 1996.)

active episode prevails. The three major categories of deformation that would be encountered are transcurrent, collisional, and extensional regimes. In fact, pure collisions are rarely observed and there is almost always a significant transcurrent component. It is generally agreed that LPO in olivine makes the primary contribution to upper mantle anisotropy. It is volumetrically most important and has single crystal shear-wave anisotropy of 18% (Mainprice, 2007). The relationships between deformation and the resulting LPO come from experiments done on olivine aggregates or from the petrographic analysis of mantle-derived rocks. It is generally thought that deformation in the dislocation creep regime is required to produce LPO, whereas diffusion creep does not produce LPO and, in fact, efficiently wipes out any preexisting fabric (Karato and Wu 1993).

12.5.3 Anisotropy in the asthenosphere

In the case of asthenospheric anisotropy, the asthenosphere serves as a decoupling zone of the differential motion between the tectonic plate and the mantle. This model is referred to as simple asthenospheric flow (Bormann et al., 1993, 1996; Silver, 1996). A special case of this model arises if we assume the sub-asthenospheric mantle as stationary.

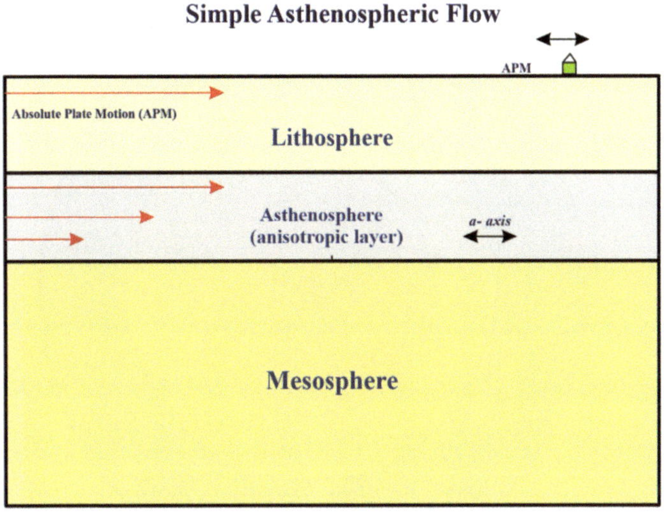

Figure 12.7 Shearing of the lithosphere over the asthenosphere helps aligning the minerals parallel to the shearing direction. (Silver, 1996.)

Then, the lineation is oriented along with the absolute plate motion (APM) direction of the plate. There is however growing evidence that the mantle is moving in a hotspot reference frame (Silver and Holt, 2002; Tarduno et al., 2003). The mantle flow can be deviated by obstacles like lithospheric keels (Bormann et al., 1996; Fouch et al., 2000), cratons (Heintz and Kennett, 2006) or subduction slabs (Margheriti et al., 2003). To completely reorient olivine aggregates, strains of order unity suffice (Nicolas et al., 1973; Mainprice and Silver, 1993). For tectonic plate speeds in the order of 1–10cm/yr and asthenospheric thickness of the order 100 km, such strains would be produced in only a few million years (Figure 12.7).

12.6 SEISMIC PHASES USED TO CHARACTERIZE CRUSTAL AND UPPER MANTLE ANISOTROPY

12.6.1 Direct *S* waves

Use of *S* wave of local earthquakes for splitting analysis with known focal depth provides constraints on the depth of the anisotropic layer. So, these earthquakes are very useful to study the anisotropic structures of the upper crustal rocks. The travel path of the direct *S* wave is shown in Figure 12.8.

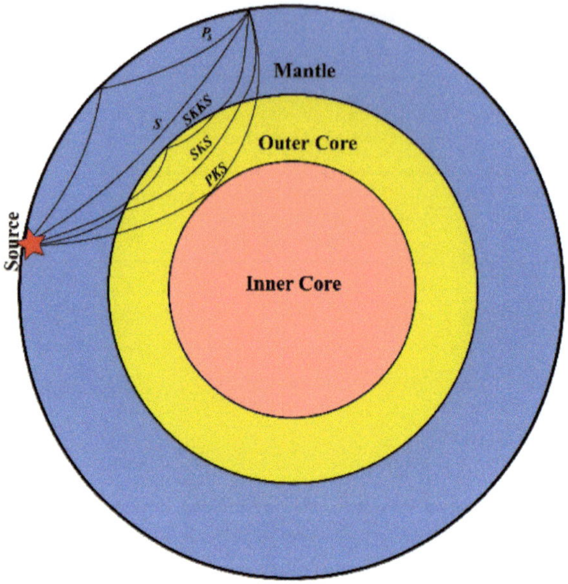

Figure 12.8 The figure is showing the travel paths of different seismic phases used for this study. (Modified after Stein and Wysession, 2003.)

12.6.2 *Ps* phase

For the investigation of the average anisotropy of the entire crust, splitting analysis of *P*-to-*S* Moho converted *Ps* phase is an effective tool. The *Ps* phase can be clearly identified in the receiver functions of teleseismic earthquakes (Paul et al., 2017). Likewise, the source of the *P*-to-*S* (*Ps*) phase is also well defined which originates at the crust-mantle boundary and reaches the recording station travelling almost vertically through the crust beneath the recording site bearing the anisotropic information of the entire crust. Another advantage of using the *Ps* phase from receiver functions is that it is free from the effect of source and instrument response and propagating path.

12.6.3 *SKS* phase

The core refracted *SKS* phase is the most commonly used for seismic anisotropy study in the mantle (Savage, 1999; Silver and Chan, 1991). The *SKS* phase leaves the source as an *S* wave, converts to *P* wave in the outer core, and, finally, propagates again as an *S* wave from the core-mantle boundary (CMB) to the surface. The *P*-to-*S* conversion at CMB causes the *SKS* phase to be radially polarized (Figure 12.8). In an isotropic medium, the energy of the polarized *SKS* phase is maximum in anisotropy. The advantage of using the *SKS* phase over other shear waves is that

i. It is naturally polarized as an SV wave at the CMB so any source side effect has been neutralized and offers good lateral resolution due to its near-vertical incidence at the receiver.
ii. Any energy present on the transverse component (SH) restricts anisotropy only to the receiver side.
iii. It is simple and computationally inexpensive.
iv. It permits extraction of structural complexity and multiple layers of anisotropy, given good azimuthal coverage of data.

We can use the core refracted *SKKS* and *PKS* wave for determining the seismic anisotropy of the mantle as well.

12.7 A CASE STUDY IN THE NORTHWEST HIMALAYA AND EASTERN LADAKH-KARAKORAM ZONE

12.7.1 Previous studies in the Himalayan arc

There are many studies carried out in the Himalayan orogenic belt, based on the splitting of *SKS*, *SKKS*, and *S* waves show a clear variation of FPDs along the strike of the Himalaya (Figure 12.9). In the easternmost extremity of the Himalayan arc, near the Namche Barwa peak, significant anisotropy has been observed by Tiwari et al. (2017). They have observed E-W to NW-SE variation in the FPDs in this southeastern extremity of the Tibetan Plateau. They have explained the origin of anisotropy as a result of a combination of lithospheric deformation and asthenospheric dynamics. An earlier study by Sol et al. (2007) showed null anisotropy near the Indus Tsangpo Suture Zone (ITSZ), in the same region which might be due to insufficient data rather than the effects of the downwelling lithosphere. In northeastern India, Singh et al. (2006) have observed that the direction of *SKS* and *SKKS* anisotropy are E-W along the Himalayan foredeep, N-S along the Indo-Burma convergence, and NE-SW away from the deformation boundaries. The E-W orientations of the FPDs are parallel to the tectonic boundaries of the region and can be explained in terms of lithospheric deformation along the tectonic

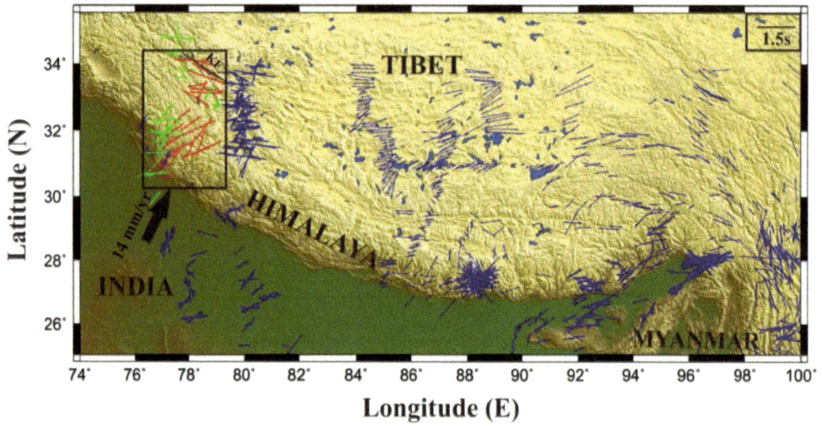

Figure 12.9 Map showing study region (black box) with the splitting measurements from previous studies in the Himalaya and surrounding regions obtained from http://splitting.gm.univmontp2.fr/DB/public/searchdatabase.html. The green bars are showing splitting measurements by Oreshin et al. (2008). The red bars show splitting measurements from the present study. The Karakoram Fault is marked in the study region. The black thick arrow shows the APM direction of the Indian plate in a no-net rotation frame following the NUVEL1A plate model (De Mets et al., 1990). The convergence rate of 14 mm/yr (Banerjee and Bürgmann, 2002) is shown along with the arrow.

boundaries. The NE-SW directions of anisotropy are explained in terms of asthenospheric flow beneath the Indian plate. The N-S orientation of anisotropy along the Indo-Burmese arc can be explained in terms of lithospheric deformation, as the subduction of the Indian plate is going on beneath the Burmese plate or it can be related to the asthenospheric flow beneath the Indian plate which is also oriented N-S in this region. A study by Hazarika et al. (2013) has shown ENE-WSE orientations of FPDs near the Tidding suture zone in the Eastern Himalayan syntaxis. This study explained the FPDs as a result of a combination of lithospheric deformation and crustal anisotropy. However, near the Tidding suture, the FPDs are parallel to the strike of the Tidding and Lohit thrust which signifies that there must be vertically coherent deformation along the thrust sheets. A similar study in the Sikkim Himalaya shows strong anisotropy with an NW-SE-oriented FPD (Singh et al., 2007) orthogonal to the APM of the Indian plate. This suggests a lithospheric deformation due to the India-Asia collision front. In the Nepal Himalaya, a NE-SW trend has been observed, in contrast to the Sikkim Himalaya. This direction is parallel to the APM of the Indian plate. This has been explained as the shearing of the Indian lithosphere over the asthenosphere (Singh et al., 2007). In the western Himalaya, Vinnik et al. (2007) observed a variation from N60E to N150E in a depth range of 160–220 km in the azimuthal anisotropy for different depths by using *P*-receiver functions. This study provides an argument to the negligible anisotropy reported by Sandvol et al. (1997). Oreshin et al. (2008) observed quite a variation in the FPDs along his seismic profile located in the Himachal and Ladakh Himalaya. The directions of FPDs are varying from NE-SW below the ITSZ to E-W above it. He explained the NE-SW orientations of FPDs in terms of the APM of the Indian plate and the E-W orientation as frozen anisotropy, due to the presently extinct subduction zone. He did not find any visible signs of mantle anisotropy along the Karakoram Fault, concluding it to be a crustal-scale strike-slip fault. Heinz et al. (2009) using almost the same station configuration for this area has reported a NE-SW-oriented anisotropy along the Himalayan foredeep and interpreted it to be due to asthenospheric flow. The transition between deformation types at the India Eurasian lithosphere may be the reason for null anisotropy further west beneath the southern Tibetan Plateau. The lack of anisotropy beneath this southern part of Tibet has been explained by the anisotropic nature of the Indian plate or due to the down welling of the Indian lithosphere creating a sub-vertical mantle flow (Sandvol et al., 1997; Singh et al, 2007).

12.7.2 Geotectonic background of the NW Himalaya and Ladakh-Karakoram zone

The northwest Himalaya and the Ladakh-Karakoram zone is the western extremity of the Himalaya-Karakoram-Tibet orogenic system that has been evolved as a result of subduction of the Tethyan oceanic crust until collision, followed by underthrusting of the Indian continental margin beneath the Eurasian Plate (Bird, 1978; Honegger et al., 1982; Searle et al., 1987; Guillot et al., 2003; Yin and Harrison, 2000). Ongoing collision and convergence between India and Asian plates, since 70 Ma, contributed to the high elevation, the formation of the Himalayan fold-thrust belt, and large crustal shortening of more than 1500 km (Molnar and Tapponnier, 1975; Yin and Harrison, 2000). Several passive seismological studies in the NW Himalaya characterize the crust and upper mantle structure and evidenced an increase in crustal thickness from south to north (e.g. Rai et al., 2006; Oreshin et al., 2008; Hazarika et al., 2014; 2017b; 2021). The NW Himalaya and eastern Ladakh-Karakoram zone cover all the geotectonic units starting from the Himalayan Frontal Thrust to the trans-Himalayan zone providing a unique opportunity to study the interaction of the Indian and Asian plates. Figure 12.10 shows a simplified geological map showing parts of the Tethyan Himalaya, Indus suture zone, and Karakoram zone (modified after Maheo′ et al., 2004; Sen et al., 2013; Hazarika et al., 2014; Hazarika et al., 2017a). The Karzok and Zildat faults are marked according to

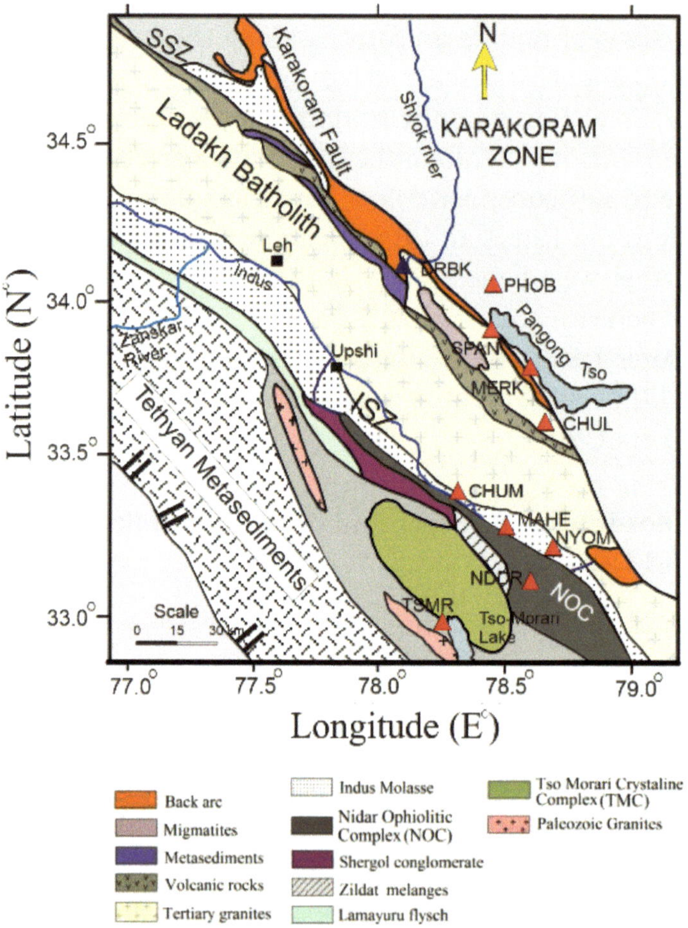

Figure 12.10 Simplified geological map showing parts of the Tethyan Himalaya, Indus suture zone (ISZ), and Karakoram zone (modified after Maheo′ et al., 2004; Epard and Steck, 2008; Hazarika et al., 2014). The seismological stations are shown in red triangles. The Shyok Suture Zone is marked by SSZ.

Epard and Steck (2008). The tectonic features shown in the map are Tso Morari Complex (TMC), Ladakh Batholith (LB), Karakoram Fault (KF), Karakoram Fault Zone (KFZ). The Indus Suture Zone (ISZ), which is considered as the surface collision boundary between the Indian and Eurasian plates, is a major tectonic unit well exposed in this part of the Trans-Himalaya. The eastern part of the Ladakh-Karakoram zone is considered as a subduction complex (Honegger et al., 1982), evidenced by its distinctive tectonic setting where an ophiolitic suite, the Nidar Ophiolitic Complex (NOC), is sandwiched between the Ultra-High Pressure (UHP) rocks of Tso-Morari Crystallines (TMC) of Indian continental affinity and the Ladakh magmatic arc (Mukherjee and Sachan 2004; Ahmed et al., 2008) and the intra-continental Karakoram Fault. Most of the current issues on geodynamic processes focus on linking the deformation/strain in the crust to deep lithospheric mantle processes (Savage, 1999; Sol et al., 2007; Wu et al., 2015). A seismic anisotropy study is an effective way to address such issues. In this study, we emphasized studying the upper crustal anisotropy of the Ladakh-Karakoram zone with the help of the *S* wave, the whole crustal anisotropy with the help of *Ps* wave, and the mantle anisotropy study along a profile passing across the NW Himalaya and the eastern Ladakh-Karakoram zone based on *SKS* splitting data. The study provides new information on the seismic anisotropy of the crust and mantle strain/deformation beneath a profile extending from the Himalayan Frontal Thrust to the Karakoram Fault, which has significant implications in understanding the geodynamic evolution of the region.

12.7.3 Crustal anisotropy in the Ladakh-Karakoram zone

Seismic anisotropy of the crust beneath the eastern Ladakh-Karakoram zone has been studied by shear-wave splitting analysis of *S* waves of local earthquakes and *P*-to-*S or Ps*-converted phases originated at the crust-mantle boundary. The *Ps* phases are clearly visible on the receiver functions of teleseismic earthquakes. The seismological stations used in this study to record local and teleseismic earthquakes are shown by red triangles plotted over the simplified geological map of the region (Figure 12.10).

We adapted the cross-correlation technique (Bowman and Ando, 1987; Piccinini et al., 2013) for shear-wave splitting study using *S* wave of local microearthquake data as well as *Ps*-converted phase of RF data. This method is based on the assumption that the radial and the transverse component of the shear wave has the same shape but are polarized in orthogonal directions with different velocities. The cross-correlation measures similarity between the slow and fast waves as a function of the time lag between the two waves. The details of the technique can be found in Bowman and Ando (1987).

The analysis procedure of *S*-wave splitting is demonstrated in Figure 12.11. The filtered north-south (N) and east-west (E) component of a local earthquake seismogram is shown in Figure 12.11a. The cross-correlation method has been applied within a small window of 0.3 s around the *S* phase. The *S* waveforms before and after correction for anisotropy as well as corresponding particle motions are shown in Figures 12.11b, c. Initially, the particle motion was elliptical which becomes linear after correction for anisotropy.

To obtain clear *Ps* phases, receiver functions are computed (Langston, 1979). The RFs are radial and transverse waveforms computed by deconvolving (Ligorria and Ammon, 1999) the vertical component waveform from the radial and tangential component waveforms results in radial and tangential RFs. An example of radial and tangential component receiver functions is shown in Figure 12.12. The cross-correlation method has been applied to a small window marked by A and F. The *Ps* waveforms before and after anisotropy correction are shown in the middle panel whereas corresponding particle motions are shown in the bottom panel.

The results (FPD and δt) of *S*-wave splitting of local earthquakes are depicted in Figure 12.13. FPDs are marked with solid black lines. The lengths of the lines are proportional to the delay time. The stations and earthquake epicenters are shown in red triangles and bigger grey dots respectively. The FPDs are plotted at the middle of the station and source path (marked with small dots) as it represents anisotropy of the medium between the event and the recording station. The splitting parameters (Φ and δt), derived from the *S* wave of local earthquakes with shallow focal depths, reveal the complex nature of anisotropy with NW-SE and NE-oriented FPDs in the upper ~22 km of the crust

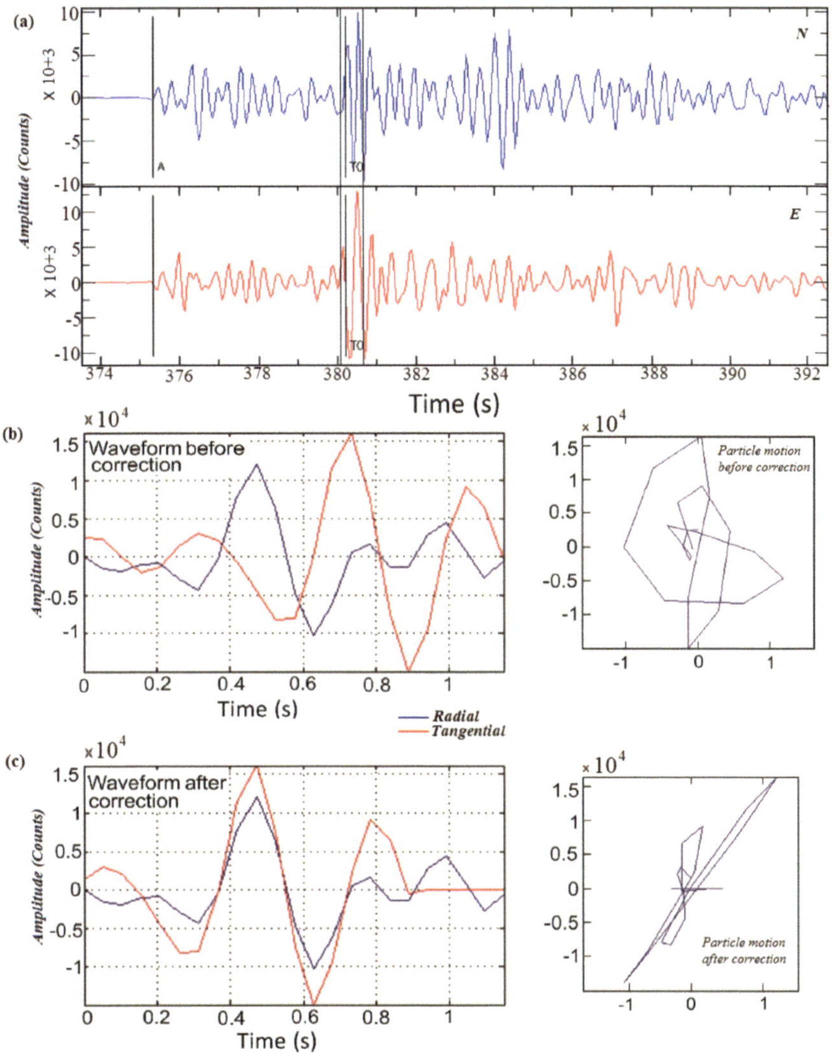

Figure 12.11 Example of S-wave splitting analysis. The NS and EW component of an event is shown in (a). The selected window of the S wave used for cross-correlation is shown in (b) with corresponding particle motion. The selected S waveform after correcting for anisotropy is shown in (c) with corresponding particle motion. The elliptical particle motion before correction signifies the presence of anisotropy whereas it becomes linear after correction.

(Figure 12.13). The observed anisotropy in the upper crust may be attributed to combined effects of existing tectonic features as well as regional tectonic stress. The maximum delay time of fast and slow waves in the upper crust is ~0.3s.

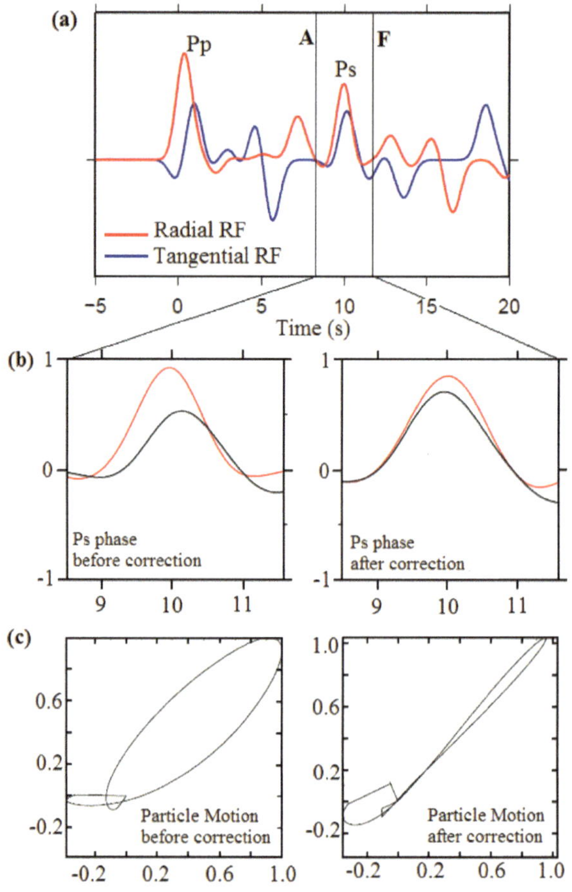

Figure 12.12 Example of *Ps*-splitting analysis. The radial and tangential components of the receiver function waveform are shown in (a). The time window for *Ps*-converted phase is shown in (b). Figure (c) shows corresponding particle motion.

Figure 12.14 shows the average FPDs for the entire crust beneath nine stations obtained through *Ps* splitting. The orientation of the bar represents FPD whereas the length of the bar is proportional to delay times. The *Ps*-splitting analysis (Figure 12.14) shows more consistent FPDs compared to *S*-wave splitting. The FPDs are parallel or sub-parallel to the Karakoram Fault and other NW-SE trending tectonic features existing in the region. The strength of anisotropy estimated for the whole crust is higher (maximum delay time δt: 0.75 s) in comparison to the upper crust. This indicates that the dominant source of anisotropy in the Trans-Himalayan crust is confined within the middle and lower crustal depths. The predominant NW-SE trending FPDs consistently

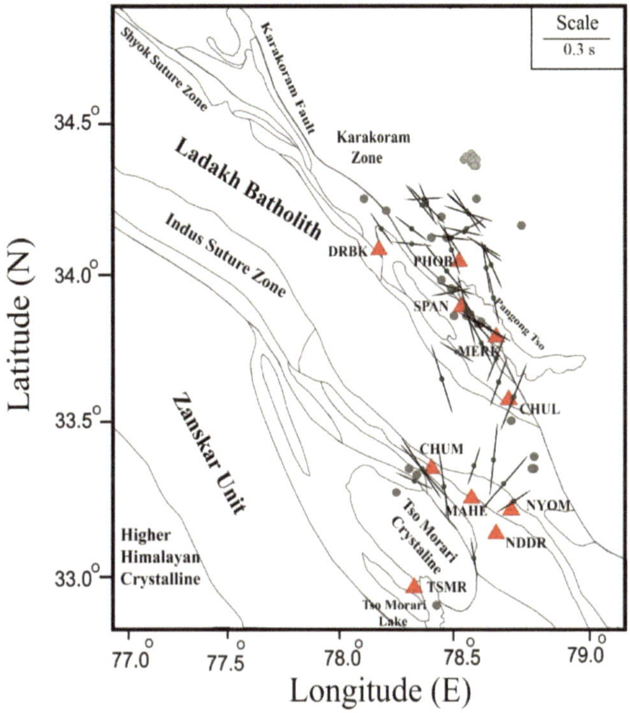

Figure 12.13 Results of S-wave splitting is depicted over the simplified tectonic features. The solid black line shows FPD. The lengths of the lines indicate the delay time. The red triangles indicate stations and big grey dots show earthquake epicenters. The FPDs are plotted at the middle of the station and source path (small dots).

observed in the upper crust as well as in the middle and lower crust near the Karakoram Fault zone support the fact that the Karakoram Fault is a crustal-scale fault that extends at least up to the lower crust. Dextral shearing of the Karakoram Fault creates shear fabric and preferential alignment of mineral grains along the strike of the fault, resulting in the observed FPDs. A similar observation in the Indus Suture Zone (ISZ) also suggests crustal-scale deformation owing to the India-Asia collision. NW-SE trend of FPDs observed in upper as well as middle and lower crust near the KFZ suggests coherent deformation throughout the crust. This also indicates that the Karakoram Fault is a crustal-scale fault that extends at least up to the lower crust. Dextral shearing of the Karakoram Fault creates shear fabric and preferential alignment of mineral grain along the fault which causes the observed FPD. The results suggest

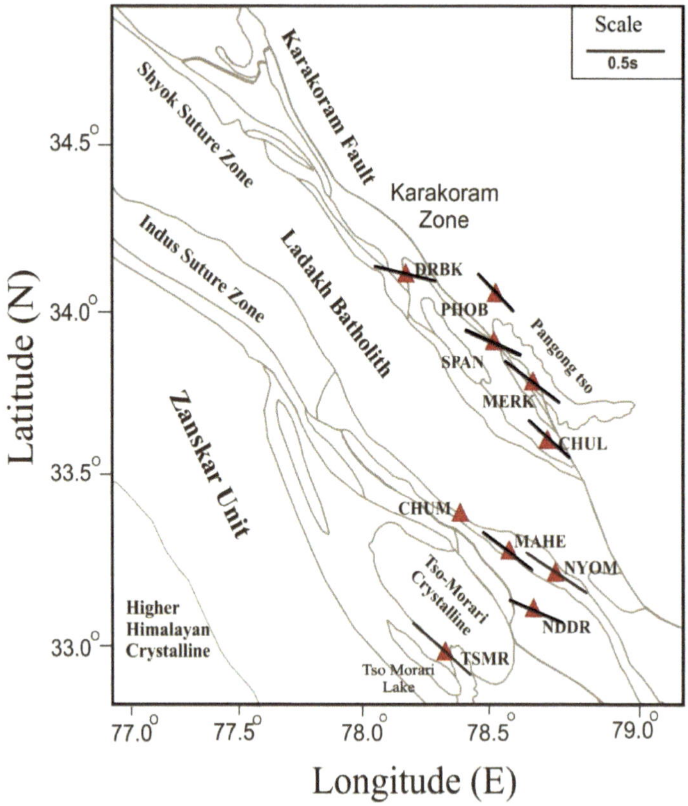

Figure 12.14 Average FPDs for the entire crust beneath nine stations (red triangles) obtained through *Ps* splitting are depicted. The orientation of the bar represents FPD, whereas the length of the bar is proportional to delay times.

the predominance of structure-induced anisotropy rather than anisotropy originated due to compressive stress (Paul et al., 2017).

12.7.4 Upper mantle anisotropy in the NW Himalaya and Ladakh-Karakoram zone

The upper mantle anisotropy beneath the northwest Himalaya and eastern part of the Ladakh-Karakoram zone has been investigated based on shear-wave splitting of core-mantle refracted *SKS* waveforms recorded at 28 broadband seismic stations (Paul et al., 2021). The SKS waveforms were obtained from teleseismic earthquakes of M ≥ 5.5 and epicentral distance 84°–120°. The transverse component minimization method of Silver and Chan (1991) has been applied for analyzing selected *SKS* waveforms to

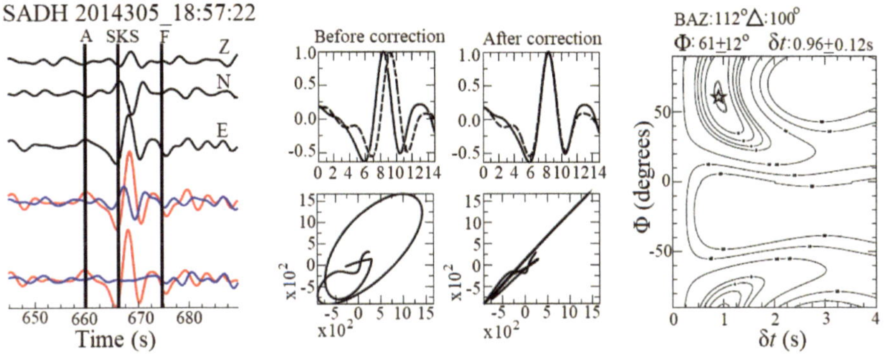

Figure 12.15 Examples of shear-wave splitting analysis at SADH station. The original waveform (Z, N, E), as well as radial (R) and transverse (T) components of the *SKS* phase before and after correction, is shown on the left panel. The fast and slow waves within the window marked by A and F (continuous and dashed lines, respectively) before and after correction for the best-calculated delay time and corresponding particle motions are presented in the middle four figures. The contour plot of energy on the transverse component as a function of δt (s) and Φ is shown on the right. The result with a 95% confidence level is shown by a star.

obtain splitting parameters (Φ, δt). In this method, a grid search is carried out over all possible values of Φ and δt by rotating and performing time-shifts of the horizontal components. The energy on the corrected transverse component (Φ, δt) is measured during the grid search to produce a contour plot. A pair of the splitting parameter is obtained that successfully corrects for the effects of anisotropy by minimizing the energy in the transverse component and produce linear particle motion after correction for anisotropy. An example of SKS-splitting analysis is depicted in Figure 12.15.

Seismic anisotropy of the upper mantle beneath the northwest Himalaya and eastern part of the Ladakh-Karakoram zone has been investigated using shear wave splitting of core-mantle refracted *SKS* waveforms from 94 teleseismic earthquakes recorded at 28 broadband seismic stations (Figure 12.16). The FPDs computed at each station are presented in the form of the rose diagrams. The longest pie slice is for the most frequent direction of anisotropy for that corresponding station. Bin size is 20 for each of the plots. The predominant FPD measurements are presented by the orientation of black bars. In addition to the splitting measurements, we observe 'nulls' measurements for a few event-station

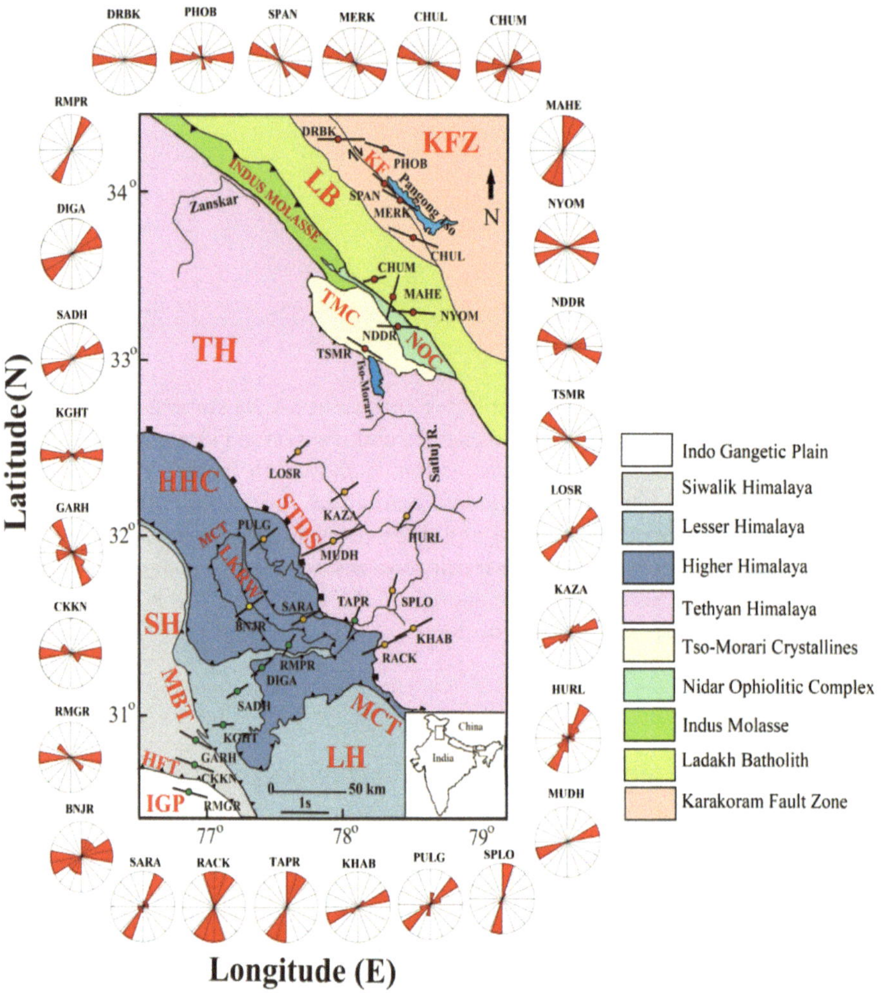

Figure 12.16 The results obtained from *SKS* splitting at each station are presented in the form of the rose diagrams. The background shows a simplified geological map. The bin size used in the rose diagram is 20 for each of the plots. The orientation and length of the black bars represent predominant FPD measurements. The green, yellow and red dots represent stations of the Lower Satluj (LSAT), Upper Satluj/Kinnaur (USAT) and Ladakh (LKZ) networks, respectively.

pairs (Figure 12.17). A 'null' measurement is that which does not exhibit any energy on the transverse component, due either to an absence of anisotropy, an initial polarization of the incoming wave parallel, or orthogonal to the fast anisotropic direction, or low signal-to-noise ratio (Savage, 1999). Thus, the analysis of null anisotropy helps in the interpretation of seismic anisotropy.

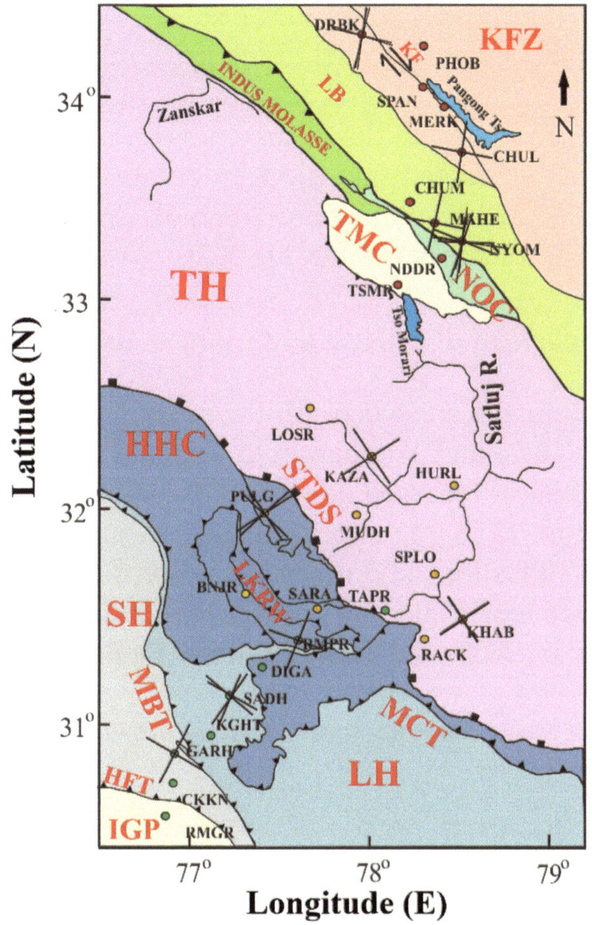

Figure 12.17 Null measurements observed at ten stations are shown by black bars.

The *SKS*-splitting parameters (Φ and δ*t*) show a significant strength of anisotropy with a wide range of delay time of split waves (~0.75–2.94 s). The distinct variations in FPDs along the study section suggest a complex source of anisotropy that cannot be explained by a simple model. In the frontal part of the Himalaya, the FPDs are mostly parallel or sub-parallel to the strike of the Himalayan orogeny suggesting deformation in the shallow lithospheric mantle under compression owing to the India-Asia collision. On the other hand, FPDs observed in the Lesser, Higher, and Tethyan Himalaya largely follow NE-oriented APM of the Indian plate which can be attributed to basal shear as the Indian plate moves over the asthenospheric mantle with a minor contribution from shallow lithospheric deformation. A complex anisotropy pattern is observed in the Indus Suture Zone. The FPDs near the Karakoram Fault Zone (KFZ) are parallel or sub-parallel to the strike of the Karakoram Fault. The strike-slip or transpressional deformation in the lithospheric mantle in the KFZ is considered as the major source of anisotropy beneath the KFZ. The study envisaged that the Karakoram Fault is a lithospheric scale fault that largely accommodates the India-Asia collision and extrusion in the Tibetan Plateau (Paul et al., 2021).

12.7.5 Comparison of crustal and mantle anisotropy

The anisotropy in the crust beneath the study region has been studied by Paul et al. (2017) and Wadhawan et al. (2022), based on shear-wave splitting of *Ps* phases recorded on receiver functions of teleseismic earthquakes. In Figure 12.18, the upper panel shows anisotropy directions in the crust. The FPDs for the crust in the western Himalayan thrust-belt and the LKZ are represented by the zone is represented by the blue lines (Wadhawan et al., 2022) and the green lines (Paul et al., 2017). Whereas,

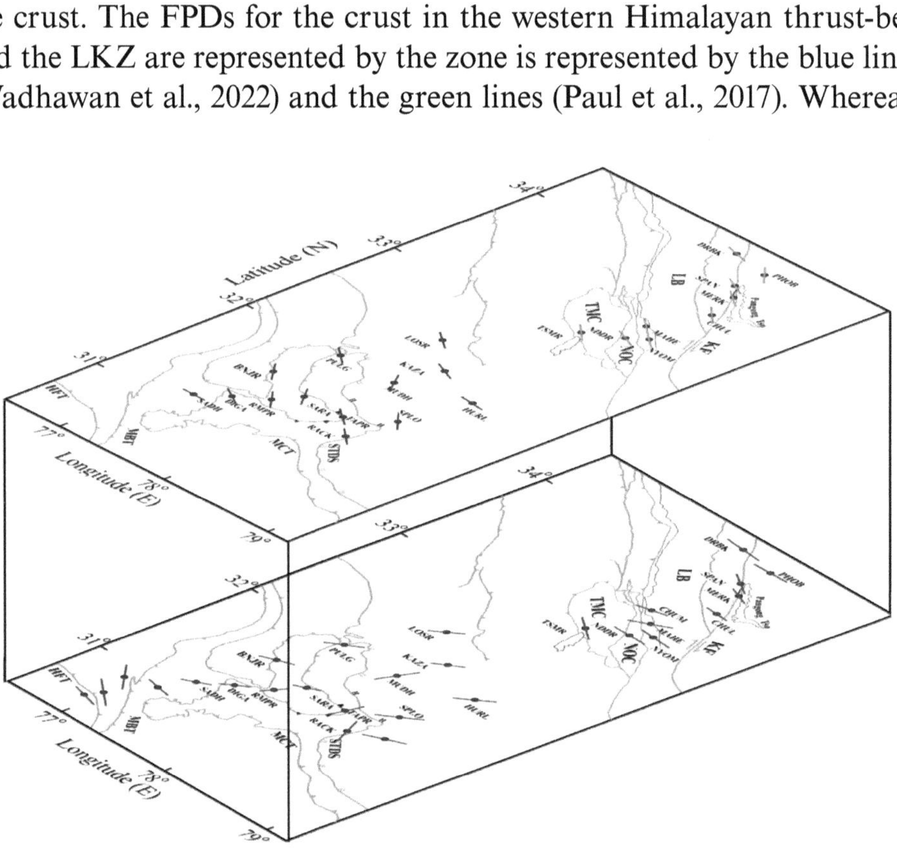

Figure 12.18 A 3-D diagram comparing the anisotropy of the crust and mantle of the northwest Himalaya and LKZ.

the grey lines in the lower panel are showing the predominant anisotropy directions in the upper mantle (Paul et al., 2021). The orientations and length of the bars show anisotropy direction and the average delay times respectively. The anisotropy directions in the crust and upper mantle are similar near the Karakoram Fault zone whereas there is a significant difference in a large part of the NW Himalaya. It can be observed that both crust and mantle anisotropy in the Trans-Himalaya particularly near the KFZ is roughly coherent and parallel/sub-parallel to the Karakoram Fault. This suggests that crust and mantle are mechanically coupled beneath the KFZ. In contrast the predominant anisotropy direction beneath the NW Himalaya shows a different trend in the crust and mantle. This may imply that the crust and mantle beneath the NW Himalaya are not mechanically coupled. The results presented in this study reveal distinct variations of anisotropy direction which cannot be explained by only one mechanism. A 3-D cartoon is shown in Figure 12.19 for an interpretation of the formation of seismic anisotropy resulting from the India–Eurasia collision and the dextral strike-slip movement of the Karakoram Fault.

12.7.6 Limitations and future scope

Shear-wave splitting is one of the most useful tools for the valuation of anisotropy underneath any tectonic setting (Savage, 1999). The two parameters of shear-wave splitting such as the first polarization direction (ϕ) and delay time (δt) give information about the direction and strength of anisotropy (Savage, 1999). The value of delay time is used to analyze the influence or thickness of mantle anisotropy but the precise depth of anisotropy cannot be determined from shear-wave splitting study (Alsina and Snieder, 1995). The core refracted phases *XKS* (*SKS*, *SKKS*, and *PKS*) minimize the vertical resolution during their path in the mantle because of the steep incident angle (Liu and Gao, 2011). The depth of anisotropy plays a huge role to depict the origin of the splitting mechanism but unfortunately, there is no research so far available to identify the source localization and depth estimation of anisotropy, which is of more importance to understand the concerned geodynamics in a more comprehensive way to solve the critical geological process and mantle deformation pattern to govern the tectonics of a particular region. Fresnel zone concept (Alsina and Snieder, 1995) was used previously to decipher the depth of anisotropy but it involves laborious

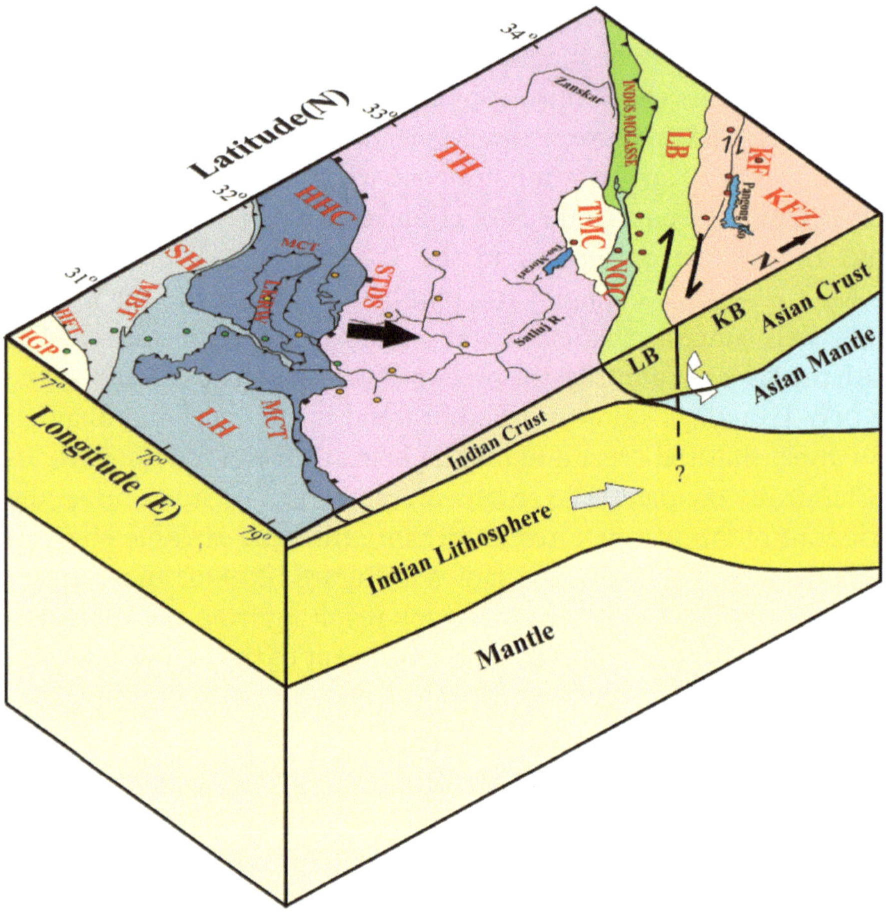

Figure 12.19 3-D cartoon showing the configuration of the underthrusting Indian lithosphere (following Xu et al., 2017 and Li et al., 2020) for interpretation of the origin of seismic anisotropy resulting from Indian plate motion in the Himalaya and dextral strike-slip motion of the Karakoram Fault.

computation of central depth which needs a pair of seismic events with a back azimuthal distance of 180 apart, which is sometimes very hard to find. To replace the drawbacks of the Fresnel zone concept, a more reliable grid search process of spatial coherency method (Gao et al., 2010; Liu and Gao, 2011; Gao and Liu, 2012) is useful to understand the crustal strain and mantle deformation but there is a depth constrain in using the *XKS* (*SKS*, *SKKS*, *PKS*, *PKKS*) phases. Due to a limited data set, the depth localization of upper mantle anisotropy could not be done which is the future scope of this study.

ACKNOWLEDGEMENTS

The authors acknowledge the support and facilities of Wadia Institute of Himalayan Geology, Dehradun, to accomplish this work. The authors are also thankful to Dr. Naresh Kumar for providing Kinnaur Network data for the study. The authors are also thankful to Dr. Rajib Biswas, the editor of the book.

REFERENCES

Ahmed, T., Tanaka, T., Sachan, H.K., Asahara, Y., Islam, R., and Khanna, P.P. 2008. Geochemical and isotopic constraints on the age and origin of the Nidar Ophiolitic Complex, Ladakh, India: Implication for the Neo-Tethyan subduction along the Indus suture zone. *Tectonophysics* 451, 206–224.

Alsina, D., and Snieder, R. 1995. Small-scale sublithospheric continental mantle deformation: Constraints from *SKS* splitting observations. *Geophys. J. Int.* 123, 431–448.

Alvizuri, C., and Tanimoto, T. 2011. Azimuthal anisotropy from array analysis of Rayleigh waves in Southern California. *Geophys. J. Int.* 186, 1135–1151.

Babuška, V., and Cara, M. 1991. Seismic anisotropy in the earth. *Mod. Approaches Geophys.* 10, 1–218.

Backus, G.E. 1962. Long-wave elastic anisotropy produced by horizontal layering. *J. Geophys. Res.* 67(11), 4427–4440.

Banerjee, P., and Bürgmann, R. 2002. Convergence across the northwest Himalaya from GPS measurements. *Geophys. Res. Lett.* 29(13), 1652. doi:10.1029/2002GL015184.

Barruol, G., Silver, P.G., and Vauchez, A. 1997. Seismic anisotropy in the eastern United States: Deep structure of a complex continental plate. *J. Geophys. Res.* 2, 8329–8348.

Barruol, G., Souriau, A., Vauchez, A., Diaz, J., Gallart, J., Tubia, J., and Cuevas, J. 1998. Lithospheric anisotropy beneath the Pyrenees from shear wave splitting. *J. Geophys. Res. Solid Earth* 103, 30039–30053.

Bird, P. 1978. Initiation of intracontinental subduction in the Himalaya. *J. Geophys. Res.* 83, 4975–4987.

Bormann, P., Burghardt, P.-T., Makeyeva, L.I., and Vinnik, L.P. 1993. Teleseismic shear-wave splitting and deformations in Central Europe. *Phys. Earth Planet. Inter.* 78, 157–166.

Bormann, P., Grünthal, G., Kind, R., and Montag, H. 1996. Upper mantle anisotropy beneath central Europe from *SKS* wave splitting: Effects of absolute plate motion and lithosphere-asthenosphere boundary topography? *J. Geodyn.* 22, 11–32.

Bowman, J.R., and Ando, M. 1987. Shear-wave splitting in the upper-mantle wedge above the Tonga subduction zone. *Geophys. J. Int.* 88, 25–41.

Cai, J., He, Y., Li, Z., Wang, B., and Guo, M. 2009. TTI/VTI anisotropy parameters estimation by focusing analysis, Part I: Theory. In SEG Technical Program Expanded Abstracts 2009 (301–305). Society of Exploration Geophysicists.

Collings, R., Rietbrock, A., Lange, D., Tilmann, F., Nippress, S., and Natawidjaja, D. 2013. Seismic anisotropy in the Sumatra subduction zone. *J. Geophys. Res.* 118(10), 5372–5390.

Crampin, S. 1984. Effective anisotropic elastic constants for wave propagation through cracked solids. *Geophys. J. Int.* 76, 135–145. https://doi.org/10.1111/j.1365-246X.1984.tb05029.x.

Crampin, S., and Lovell, J.H. 1991. A decade of shear-wave splitting in the Earth's crust: What does it mean? what use can we make of it? and what should we do next? *Geophys. J. Int.* 107, 387–407.

De Mets, C., Gordon, R.G., Argus, D.F., and Stein, S. 1990. Current plate motions. *Geophys. J. Int.* 101, 425–478. https://doi.org/10.1111/j.1365-246X.1990.tb06579.x.

Epard, J.L., and Steck, A. 2008. Structural development of the Tso Morari ultra-high pressure nappe of the Ladakh Himalaya. *Tectonophysics* 451, 242–264.

Faccenda, M., Ferreira, A.M., Tisato, N., Lithgow-Bertelloni, C., Stixrude, L., and Pennacchioni, G. 2019. Extrinsic elastic anisotropy in a compositionally heterogeneous Earth's mantle. *J. Geophys. Res. Solid Earth* 124(2), 1671–1687.

Fouch, M.J., Fischer, K.M., Parmentier, E.M., Wysession, M.E., and Clarke, T. J. 2000. Shear wave splitting, continental keels, and patterns of mantle flow. *J. Geophys. Res.* 105, 6255–6275. https://doi.org/10.1029/1999JB900372.

Gao, S.S., and Liu, K.H. 2012. AnisDep: A FORTRAN program for the estimation of the depth of anisotropy using spatial coherency of shear-wave splitting parameters. *Comp. Geosc.* 49, 330–333. https://doi.org/10.1016/j.cageo.2012.01.020.

Gao, S.S., Liu, K.H., and Abdelsalam, M.G. 2010. Seismic anisotropy beneath the Afar Depression and adjacent areas: Implications for mantle flow. *J. Geophys. Res. Solid Earth* 115(B12), 330. https://doi.org/10.1029/2009JB007141.

Gao, Y., Shi, Y., and Wang, Q. 2020. Seismic anisotropy in the southeastern margin of the Tibetan Plateau and its deep tectonic significances. *Chinese J. Geophys.* 63(3), 802–816.

Guillot, S., Garzanti, E., Baratoux, D., Marquer, D., Mahéo, G., and de Sigoyer, J. 2003. Reconstructing the total shortening history of the NW Himalaya. *Geochemistry, Geophysics, Geosystems* 4(7), 1064.

Hall, S.A., Kendall, J.M., and van der Baan, M. 2004. Some comments on the effects of lower-mantle anisotropy on SKS and SKKS phases. *Phys. Earth Planet. Inter.* 146(3–4), 469–481.

Hazarika, D., Hajra, S., Kundu, A., Bankhwal, M., Kumar, N., and Pant, C.C. 2021. Imaging the Moho and Main Himalayan Thrust beneath the Kumaon Himalaya: Constraints from receiver function analysis. *Geophys. J. Int.* 224, 858–870. doi: 10.1093/gji/ggaa478.

Hazarika, D., Paul, A., Wadhawan, M., Kumar, N., Sen, K., and Pant, C.C. 2017a. Seismotectonics of the Trans-Himalayan, eastern Ladakh, India: Constraints from moment tensor solutions of local earthquake data. *Tectonophysics* 698, 38–46. https://doi.org/10.1016/j.tecto.2017.01.001.

Hazarika, D., Sen, K., and Kumar, N. 2014. Characterizing the intra-crustal low velocity zone beneath northwest India-Asia collision zone. *Geophys. J. Int.* 199, 1338–1353. https://doi.org/10.1093/gji/ggu328.

Hazarika, D., Wadhawan, M., Paul, A., Kumar, N., and Borah, K. 2017b. Geometry of the Main Himalayan Thrust and Moho beneath Satluj valley, Northwest Himalaya: Constraints from receiver function analysis. *J. Geophys. Res. Solid Earth* 122, 2929–2945. https://doi.org/10.1002/2016JB013783.

Hazarika, D., Yadav, D.K., Sriram, V., and Rai, A. 2013. Upper mantle anisotropy beneath northeast India – Asia collision zone from shear-wave splitting analysis. *Int. J. Earth Sci.* 102, 2061–2076. https://doi.org/10.1007/s00531-013-0922-4.

Heintz, M., and Kennett, B.L.N. 2006. The apparently isotropic Australian upper mantle. *Geophys. Res. Lett.* 33, L15319.

Heintz, M., Kumar, V.P., Gaur, V.K., Priestley, K., Rai, S.S., and Prakasam, K.S. 2009. Anisotropy of the Indian continental lithospheric mantle. *Geophys. J. Int.* 179, 1341–1360. doi:10.1111/j.1365-246X.2009.04395.x.

Honegger, K., Dietrich, V., Frank, W., Gansser, A., Thöni, M., and Trommsdorf, V. 1982. Magmatism and metamorphism in the Ladakh Himalayas (the Indus-Tsangpo suture zone). *Earth Planet. Sci. Lett.* 60(2), 253–292. https://doi.org/10.1016/0012-821X(82)90007-3.

Ismail, W.B., and Mainprice, D. 1998. An olivine fabric database: An overview of upper mantle fabrics and seismic anisotropy. *Tectonophysics* 296, 145–157.

James, D.E., and Assumpção, M. 1996. Tectonic implications of S-wave anisotropy beneath SE Brazil. *Geophys. J. Int.* 126, 1–10.

Karato, S., and Wu, P. 1993. Rheology of the upper mantle: A synthesis. *Science* 260, 771–778.

Kendall, J. 1994. Teleseismic arrivals at a mid-ocean ridge: Effects of mantle melt and anisotropy. *Geophys. Res. Lett.* 21, 301–304.

Kern, H., Ivankina, T.I., Nikitin, A.N., Lokajíček, T., and Pros, Z. 2008. The effect of oriented microcracks and crystallographic and shape preferred orientation on bulk elastic anisotropy of a foliated biotite gneiss from Outokumpu. *Tectonophysics* 457(3–4), 143–149.

Kern, H., and Richter, A. 1981. Temperature derivatives of compressional and shear wave velocities in crustal and mantle rocks at 5 kbar confining pressure. *J. Geophys.* 49, 47–56.

Lacassin, R., Valli, F., Arnaud, N., Hervé Leloup, P., Paquette, J.L., Haibing, L., Tapponnier, P., Chevalier, M.-L., Guillot, S., Maheo, G., and Xu, Z. 2004. Large-scale geometry, offset and kinematic evolution of the Karakorum fault. *Tibet. Earth Planet. Sci. Lett.* 219, 255–269.

Langston, C.A. 1979. Structure under Mount Rainer, Washington, infer from teleseismic body waves. *J. Geophys. Res.* 84, 4749–4762.

Li, L., Murphy, M.A., and Gao, R. 2020. Subduction of the Indian plate and the nature of the crust beneath western Tibet: Insights from seismic imaging. *J. Geophys. Res. Solid Earth* 125(8), e2020JB019684. https://doi.org/10.1029/2020JB019684 e2020JB019684.

Ligorria, J.P., and Ammon, C.J. 1999. Iterative deconvolution and receiver function estimation. *Bull. Seismol. Soc. Am.* 89, 1395–1400.

Liu, J., Wu, J., Wang, W., Fang, L., and Chang, K. 2020. Seismic anisotropy beneath the eastern margin of the Tibetan Plateau from SKS splitting observations. *Tectonophysics* 785, 228430.

Liu, K.H., and Gao, S.S. 2011. Estimation of the depth of anisotropy using spatial coherency of shear wave splitting parameters. *Bull. Seismol. Soc. Am.* 101, 2153–2161.

Long, M.D., and Becker, T.W. 2010. Mantle dynamics and seismic anisotropy. *Earth Planet. Sci. Lett.* 297(3–4), 341–354.

Long, M.D. 2013. Constraints on subduction geodynamics from seismic anisotropy. *Rev. Geophys.* 51(1), 76–112.

Love, A.E.H. 2013. *A Treatise on the Mathematical Theory of Elasticity*. Cambridge, England: Cambridge University Press.

Maggi, A., Debayle, E., Priestley, K., and Barruol, G. 2006. Azimuthal anisotropy of the Pacific region. *Earth Planet. Sci. Lett.* 250(1–2), 53–71.

Mahéo, G., Bertrand, H., Guillot, S., Villa, I.M., Keller, F., and Capiez, P. 2004. The south Ladakh ophiolites (NW Himalaya, India): An intra oceanic tholeiitic origin with implication for the closure of the Neo-Tethys. *Chem. Geol.* 203, 273–303. https://doi.org/10.1016/j.chemgeo.2003.10.007.

Mainprice, D., Barruol, G., and Ismail, W.B. 2000. The seismic anisotropy of the Earth's mantle: From single crystal to polycrystal. In: Karato, S.I. (Ed.), *Earth's Deep Interior: Mineral Physics and Tomography from the Atomic to the Global Scale*. Washington, DC: American Geophysical Union, 237–264. doi:10.1029/GM117.

Mainprice, D., Le Page, Y., Rodgers, J., and Jouanna, P. 2007. Predicted elastic properties of the hydrous D phase at mantle pressures: Implications for the anisotropy of subducted slabs near 670-km discontinuity and in the lower mantle. *Earth Planet. Sci. Lett.* 259, 283–296.

Mainprice, D. 1997. Modelling the anisotropic seismic properties of partially molten rocks found at mid-ocean ridges. *Tectonophysics* 279(1–4), 161–179.

Mainprice, D., and Silver, P.G. 1993. Interpretation of *SKS*-waves using samples from the subcontinental lithosphere. *Phys. Earth Planet. Inter.* 78, 257–280.

Margheriti, L., Nostro, C., Cocco, M., and Amato, A. 1996. Seismic anisotropy beneath the Northern Apennines (Italy) and its tectonic implications. *Geophys. Res. Lett.* 23(20), 2721–2724.

Mc Namara, D.E., and Owens, T.J. 1993. Azimuthal shear wave velocity anisotropy in the Basin and Range province using Moho Ps converted phases. *J. Geophys. Res.* 98, 12003–12017.

Molnar, P.H., and Tapponnier, P. 1975. Cenozoic tectonics of Asia: Effects of continental collision. *Science* 189, 144–154.

Mukherjee, B.K., and Sachan, H.K. 2004. Discovery of coesite from Indian Himalaya: A record of ultra-high pressure metamorphism in Indian continental crust. *Curr. Sci.* 81, 1358–1361.

Nicolas, A., Boudier, F., and Boullier, A.M. 1973. Mechanisms of flow in naturally and experimentally deformed peridotites. *Am. J. Sci.* 273, 853–876.

Nicolas, A., and Christensen, N.I. 1987. Formation of anisotropy in upper mantle peridotites-A review. *Composition, Structure and Dynamics of the Lithosphere-Asthenosphere System*, vol. 16, Washington, DC: American Geophysical Union, 111–123.

Oreshin, S., Kiselev, S., Vinnik, L., Surya Prakasam, K., Rai, S.S., Makeyeva, L., and Savvin, Y. 2008. Crust and mantle beneath western Himalaya, Ladakh and western Tibet from integrated seismic data. *Earth Planet. Sci. Lett.* 271, 75–87. https://doi.org/10.1016/j.epsl.2008.03.048.

Park, J., and Levin, V. 2002. Seismic anisotropy: Tracing plate dynamics in the mantle. *Science*, 296(5567), 485–489.

Paul, A., Hazarika, D., Kumar, N., and Wadhawan, M. 2021. Upper mantle anisotropy in the northwest Himalaya and Ladakh Karakoram zone based on SKS splitting analysis. *J. Geodyn.* 144, 101817. https://doi.org/10.1016/j.jog.2021.101817.

Paul, A., Hazarika, D., and Wadhawan, M. 2017. Shear wave splitting and crustal anisotropy in the Eastern Ladakh-Karakoram zone, northwest Himalaya. *J. Asian Earth Sci.* 140, 122–134. https://doi.org/10.1016/j.jseaes.2017.04.010.

Piccinini, D., Pastori, M., and Marghieriti, M. 2013. ANISOMAT+: An automatic tool to retrieve seismic anisotropy from local earthquakes. *Comput. Geosci.* 56, 62–68. http://dx.doi.org/10.1016/j.cageo.2013.01.012.

Rai, S.S., Priestley, K., Gaur, V.K., Mitra, S., Singh, M.P., and Searle, M. 2006. Configuration of the Indian Moho beneath the NW Himalaya and Ladakh. *J. Geophys. Res.* 33, L15308. http://dx.doi.org/10.1029/2006GL026076.

Sandvol, E., Ni, J., Kind, R., and Zhao, W. 1997. Seismic anisotropy beneath the southern Himalayas-Tibet collision zone. *J. Geophys. Res.* 102, 17813–17823. doi:10.1029/97JB01424.

Savage, M.K. 1999. Seismic anisotropy and mantle deformation: what have we learnt from shear wave splitting? *Rev. Geophys.* 37, 65–106. https://doi.org/10.1029/98RG02075.

Searle, M.P., Windley, B.F., Coward, M.P., Cooper, D.J.W., Rex, A.J., Rex, D., Tingdong, L., Xuchang, X., Jan, M.Q., and Thakur, V.C. 1987. The closing of Tethys and the tectonics of the Himalaya. *Geol. Soc. Am. Bull.* 98, 678–701.

Sen, K., and Collins, A.S. 2013. Dextral transpression and late Eocene magmatism in the trans-Himalayan Ladakh Batholith (North India): Implications for tectono-magmatic evolution of the Indo-Eurasian collisional arc. *Int. J. Earth Sci.* 102, 1895–1909.

Shearer, P.M. 2009. *Introduction to Seismology* (second edition). Cambridge, England: Cambridge University Press.

Silver, P.G. 1996. Seismic anisotropy beneath the continents: Probing the depths of geology. *Ann. Rev. Earth Planet Sci.* 24, 385–432.

Silver, P.G., and Chan, W.W. 1991. Shear wave splitting and subcontinental mantle deformation. *J. Geophys. Res.* 96, 16429–16454. https://doi.org/10.1029/91jb00899.

Silver, P.G., and Holt, W.E. 2002. The mantle flow field beneath western North America. *Science* 295, 1054–1057.

Singh, A., Kumar, M.R., and Raju, P.S. 2007. Mantle deformation in Sikkim and adjoining Himalaya: Evidences for a complex flow pattern. *Phys. Earth Planet Inter.* 164, 232–241. doi:10.1016/j.pepi.2007.07.003.

Singh, A., Kumar, M.R., Raju, P.S., and Ramesh, D.S. 2006. Shear wave anisotropy of the northeast Indian lithosphere. *Geophys. Res. Lett.* 33, L16302. doi.org/10.1029/2006GL026106.

Sivaram, K., Kumar, V.P., Gupta, S., Prasad, B.N.V., and Kumar, S. 2022. Upper mantle seismic anisotropy beneath the Deccan Volcanic Province and the adjacent Eastern Dharwar Craton in south Indian shield from shear wave splitting analysis. *Phys. Earth Planet. Inter.* 322, 106829.

Sol, S., Meltzer, A., Bürgmann, R., van der Hilst, R.D., King, R., Chen, Z., Koons, P.O., Lev, E., Liu, Y.P., Zeitler, P.K., Zhang, X., Zhang, J., and Zurek, B. 2007. Geodynamics of the southern Tibetan Plateau from seismic anisotropy and geodesy. *Geology* 35, 563–566. https://doi.org/10.1130/G23408A.1.

Stein, S., Wysession, M., and Houston, H. 2003. An introduction to seismology, earthquakes, and earth structure. *Physics Today* 56(10), 65–72.

Tarduno, J.A., Duncan, R.A., Scholl, D.W., Cottrell, R.D., Steinberger, B., Thordarson, T., Kerr, B.C., Neal, C.R., Frey, F.A., and Torii, M. 2003. The Emperor Seamounts: Southward motion of the Hawaiian hotspot plume in Earth's mantle. *Science* 301, 1064–1069.

Tiwari, A., Singh, A., Eken, T., and Singh, C. 2017. Seismic anisotropy inferred from direct *S*-wave-derived splitting measurements and its geodynamic implications beneath southeastern Tibetan Plateau. *Solid Earth, EGU* 8, 435–452.

Vauchez, A., and Nicolas, A. 1991. Mountain building: Strike-parallel motion and mantle anisotropy. *Tectonophysics* 185, 183–201. https://doi.org/10.1016/0040-1951(91)90443-V.

Vinnik, L., and Farra, V. 2007. Low S velocity atop the 410-km discontinuity and mantle plumes. *Earth Planet. Sci. Lett.* 262, 398–412.

Vinnik, L.P., Makeyeva, L.I., Milev, A., and Usenko, A.Y. 1992. Global patterns of azimuthal anisotropy and deformations in the continental mantle. *Geophys. J. Int.* 111(3), 433–447.

Wadhawan, M., Hazarika, D., Paul, A., Kumar, N., Gupta, V., and Agarwal, M. 2022. Seismic anisotropy and crustal deformation in the Satluj valley, Northwest Himalaya revealed by the splitting analysis of Moho converted Ps phases. *J. Asian Earth Sci.* (Under Review).

Walker, K.T., Nyblade, A.A., Klemperer, S.L., Bokelmann, G.H.R., and Owens, T.J. 2004. On the relationship between extension and anisotropy: Constraints from shear wave splitting across the East African Plateau. *J. Geophys. Res. Solid Earth* 109, B08302. https://doi.org/10.1029/2003JB002866.

Wei, W., Zhao, D., Xu, J., Zhou, B., and Shi Y. 2016. Depth variations of P-wave azimuthal anisotropy beneath Mainland China. *Sci. Rep.* 6, 29614.

Weinberg, R.F., and Mark, G. 2008. Magma migration, folding and disaggregation of migmatites in the Karakoram Shear Zone, Ladakh, NW India. *Geol. Soc. Am. Bull.* 120, 994–1009.

Weinberg, R.F., Mark, G., and Reichardt, H. 2009. Magma ponding in the Karakoram shear zone, Ladakh, NW India. *Geol. Soc. Am. Bull.* 121, 278–285.

Wu, J., Zhang, Z., Kong, F., Yang, B., Yu, Y., Liu, K., and Gao, S. 2015. Complex seismic anisotropy beneath western Tibet and its geodynamic implications. *Earth Planet. Sci. Lett.* 413, 167–175. doi: 10.1016/j.epsl.2015.01.002.

Xu, Q., Zhao, J., Yuan, X., Liu, H., and Pei, S. 2017. Detailed configuration of the underthrusting Indian lithosphere beneath western Tibet revealed by receiver function images. *J. Geophys. Res. Solid Earth* 122, 8257–8269. https://doi.org/10.1002/2017JB014490.

Yin, A., and Harrison, T.M. 2000. Geologic evolution of the Himalayan-Tibetan orogen. Annual Review of Earth Planet. *Sci. Lett.* 28, 211–280. doi:10.1146/annurev.earth.28.1.211.

Zhao, D., Yu, S., and Liu, X. 2016. Seismic anisotropy tomography: New insight into subduction dynamics. *Gondwana Res.* 33, 24–43.

Zheng, T., Gao, S.S., Ding, Z., Liu, K.H., Chang, L., Fan, X., Kong, F., and Yu, Y. 2021. Crustal azimuthal anisotropy and deformation beneath the north-eastern Tibetan Plateau and adjacent areas: Insights from receiver function analysis. *Tectonophysics* 816, 229014.

Chapter 13

Lessons learned from earthquakes

Mehdi Zaré

13.1 INTRODUCTION

Moderate to large magnitude earthquakes may lead to large damages. Field observations performed after many earthquakes in Iran indicate that there has been a major difference between the *observed* and the *expected* ground motions and intensities. A large part this difference about ground motions goes back to the lack of enough data or hypo chapter (e.g. source geometry, faulting probabilities) as it was mentioned in Chapters 1 and 5. In many cases, expected intensities are intensified by the unknown site effects, such as soil amplifications as well as topography conditions.

In this chapter, a combination of lessons learned from in-situ earthquakes damages and subsequent phenomena observed during reconnaissance studies in different earthquake-affected areas are presented. In this regard, selected topics such as the data acquisition in earthquake recognitions, the analysis of recorded strong motions, the determination of physical damage, the near-field effects associated with directivity and fling-step, and the importance of these analysis in hazard zoning provide some insights into the seismic hazard assessment. These studies are the results of field investigations that I performed following the 14 important recent earthquakes in Iran and other countries in the Middle East. The field studies by the author start from the 2003 Bam (Iran) earthquake, continue with the 2003 Boumerdès-Zemmouri (Algeria), 2005 Balakot (Pakistan), 2009 Padang (Indonesia), 2009 Southeastern Tehran (Iran), 2010 Kuh-Zar (Iran), 2011 Tohoku (Japan), 2011 Van (Turkey), 2012 Varzeghan (Iran) twin events, 2013 Shonbeh (Iran), 2013 Saravan (Iran), 2013 Bashagard (Iran), 2014 Mormori (Iran) earthquakes and finally ends with the 2015 Gorkha (Nepal) earthquake. The chapter shows a syn chapter of

DOI: 10.1201/9781003177692-13

the main observations and comparison of local seismic hazard analysis (SHA) and scenarios between different regions.

13.2 EARTHQUAKE INTENSITY AND DAMAGE DISTRIBUTION

In Table 13.1, a summary of the visited earthquakes is shown.

The 25 April 2015, Nepal earthquake was practically the most recent destructive event, with widespread damages to many buildings, infrastructures and historic monuments mostly within the central part of the country. In all the visited areas, there were many low-resilient multistorey masonry or concrete buildings which were fully collapsed, tilted or severely damaged. Most of the destructions had occurred in mountainous villages especially in Barpak, Laprak and Saurpani villages in Gorkha district as well as Langtang village, 60 kilometer distance northeast of Kathmandu. In Kathmandu, violent damages were mostly observed in the central and western parts of the city. Around the Dhading New Bus Park in Gongabu Ring Road northwest of Kathmandu, business and shopping centers and residential complexes sustained major damages (Figure 13.1). Based on our observations, the earthquake intensity

Table 13.1 List of the visited earthquakes by the author

No	*Earthquake*	*Date*	M_w	*Depth (km)*	*I0*	*No. Death*	*No. Injured*
1	Manjil, Iran	1990/06/21	7.3	19	X	35,000	60,000
2	Boumerdès, Algeria	2003/05/21	6.8	12	X	2,266	10,261
3	Bam, Iran	2003/12/26	6.5	8	IX	33,000	50,000
4	Balakot, Pakistan	2005/10/08	7.6	15	VIII	87,000	70,000
5	Padang, Indonesia	2009/09/30	7.6	90	VII	1,115	2,902
6	Kuh-Zar, Iran	2010/08/27	5.7	15	VII	4	40
7	Tohoku, Japan	2011/03/11	9.1	29	IX	15,894	6,152
8	Van, Turkey	2011/10/23	7.1	7	VIII	644	4,412
9	Varzeghan, Iran	2012/08/11	6.3	12	VIII+	306	3,037
10	Shonbeh, Iran	2013/04/09	6.3	10	VII	37	850
11	Saravan, Iran	2013/04/16	7.7	82	VI	35	117
12	Bashagard, Iran	2013/05/11	6.2	25	VI+	1	-
13	Mormori, Iran	2014/08/18	6.2	10	VII	0	250
14	Gorkha, Nepal	2015/04/25	7.8	8	IX	8,964	21,952

Figure 13.1 Severe damages to the building in the Dhading New Bus Park in Gongabu Ring Road northwest of Kathmandu.

was estimated to be about VIII, VII and V on the EMS98 intensity scale for the epicentral region Kathmandu as well as Bhaktapur and Pokhara, respectively.

In 2014, M_W 6.2 earthquake occurred in Mormori (Ilam), southwest of Iran. Based on the fields investigations (Zaré et al., 2014), the most damages were observed in Mormori (a population of about 4,000) and the villages of Abanar (Figure 13.2) and Siahgel, and there were also partial damages to historical and archaeological buildings. There were also some damages to the residential complexes of "Maskan-e Mehr" (social apartments built in the last decade in Iran for low-income people) in Abdanan (Figure 13.3). The shake maps were developed for this event in which a maximum intensity of VII–VIII (EMS98) are estimated for the epicenter. The maximum acceleration between 100 and 200 gals are estimated for the epicenter.

Figure 13.2 Damage to the buildings in Abanar: collapse of walls, figures and object and plaster fall.

In 16 April 2013, Saravan earthquake of M_W 7.7 occurred in a *sparsely populated area* with about 35 km distance to the city of Gosht (Figure 13.4).

The shock was felt strongly in Gosht (assessed VI+, 35 km southwest of epicenter), Saravan (VI, 60 km S) and Zahedan (VI, 190 km NW) and less strongly in Masqat (IV, 650 km S) and Persian Gulf Islands

Figure 13.3 Fissures created in "Mehr" Apartments (Maskan-e Mehr) in Abdanan (38 km to the epicenter); constructed about one year ago and suffered extensive damages due to an intensity VI (EMS98) on 18 August 2014 Mormori earthquake.

Figure 13.4 The damages in the epicentral region of Gosht, Saravan Earthquake, 16 April 2013, M_w 7.8 earthquake. (Photo by AFP.)

> (IV, >700 km W) as well as the strong shaking felt in the high-rise buildings of Dubai, Doha, Manama and even in Kuwait.
>
> (Zaré et al., 2013a)

The M_W 6.4 Shonbeh (Bushehr) earthquake was the next important event that happened in 2013. In Zaré et al. (2013a), we show that

> Most of the buildings damaged in the 2013 M_W 6.4 Shonbeh (Bushehr), Iran earthquake were adobe buildings, made of heavy material and constructed in the last 50 years (Figure 13.5). The walls collapsed mostly in east-west or NNW-SSE directions. The shock was felt strongly in Shonbeh (assessed at VII), 18 km SW of the epicenter; Khormoj (assessed at VI+), 22 km NE; and Bushehr (assessed at V),

Figure 13.5 Buildings damaged in the 2013 M_w 6.4 Shonbeh (Bushehr), Iran, earthquake.

> 86 km NW. The maximum recorded acceleration in this event was obtained in Khormoj (70 gals); the mainshock was recorded in Khormoj (the records are available at the BHRC web site: https://www.bhrc.ac.ir). In Bushehr (80 km NW of the epicenter), where the maximum acceleration was estimated at around 50 gals, there was no damage reported to a nuclear power plant.

On 11 August 2012, two relatively large shallow that struck Varzeghan region in NW Iran, causing about 306 deaths, 3,037 injuries and 30,000 homeless people. Following the earthquakes, reconnaissance team was dispatched to the area by IIEES (Zaré et al., 2012). According to Vasheghani Farahani and Zaré (2014),

> the maximum intensity was estimated to be VIII+ on the EMS98 scale in Varzeghan and affected villages, VII+ in Ahar city and VI+ in Heris city. The intensity of the earthquake in Tabriz was V.
>
> Based on observations in the region, the main failure mode was the collapse of heavy roofs. Masonry buildings, for example in Figure 13.6a, b (very weak mortar), suffered very heavy structural damage with a damage grade (based on Grunthal, 1998) of five (destruction), total or

Figure 13.6 (a and b) Examples of total collapse in very weak buildings (c and d). Examples of demolished buildings in Varzeghan (left-c) and Baj e Baj (right-d). (Zaré et al., 2012.)

near total collapse. In Baj e Baj, most of the buildings were made of bricks and arched wood beam roofs, adobe and low-quality masonry materials. In types of structure with simple stone masonry such as in Figure 13.6c, d, some parts of the bearing walls collapse of the roof and floor slabs. This kind of heavy structural damage is of damage grade 4. In Heris city, a hospital was built on a hill slope one year before the 11 August 2012 earthquakes. Although, the city was located about 20 km away from the earthquake epicenter, the hospital was rendered out of service. This hospital (the main health center near the epicentral zone) was located 73 kilometers east of Tabriz, the capital city of the east Azerbaijan province (Figure 13.7a, b). Moreover, there were some structures with large diagonal cracks in the walls or some cracks in the exterior wall in the twin Varzeghan earthquakes region. Figure 13.7c shows an example of these buildings that exhibited moderate structure damage of grade 3. In type of structure with many vertical cracks, such as Figure 13.7d, slight structural damage resulted and the damage was of grade two.

By the occurrence of the 27 August 2010 M_W 5.7 Kuh-Zar earthquake (Iran), more than 80% of the rural houses in Iran are composed of adobe or stone masonry and are constructed using only local materials and

Figure 13.7 (a and b). Examples of Heris Hospital (c and d). Example of diagonal cracks in walls, and of slight structural damage in the region, respectively. (Zaré, 2012.)

unskilled labor (Maheri et al., 2005). In the Kuh-Zar region, the greatest intensity of damages was due to the traditional construction, very poor clay materials, single-storey houses of weak masonry, non-reinforced adobe (Figure 13.8a–e), overloaded construction, and flat wooden and steel-beamed roofs (Figure 13.8e), a number of which were damaged in the earthquake (Shahvar and Zaré, 2012).

An M_W 7.6 earthquake hit the western Sumatra and Padang city on 30 September 2009. A reconnaissance team was dispatched to the area by IIEES. There was no ground motion record available to estimate the earthquake peak ground acceleration or the response spectra. The damage observation covers building and houses, water supply network and

Figure 13.8 Field photographs: (a) Coseismic fracture near the epicenter (Foroutan et al., 2010). (b) Crack from the 2010 Kuh-Zar earthquake (near the epicenter), cracks strike E to W. (c) Fissuring in Kelu village, 5 kilometers from the epicenter, crack strike is N70. (d) Typical adobe-brick house at Kelu; the non-load-bearing walls collapse in an earthquake first. (e) Collapse started in the corner, where it was easier to initiate instability.

Compressive failure of RC column — Soft storey response in RC building

Lack of correct practice in masonry buildings — Gantiang mosque, city's oldest mosque with heavy damage

Airport bridge in Padang city — Railroad bridge on the way from Pariaman to Padang — The Padang city bus terminal totally collapsed

Figure 13.9 Different examples of damages to buildings and transportation facilities.

transportation system. A quick survey of hospitals was also performed during the reconnaissance trip. The buildings in the region can be categorized mainly as masonry and RC buildings. Low-quality material and incorrect construction methods were identified as the main reasons of damage as shown in the following pictures. Soft storey effect was observed in several RC buildings in the Padang City. Considerable damage to historical buildings was observed (Figure 13.9).

Following the 8 October 2005 Pakistan earthquake, a visit to the earthquake area was performed by Mehdi Zaré.

The visits took place less than 3 weeks after the epicentral mainshock. The maximum intensity was estimated to be around Balakot and

Figure 13.10 About 1 m displacement in the abutment of Balakot southern bridge due to the earthquake mainshock on 8 October 2005.

along Muzaffarabad fault, with a width of 10–15 km around the fault. The greatest damage was observed in Balakot representing complete destruction (i.e., more than 1 m dislocation of the concrete bridge over Balakot River; Figure 13.10). Balakot region is as-signed intensity X [according to The European Macroseismic Scale (EMS-98) scale]. All buildings in this valley city were destroyed totally. The city of Muzaffarabad is located on a river terrace. Based on damages seen in Ijaz Golab military hospital, an intensity of IX on EMS-98 scale is estimated. The retaining walls were collapsed in some roads (Figure 13.11). In Ijaz Golab hospital, the northern block fell down due to the first floor collapse (soft storey), but other parts of the hospital show less damage. The daily life had returned to normal 18 days after the event when IIEES reconnaissance team visited the epicentral region. Intensity in Abbottabad was estimated to be VIII (EMS-98 scale) derived from a hotel col-lapse. Due to the observed damage of residential buildings, VIII intensity is assigned to Mansehra (Figure 13.12). No damage was observed on the dam and the facilities upon the inspection of Tarbela earth-rock fill dam. To the SW of the epicentre, the landslides were

Figure 13.11 Muzaffarabad: The retaining wall collapse due to Balakot earthquake mainshock on 8 October 2005.

Figure 13.12 Destruction at Mansehra.

Figure 13.13 Margala Tower collapse in Islamabad. (Associated Press, October 2005.)

> observed in the region between Punjab and Jammu and Kashmir (India). This region is assigned an intensity of about VI. While there was not as much damage in Islamabad, a Margala tower collapsed in the north-west of Islamabad (Figure 13.13) because of its struc-tural engineering and also the type of soil (grain alluvium with clay matrix).
> (Zaré and Karimi-Paridari, 2009)

Another destructive earthquake during last decades in Iran was the Bam (SE Iran) earthquake of 26 December 2003, M_W 6.5 that was explained previously in Chapter 5. The macroseismic intensity of the earthquake is estimated to be I0 = IX (in the EMS98 scale), where the strong motions and damaging effects seems to be attenuated very fast especially in the fault-normal direction. The intensity levels are estimated to be VIII in Baravat, VII in New-Arg (Arg-e Jadid) and the airport area. The intensity level was estimated to be around IV–V in Kerman and Mahan.

In 2009, M_W 7.6, Padang, Sumatra, Indonesia, site effect had an essential role in amplifying the ground shaking in the epicentral area, especially in the city of Padang. This idea comes from the most severe damages observed in the engineered constructed building in this city, while the epicentral distances for this city is greater than the central part of the Pariaman, for example. A preliminary Vs30 map prepared by Petresen (2007) based on assigning a shear-wave velocity to geological units. The soils in Padang City consist mainly of Quaternary alluvium (includes some swamps, bay and estuary mud) and have a very low shear-wave velocity in upper 30 meters (180–360 m/s). Regarding this range of velocity, we can expect the ground motion amplification for the frequencies ranging from 1.5 to 4 Hz.

13.3 EARTHQUAKE-INDUCED LANDSLIDES

Earthquake-induced landslides are common events that occur after large earthquakes, moving medium to huge debris and blocking roads and lifelines. Based on the performed reconnaissance surveys by the author, large earthquake-induced landslides were observed in mountainous areas during the 1990 Manjil-Iran, 2005 Balakot-Pakistan, 2009 Padang-Indonesia, 2013 Goharan-Iran and 2015 Gorkha-Nepal earthquakes some of which are pointed out briefly in the following.

In Zaré et al. (2008), we show that

> the 2005 Balakot, Pakistan earthquake occurrence accompanied by landslides made a lot of dust especially in Darebagh, Muzaffarabad, and Balakot. Such phenomena with huge stones falling destroyed the roads in different places (Figure 13.14) and blocked the traffic, leading to extreme congestion. Numerous landslides occurred in the meisoseismal region, and most of them were associated with the fault scarp where the rupture reached the surface. The landslides mostly occurred from the eastern parts of Darebagh to Muzaffarabad and Balakot (Figure 13.15) which damaged or blocked roads and also induced some damages to the transport facilities such as tunnels, having no halter or cover, and bridges in the south of Muzaffarabad. Some vertical and horizontal cracks and fissures have been observed on the south Muzaffarabad Bridge. The bridge was barricaded, and the road was closed in the first days after the event. By reopening this tunnel, the connection between Islamabad and Muzaffarabad was resumed.

Relief Web reported that over 1,000 landslides were triggered by the 2009 Padang, Indonesia, earthquake. Based on the reports, 400–600 death of the total of more than 1,200 death caused by earthquake were killed by landslides. Three villages were completely demolished by landslide and

Figure 13.14 Islamabad–Muzaffarabad road destroyed by Balakot earthquake on 8 October 2005.

Figure 13.15 An example of landslide south of Balakot along the earthquake fault rupture.

Figure 13.16 Two giant landslides happened on Tandikat regency along the Lembah Tiga valley. Upper part of the photo: Cumanak landslide wiped and buried the village of Cumanak and killed nearly 200–300 peoples gathered for a wedding ceremony. Lower part of the photo: A landslide adjacent to the Lubuk Laweh landslide cut a road.

many others were damaged (Figure 13.16). Many roads were damaged or blocked by landslides and resulted in difficulties in rescue efforts. The most destructive landslides happened during this earthquake are concentrated in the Tandikat regency.

Such phenomenon was also observed as a major rock slide about 5 km north of Goharan during 11 May 2013 Bashagard earthquake of Iran (Figure 13.17).

Following the 2015 Gorkha, Nepal, earthquake, many landslides, avalanches, ground fissures and fault ruptures happened. Based on our observations, the earthquake triggered some new landslides in the form of debris falls and rock slides along the intercity road from Kathmandu westward to Gorkha and Pokhara. At about 62 km west of Kathmandu, a relatively large landslide could be observed with an approximate size of 100m width × 50m length (Figure 13.18). Because of the continued

Figure 13.17 A rock slide in the Goharan to Irar road, during the 11 May 2013 earthquake.

Figure 13.18 Observed debris slide with about 100m width and 50m length 62 km west of Kathmandu.

aftershocks occurred throughout Nepal in the first weeks, the country also had a continued risk of landslides. The earthquake also sparked an avalanche on Mount Everest and a huge avalanche in the Langtang valley, blocking mountain routes, killing several individuals and missing hundreds. In Pokhara City, there were some structure damages mostly due to earthquake-induced landslides. It is notable that the city of Pokhara is located aside the Phewa Lake in 207 km west of Kathmandu. Evidently, the Phewa Lake seems to be formed by a landslide dam (possibly triggered by the 1833 or 1934 earthquakes) (Figure 13.19).

Figure 13.19 The Phewa Lake that seems to be formed by a landslide dam (possibly triggered by a previous earthquakes). Red arrow shows the direction of the probable earthquake-induced landslide blocking the lake.

13.4 FAULT RUPTURES, SURFACE FISSURES AND GEOTECHNICAL FEATURES

One of the superficial phenomena caused by the earthquake are surface fissures and cracks as well as fault rupture. In the following, some of the observed fault ruptures and surface fissures are pointed out briefly.

A typical ground fissure was observed at a distance about 9 km southeast of Kathmandu toward Bhaktapur with a NE-SW direction and 2 metre vertical displacement (Figure 13.20) as an aftermath of the 2015 Gorkha, Nepal, earthquake.

Such surface fissures were found nearby Gosht and in the village of Gol-Pichak during the 2013, M_W 7.8, Gosht (Saravan) earthquake (Figure 13.21).

In Zaré et al. (2013b), we show that

> according to the existing faults in the region of the 2013 Bashagard, SE of Iran, the Manujan fault with a right lateral strike-slip movement and a NW-SE trend having a slip towards east was expected to be the causative fault. The location of the epicentre and the fault slip towards east might be a justification for a depth between 20 to 25 km for this earthquake. The Surface fissures, is observed around

Figure 13.20 Ground surface fissures with azimuth of 19 208° and 2m vertical displacement occurred in the road between Bhaktapur and Kathmandu.

Figure 13.21 The surface fissure in Gol-Pichak village (nearby Gosht) created during the mainshock of 16 April 2013 M_w 7.8 earthquake.

the road from Goharan to Irar (kilometre 5, north of Goharan, and just at the end of this road (Figure 13.22). However it does not present the systematic evidences of surface fault rupturing, it continues in an en echelon NE-SW direction near Irar (Figure 13.23), along about 5 kms, and the fissure segments has mostly the azimuths of N320 to N330.

Figure 13.22 The tension cracks in the Goharan to Irar road, created during the mainshock of the 11 May 2013 earthquake.

Figure 13.23 Surface fissures in the epicenter of 11 May 2013, M_w 6.2 Iran earthquake (in the village of Irar).

The 2012 Varzeghan earthquakes caused ground surface rupture. The causative fault was assessed to be the South Ahar fault, having an east-west trend and a length of about 60 km. The surface rupture of the Varzeghan fault was surveyed along about 4 km of ground surface by the investigating team and is located 13 km south of Varzeghan. Its azimuth is N 245–260 degrees, as shown in Figure 13.24.

Figure 13.24 Surface rupture on the Varzeghan twin earthquakes fault. (Zaré, 2012.)

Figure 13.25 The surface fissures along the Bam fault (near Baravat) in the prone area of the 26 December 2003 earthquake. (a) the fissures nearby Baravat, (b) the fissure that ruptured the road from Bam to Baravat and (c) the fissures in the foot of the fault scarp of the Bam fault.

Such a situation was also observed in the 2003 Bam earthquake in which, according to Eshghi and Zaré (2003),

> The Bam fault with a near north-south direction passes from the vicinity of the city of Bam (less that 1 km distance to the east of Bam, and between the cities of Bam and Baravat. The surface fissures created after the Bam earthquake are observed around the Bam fault between the cities on Bam and Baravat (Figure 13.25). The fissures are created in the form of the sinkholes in the city of Baravat (Figure 13.26).

The 2014 Mormori (Ilam) SW Iran's earthquake caused road blocks, fissures and subsidence in the road (32 kilometers from Darreh-Shahr; Figure 13.27 and 13.28), and followed up by the debris slides during the aftershocks.

In Zaré et al. (2013a), we show that

> the 2013, Mw 6.4 Shonbeh earthquake had some geotechnical features. Near Shonbeh (4 km southeast of the city) there is clear

Figure 13.26 The sinkholes created during the 26 December 2006 Bam earthquake in Baravat.

Figure 13.27 Darreh-Shahr to Abdanan Road (20 km from Darreh-Shahr) caused the road block (about 52 km to the epicenter).

Figure 13.28 Road from Darreh-Shahr to Abdanan (32 km from Darreh-Shahr): Fissures and land subsidence caused by the 18 August 2014 earthquake (about 42 km distance to epicenter).

evidence of liquefaction along the Mond River in the Dashte-Zaal ('white plain'), (Figure 13.29). Some surface tension cracks were induced by liquefaction in the Mond River terrace (Figure 13.30). These features could be seen along about 3 km of the terrace. Surface fissures along the reported causative fault could be observed south of Shonbeh, on the road to Baghan and Bandar-Daiyer (Figures 13.31 and 13.32). These en-echelon, non-continuous segmented fissures do not show the displacement consistent with the fault mechanism (compressional with a strike slip movement); however, it is mostly a vertical displacement in most of segment, with some surface strike slip movements (not systematic in the same direction). The segments were located about 1 km distant from the Jashak Salt Diapir and it may be that the salt layers are effective in conducting the rupture to the alluvium surface.

Figure 13.29 Close-up view of liquefaction in the Dashte Zaal area.

Figure 13.30 Extension cracks and joints in young alluvium river bank deposit in the Dashte Zaal.

Figure 13.31 Surface fissure (N300–320) in young alluvium deposit, east area of Jashk Salt-Diapir 8 km of this fissure could be seen in the field.

Figure 13.32 Closer view of surface fissure (N300–320) in young deposits.

13.5 CONCLUSIONS AND OUTLOOK

Large earthquake-induced landslides were observed in mountainous areas during the 1990 Manjil-Iran, 2005 Balakot-Pakistan, 2009 Padang-Indonesia, 2013 Goharan-Iran and 2015 Gorkha-Nepal earthquakes. The damage observation covers building and houses, water supply network and transportation system. Based on the survey in the epicentral zones of the major studied earthquakes, we may conclude that in the masonry and RC buildings, low-quality material and incorrect construction methods were identified as the main reasons of damage.

REFERENCES

Eshghi, S. and Zaré, M. (2003). Bam (SE Iran) earthquake of 26 December 2003, Mw6. 5: A preliminary reconnaissance report, 7 pages, available on *IIEES*. http://www.iiees.ac.ir.

Shahvar, M. P. and Zaré, M. (2013). The 27 August 2010 Mw 5.7 Kuh-Zar earthquake (Iran): Field investigation and strong-motion evidence. *Natural Hazards*. doi: 10.1007/s11069-012-0507-8.

Talebi, M., Zaré, M., Madahi-zadeh, A. R. and Bali-Lashak. (2015). Spatial-temporal analysis of seismicity before the 2012 Varzeghan, Iran, Mw 6.5 earthquake. *Turkish Journal of Earth Sciences*, 24 (24): 1–13. doi: 10.3906/yer-1410-13.

Vasheghani Farahani, J. and Zaré, M. (May/June 2011). The southeastern Tehran earthquake of 17 October 2009 (Mw = 4.0). *Seismological Research Letters*, 82 (3): 404–412. doi: 10.1785/gssrl.82.3.404.

Vasheghani-Farhani, J. and Zaré, M. (2014). Seismological aspects of the Varzeghan twin earthquakes on 11 August 2012 (Mw 6.3 and Mw 6.1), in East Azerbaijan province, NW Iran. *Episodes*, 37 (2): 9–104.

Zaré, M. (2015). "Le séisme de Gorkha, Népal, 2015; Rapport de Reconnaissance et distribution des dégâts", *9ème Colloque National AFPS 2007* – Ecole Centrale Paris, 8 pages.

Zaré, M. and Afrouz, S. Ghaychi. (June 2012). Crisis management of Tohoku; Japan earthquake and tsunami, 11 March 2011. *Iranian Journal of Public Health*, 41 (6): 12–20.

Zaré, M. (2003). "A preliminary reconnaissance report on the Algeria earthquake of 21 May 2003, Mw6. 8". http://www.iiees.ac.ir.

Zaré, M. and Nazmazar, B. (February 2013). Van, Turkey Earthquake of 23 October 2011, Mw 7.2; An overview on disaster management. *Iranian Journal of Public Health*, 42 (2): 134–144.

Zaré, M., Ansari A., Ashkpour-Motlagh, S. H., Esmaeili, B. and Shahvar, M. (2013a). "The Mw6.4 Shonbeh (Bushehr), Iran Earthquake, April 2013", published by *EERRI*, 7 pages.

Zaré, M., Ansari, A., Heydari, H., Shahvar, M., Daneshdust, M., Mahdian, M., Sinaiean, F., Farzanegan, E. and Mirzaei-Alavijeh, H. (2013b). "A Reconnaissance Report on two Iran, Makran Earthquakes: Iran, Makran Earthquakes; 16 April 2013, Mw7.8, Gosht (Saravan) and 11 May 2013 Irar (Goharan), Bashagard, SE of Iran", report published by *EERI*, 12 pages.

Zaré, M., Farzanegan, E., Shahvar, M., Kamali, E. and Saeidi A. (2014). "Mormori (Ilam) SW Iran's Earthquake of 18 August 2014, Mw6.2: A Preliminary Reconnaissance Report", published by *EERRI*, 22 pages.

Zaré, M., Haghshenas, E. and Kalantari, A., (2009). "Earthquake of 30 September 2009, Mw7.6, Padang, Sumatra, Indonesia: A Preliminary Reconnaissance Report", published by IIEES and the collaboration of Center of Technology for Natural Resources Inventory (PTISDA), Jakarta – Indonesia, 6 pages.

Zaré, M. and Karimi-Paridari, S. (2009). An investigation on Balakot, Muzaffarabad (Pakistan) earthquake, 8 Oct. 2005, Mw 7.6; geological aspects and intensity distribution. *Journal of Seismology*, 13: 327–337. doi: 10.1007/s10950-008-9120-4.

Zaré, M., Karimi-Paridari, S. and Sabzali, S. (2008). Spectral attenuation of strong motions for near source data in Iran. *Journal of Seismology and Earthquake Engineering*, 10(3): 146–152.

Chapter 14

Integrated geotechnical and geophysical approaches to assess integrity of tank foundations for a crude oil terminal project in the Middle East

C.R. Parthasarathy, Prashanth Talkad, and Sreevalsa Kolathayar

14.1 INTRODUCTION

For important infrastructure projects like crude oil terminals, the foundation design of the storage tanks is very critical. For such sensitive projects, a conventional geotechnical investigation may not be sufficient to make design recommendations. There is a need to integrate geophysical tests along with geotechnical investigations to get reliable information of the subsurface profile. The very low-frequency electromagnetic (VLF-EM) measurement survey is a popular passive technique using the radiation from radio transmitters functioning in the very low-frequency band of 15–30 kHz as the primary electromagnetic field (Paterson and Ronka, 1971). Several studies in the past employed VLF-EM techniques to investigate the ground for diverse applications like identification of fractures, groundwater monitoring, foundation design, and so forth (Saydam, 1981; McNeill and Labson, 1991; Jeng, et al., 2004; Sharma and Baranwal, 2005; Santos et al., 2006; Babu et al., 2007; Gnaneshwar et al., 2011; Singh and Sharma, 2016; Taiwo et al., 2016; Ebrahimi et al., 2019). Electrical resistivity is another popular non-destructive technique to detect geological discontinuities and anomalies in the ground that has been used widely for different studies in the past (Herman, 2001; Schoor, 2002; Buck, 2003; Chandra et al., 2010; Papadopoulos et al., 2010; Gambetta et al., 2011).

In the present study, it is proposed to develop a crude oil terminal project at RasIssa, Hodeidah, Yemen. The project comprised of installing four storage tanks in the region. A VLF-EM measurement survey was carried

DOI: 10.1201/9781003177692-14

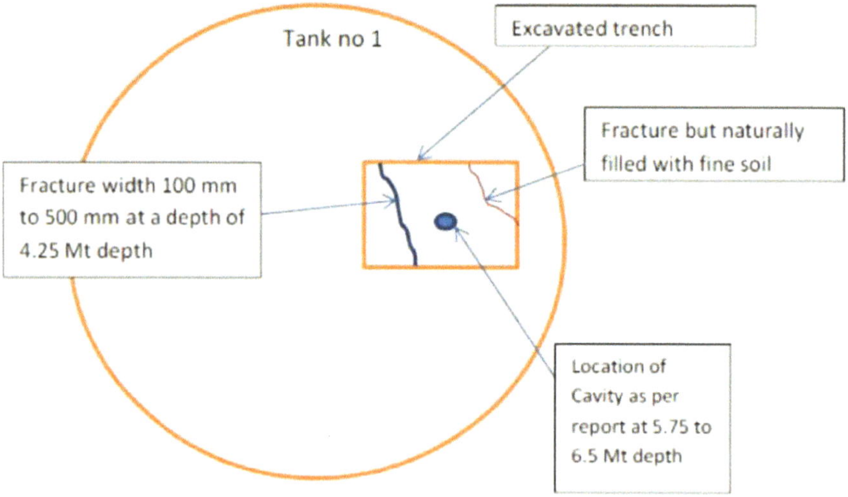

Figure 14.1 Cracks identified in Tank 1.

out at the site, which indicated presence of two intermediate scales, near-surface highly fractured/cavernous structures (appx. 2.0 m thick) at depth ranges between 5.75 m and 6.5 m below the existing natural ground level at both Tank 1 and Tank 4 sites. Excavation is done up to the crack depth and fills it back with controlled compaction. Inspection of excavation at Tank 1 suggested that the reported crack was absent instead, two cracks were noticed (as shown in Figure 14.1). One crack of 100–500 mm wide indicated by blue line encountered at about 4.25 m below the ground level. This extends throughout the cross-section of the tank. The second crack is indicated by the red line in Figure 14.1.

14.1.1 Study area

The strata at the proposed tank location consisted of weak, highly weathered to disintegrated (to soil-like) coral reef and medium dense to very dense silty/clayey sand materials up to the depth of investigation. Groundwater is reported to have encountered at a depth ranging between 4.9 m and 10.0 m (about 5.0 m in general as per the bore logs). For sandy material, the *N* values varied between 9 and 50 in the top 20 m and increasing with depth (except at a very few borehole locations). Inspection of the geotechnical report suggests the stratigraphy is good enough to withstand the pressure due to the loading from the tank.

The visible cracks were minor in volume vis-à-vis the stress distributed area due to the size of the tank foundation. No cracks were visible in the partial excavation at Tank 4. The extract of *N* values is presented in Figures 14.2 to 14.6 for the four tank foundations.

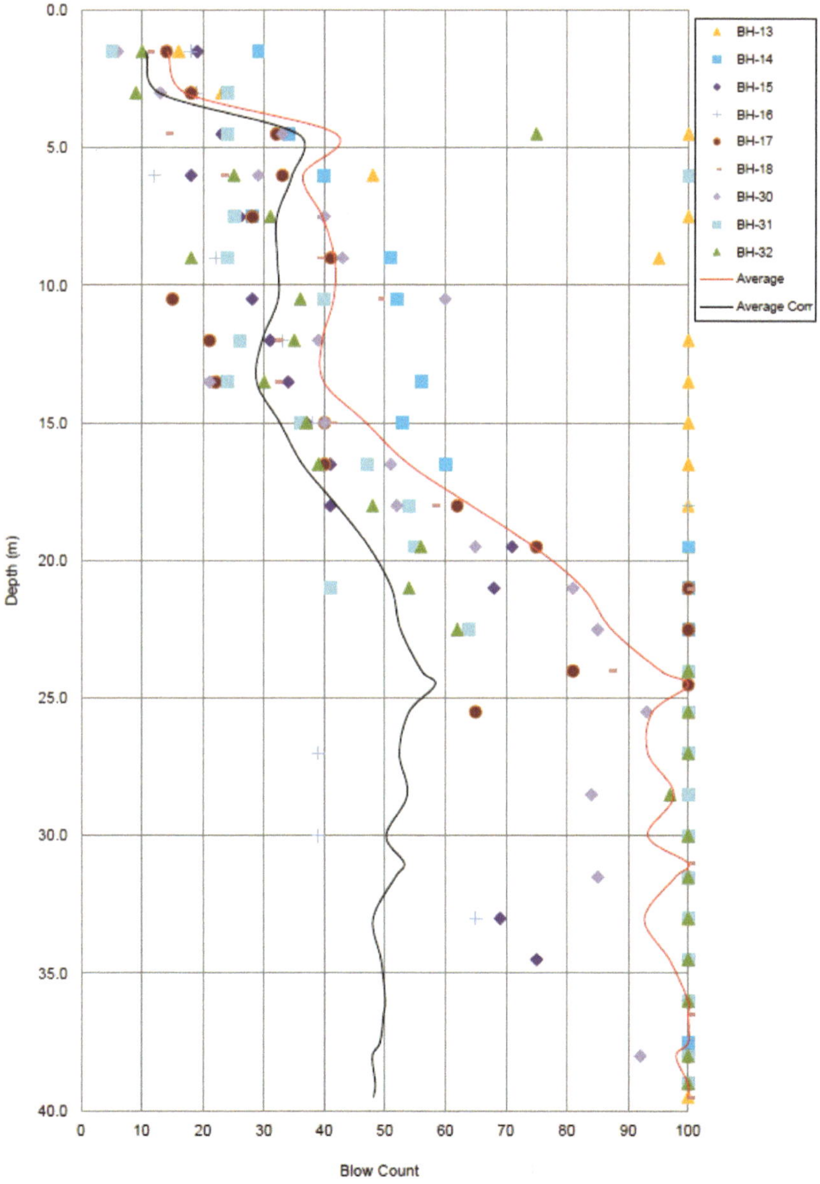

Figure 14.2 Graphical illustration of standard penetration test (SPT) results for the area of Tank 1.

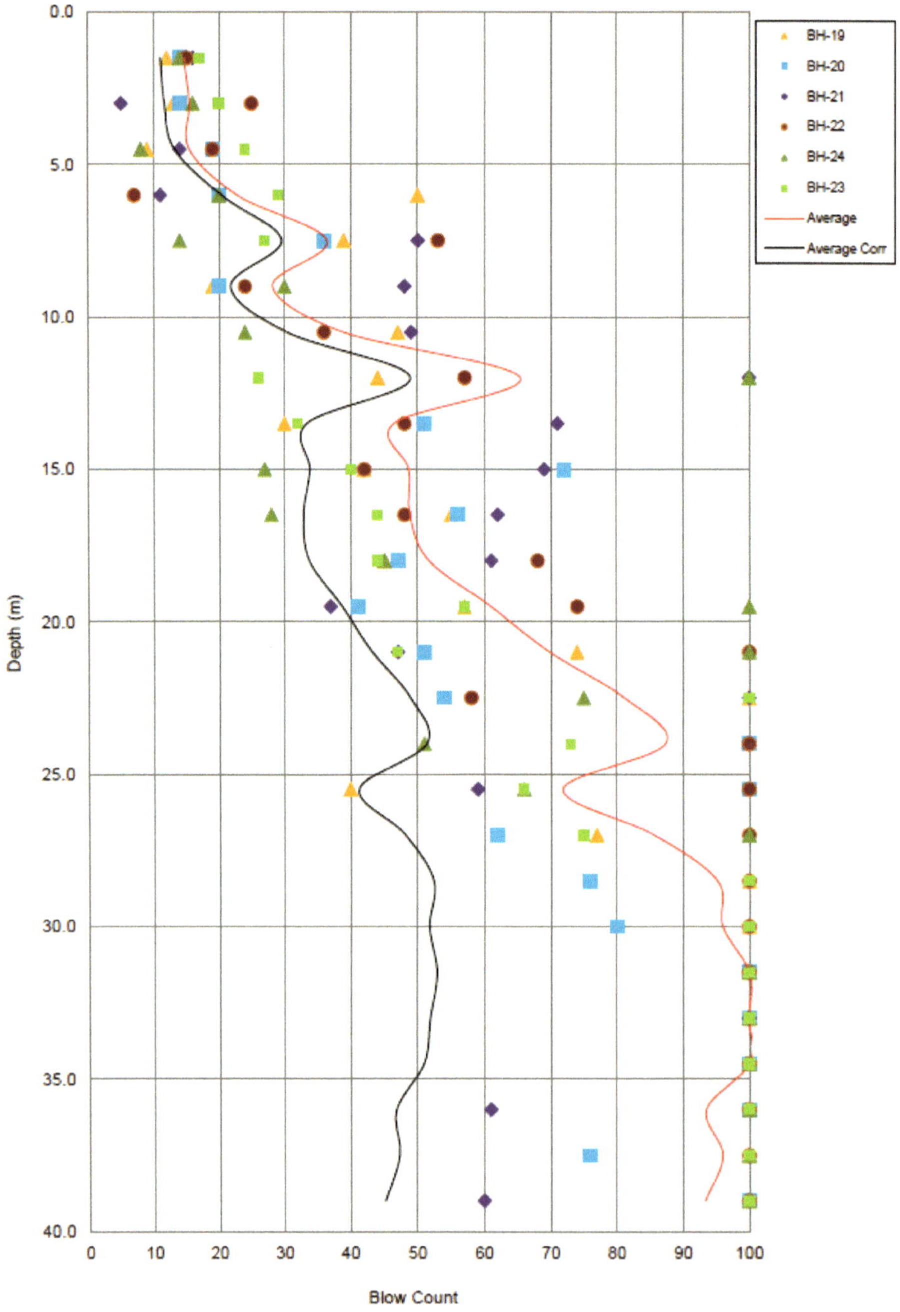

Figure 14.3 Graphical illustration of SPT results for the area of Tank 2.

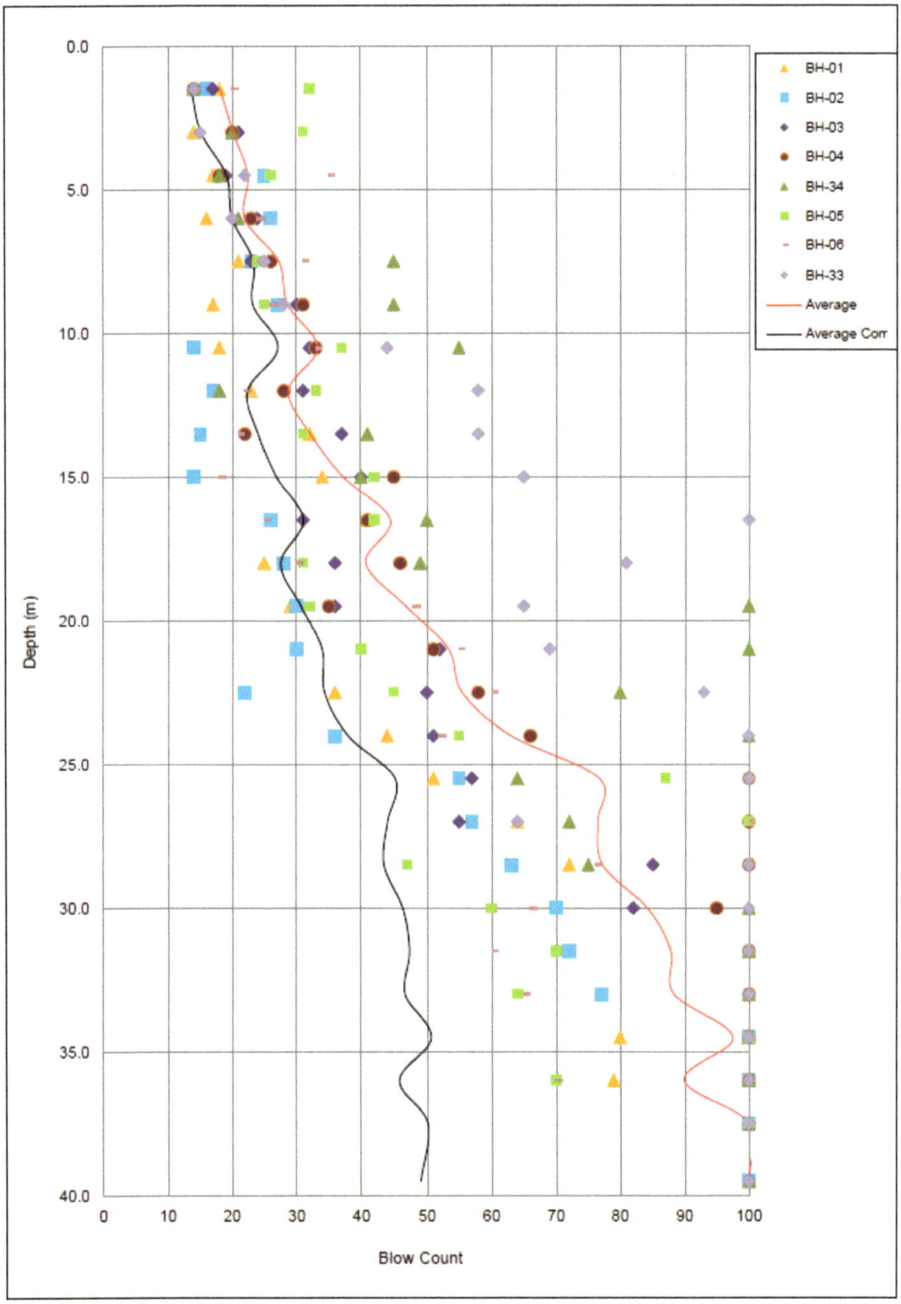

Figure 14.4 Graphical illustration of SPT results for the area of Tank 3.

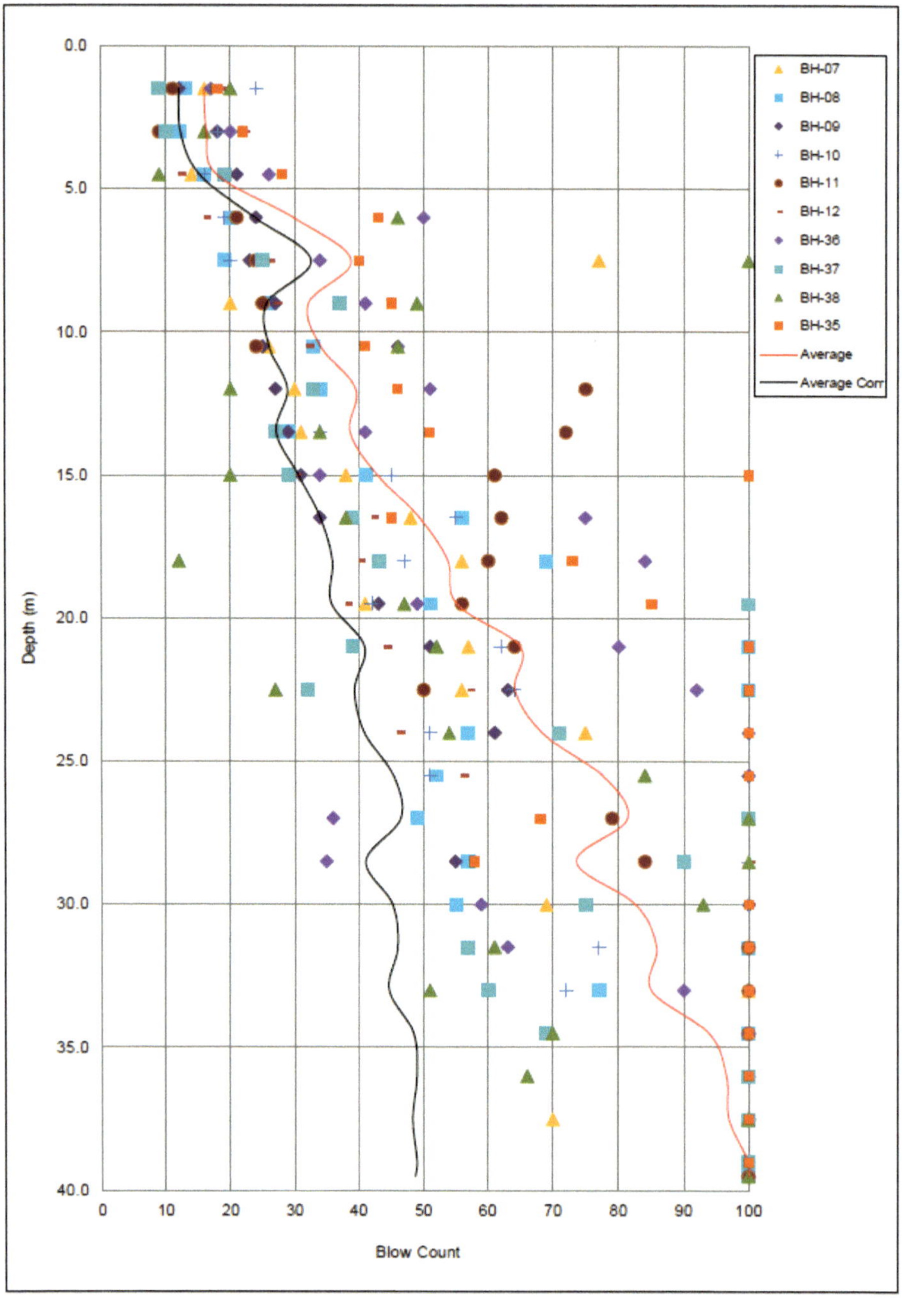

Figure 14.5 Graphical illustration of SPT results for the area of Tank 4.

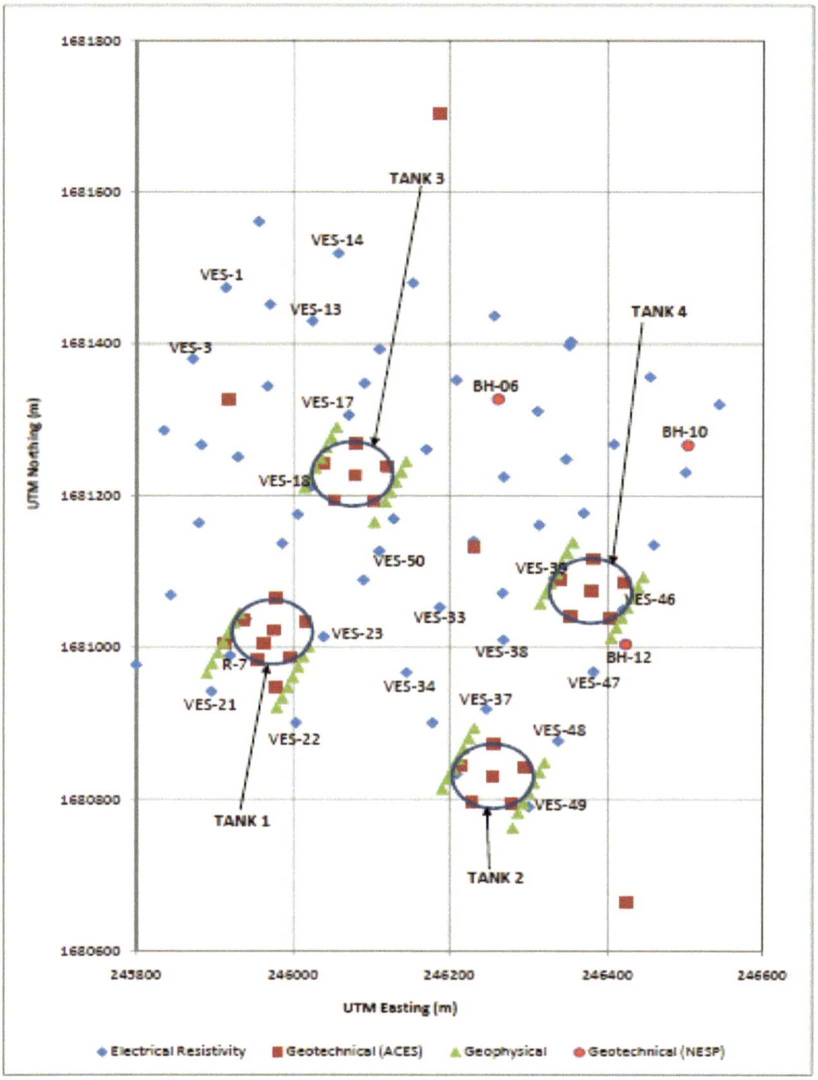

Figure 14.6 The coordinates of survey locations.

14.2 ELECTRICAL RESISTIVITY TEST

Two-dimensional electrical resistivity mapping was performed by surveying 12 profiles each of 150 m length over each tank location. The station interval was kept at 5 m, and data were acquired using the Wenner-Schlumberger method for continuous subsurface profiling (Figure 14.7). The concern behind the study was to identify subsurface

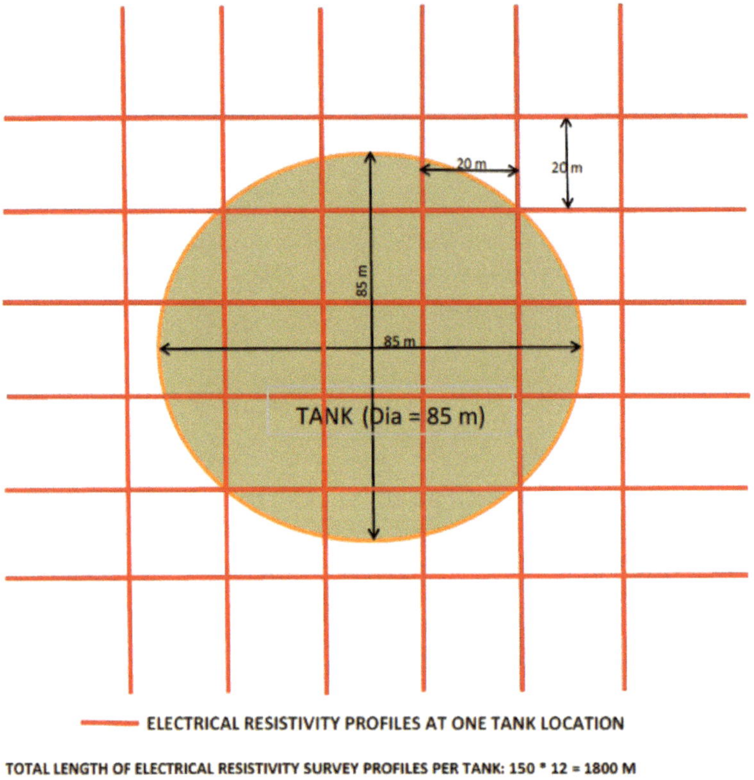

Figure 14.7 Schematic layout of resistivity profiles.

anomalies containing loose formations and cavernous zones. Such anomalous features were identified based on their relatively higher resistive properties as compared to their surrounding environment. Cracks, voids, and fractures, usually show up as spots of high resistivity values.

Wenner - Schlumberger

C1 P1 P2 C2

•←na→•←a→•←na→•

$$k = \pi n(n+1)a$$

C1 and C2 are current electrodes, P1 and P2 are potential electrodes, *a* is the electrode separation, and *k* is the geometrical factor.

The survey was conducted using the roll-along method to extend area coverage.

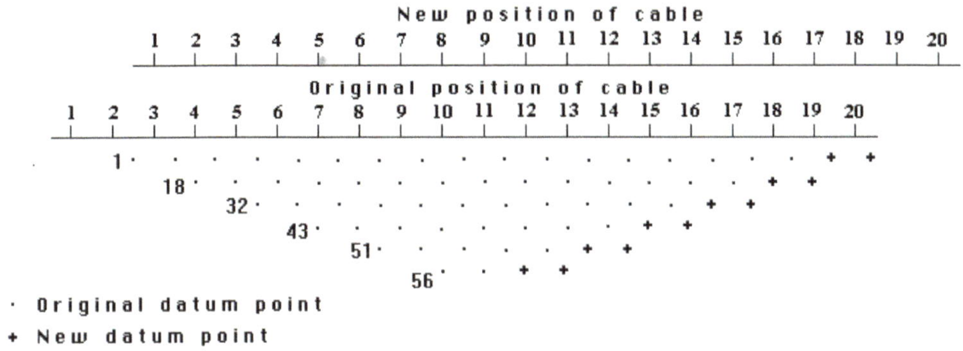

Figure 14.8 Resistivity profile imaging.

Each resistivity profile will be imaged in a similar way, as illustrated in Figure 14.8.

14.3 TANK LOCATIONS

A total of 48 resistivity profiles, each of length 150 m, were mapped over four tank locations – Tank 1–Tank 4. The project area is in the general flat. The ground elevations of the tanks vary between 5.5 m and 7.5 m above mean sea level (MSL). The location of the tanks and their ground elevations are shown in Figure 14.9.

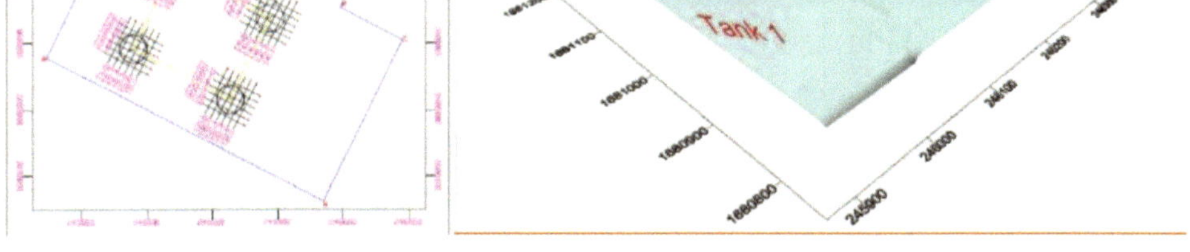

Figure 14.9 Tank locations and elevations.

14.4 SUMMARY OF THE COLLECTED INFORMATION

The coordinates of survey locations are plotted as presented in Figure 14.6.

- The electrical resistivity test states that there are no major cavities/joints/fractures at shallow depth.
- The geotechnical investigation has indicated a loss of drill fluid at few boreholes particularly in the region of BH-6, 10 and 12, indicating some localized cavities possibly insignificant.
- The low-frequency VLF-EM test survey has indicated the presence of two intermediate scales, near-surface highly fractured/cavernous structures (appx. 2.0 m thick) at depth ranges between 5.75 m and 6.5 m below the existing natural ground level at both Tank 1 and Tank 4 sites.
- A site-specific and detailed geotechnical investigation indicate no loss of drill fluid and thus no cavities in the investigated area.

To mitigate the cracks identified during VLF-EM test, excavation is done up to the crack level, and backfill with controlled compaction. During the excavation in Tank 1, the VLF-EM-identified crack was absent, but a crack of 0.1 m–0.5 m wide was identified at about 4.25 m below the ground level.

14.4.1 Tank foundation

The size of the tank is proposed as 85.0 m in diameter and 18.0 m height with a ring beam foundation. A generic ground model was developed to verify the effect of openings/cracks on the settlement of the foundations, and numerical analysis was performed in Plaxis 2D software. The tank foundation was modelled numerically using commercially available finite element software Plaxis 2D. The finite element analysis was performed for plain strain condition with two translation degrees of freedom in x and y direction. A 15-node triangular element was used to model the soil. The foundation was modelled using beam element. The triangular elements used in the model have 12 gauss points for numerical integration and fourth-order interpolation for displacement. The beam element has two translational and one rotational degree of freedom per node. Interface elements were also used to model foundation soil and soil-cavity interaction.

Table 14.1 Ground model

Layer	*Thickness of layer (m)*	*Cohesion (kPa)*	*Internal frictional angle (deg)*	*Unit Wt (kN/m3)*	*Young's modulus (kPa)*
Layer – 1	0.0–15.0	5	28	18	30000
Layer – 2	15.0–significant depth	1	42	20	75000

The analysis indicated that for an 85.0 m diameter foundation, the settlement was in the order of about 100 mm for pressure of 20T/m2. This is also verified by analytical method for a circular foundation based on the solutions by Selvadurai (1996), which is about 80 mm. Further two openings were modelled. One opening of 0.5 m wide, 2.0 m deep was at a depth of 4.5 m, and the second opening of 0.7 m wide, 2.0 m deep was at a depth of 5.0 m below the ground level. These two openings were representative of the cracks detected by the VLM method and the physical observation during excavations. Numerical analysis indicated that the presence of these two cracks within the stressed zone showed an insignificant contribution to the settlement of the tank foundation. Parameters of the ground model are listed in Table 14.1.

14.5 RESULTS AND DISCUSSIONS

The interpretation of the resistivity distribution across the surveyed area is based on the following assumptions:

Resistivity Values	**Probable lithology/feature**
Less than 5.0 Ω-m	clay or clayey silt, mud deposits, sandy silt/shale/claystone with saline water
5.1 to 10.0 Ω-m	cemented sand/consolidated clayey silts with saline water
10.1 to 30.0 Ω-m	weathered/fractured sandstone/limestone/coral reef with saline water (referred in this report as containing major crack, major fracture, or major cavity)
Above 30.1 Ω-m	fractures and cavities that may be partially saturated with saline water (referred in this report as containing major crack, major fracture, or major cavity)

An attempt has been made to infer fractures and cavities from the resistivity profiles and estimate the location and extent of such features.

14.5.1 Tank 1

Nearer to the ground surface, cavernous features (characterized by higher resistivity values of over 30 Ω-m) are minor and smaller in size as compared to the surrounding weathered or less compacted formations (characterized by medium resistivity values in the range of 10–30 Ω-m). Minor cavities are inferred at this level inside the tank boundary. At lower levels, cavernous features are more prominent looking as if these are cracks developed in otherwise compact rocks, partially charged with saline water.

14.5.2 Tank 2

Resistivity distribution for Tank 2 is mapped at different elevation levels: 2.42 m, –8.24 m, and –18.28 m above MSL. The mapping at EL 2.42 m shows two locations of high resistivity values inside the tank boundary in contrast to the general low resistive environment of the subsurface. However, these locations are not prominent and might constitute minor cavities of very small sizes. At lower levels, no resistivity anomalies are visible under Tank 2 location.

14.5.3 Tank 3

Resistivity distribution for Tank 3 is mapped at elevations –0.86 m, –5.04 m, –11.75 m, –16.22 m, –21.56 m, and –26.26 m above MSL. The mapping at EL –0.86 m shows the zone of higher resistivity in the South to the Southwest side of the tank location in contrast to the general low resistive environment of the subsurface. The high resistivity zone indicates the presence of a fracture/cavernous zone. The resistivity distribution at an elevation level of –5.04 m also depicts similar anomaly features. The resistivity distribution at elevation levels –11.75 m and –16.22 m has a single high resistivity anomaly on the northern side of the tank location outside the tank boundary which appears to be a major crack propagating at both elevations.

14.5.4 Tank 4

Resistivity distribution for Tank 4 is mapped at elevations 1.94 m, –2.25 m, –5.13 m, –8.72 m, and –18.77 m above MSL. The mapping at EL 1.94 m shows areas of medium resistivity in the northern and western part of the tank location which may be attributed to the presence of weathered and un-compacted formations which may be prone to house cracks. The resistivity distribution at other elevation levels –2.25 m, –5.13 m, –8.72 m, and –18.77 m above MSL depict high-resistivity anomalies indicating the presence of caverns of bigger sizes. Such features appearing within the tank boundary are of real concern.

14.6 CONCLUSIONS AND RECOMMENDATIONS

This chapter presented a case study of integrating geotechnical and geophysical approaches in the design of 4 Nos of foundations for a crude oil terminal project in the Middle East. The summary of the review of past geotechnical data, observations during the field visit, challenges associated, and conclusions and recommendations for the progress of the work are discussed. A low-frequency VLF-EM measurement survey was carried out at the site, which indicated the presence of two intermediate scales, near-surface highly fractured/cavernous structures of 2.0 m thickness at depth ranges between 5.75 m and 6.5 m below the existing natural ground level. This necessitated remedial measures, and, accordingly, it was proposed to excavate up to the crack depth and fill it back with controlled compaction. Inspection of excavation suggested that the reported crack was absent; instead, two cracks were noticed. The absence of cavity detected in VLF-EM and observation of fresh cracks in excavation alerted to relook at the integrity of the tank foundation. Extensive data viz., foundation details/drawings of crude oil storage tank, geotechnical investigation reports, electromagnetic test report, electrical restively report, restoration scheme reports, and geological information reports, were reviewed in detail and their indications are discussed in the chapter.

The strata at the proposed tank location consisted of weak, highly weathered to disintegrated coral reef and medium dense to very dense silty/clayey sand materials up to the depth of investigation. A generic ground model was developed to verify the effect of openings/cracks on the settlement of the foundations, and numerical analysis was performed in Plaxis 2D software. The tank foundation was modelled numerically using commercially available finite element software PLAXIS 2D. The finite element analysis was performed for plain strain condition with two translation degrees of freedom in x and y direction. The analysis indicated that for an 85.0 m diameter foundation, the settlement was in the order of about 100 mm for pressure of 20T/m2. This is also verified by the conventional analytical method for the circular foundation. Numerical analysis indicated that the presence of two cracks identified within the stressed zone showed an insignificant contribution to the settlement of the tank foundation. However, the concern of fresh cracks during excavation prompted the stakeholders to take appropriate measure. Accordingly, it was recommended to conduct a detailed cavity mapping with geophysical methods, and the survey was carried out. In one of the tank foundations (Tank 4), interpretation of geophysical methods revealed the cavity of about 35% within the stressed zone and, accordingly, planned to shift the tank location to avoid transitional tilt. The other tank locations have a minor indication of cavities not jeopardizing the safety of foundation, and thus the planned foundation system was adopted. The proper geophysical survey is imperative for identifying the geohazards, which will be detrimental to the safety of the foundation, as highlighted in this case study.

- The fracture/caverns found at 4.25 m deep at Tank 1 is significantly smaller, and it will not have any effect on the settlement of tank foundation due to its size which is verified in Plaxis analysis also.
- The excavations done for Tank 1 shall be backfilled (controlled and in layers).
- At all the tank locations, ring beam foundations as already planned shall be constructed, without any excavations for crack detections.
- Based on the review of the available information, though major cavities/cracks is unlikely to be present at the tank locations, a detailed geophysical investigation comprising of electrical resistivity mapping as per ASTM D 6431-99 is recommended at all the tank locations to rule out the presence of major caverns/cavities. The mapping shall cover an area of 100.0 m × 100.0 m for each of the tank locations. The investigation shall be focussed in the top 15.0 m–20.0 m, which is the stressed zone and depth of interest. Below 20.0 m, stratigraphy is essentially incompressible.
- During the confirmatory mapping, the approved grouting technique method shall be adopted as a remedial measure.

- The tank foundations shall be monitored, and levels were taken at various locations periodically during construction and hydro-testing to measure the settlement of tanks.

REFERENCES

ASCE (American Society of Civil Engineers). (2014) *Minimum Design Loads for Buildings and Other Structures*, Standard ASCE/SEI 7-10. Third printing. Reston, VA: ASCE.

Buck, S. (2003). Searching for graves using geophysical technology: Field tests with ground penetrating radar, magnetometry, and electrical resistivity. *Journal of Forensic Science*, 48(1), 1–7.

Chandra, S., Dewandel, B., Dutta, S., & Ahmed, S. (2010). Geophysical model of geological discontinuities in a granitic aquifer: Analyzing small scale variability of electrical resistivity for groundwater occurrences. *Journal of Applied Geophysics*, 71(4), 137–148.

Ebrahimi, A., Sundararajan, N., & Babu, V. R. (2019). A comparative study for the source depth estimation of very low frequency electromagnetic (VLF-EM) signals. *Journal of Applied Geophysics*, 162, 174–183.

Gambetta, M., Armadillo, E., Carmisciano, C., Stefanelli, P., Cocchi, L., & Tontini, F. C. (2011). Determining geophysical properties of a near-surface cave through integrated microgravity vertical gradient and electrical resistivity tomography measurements. *Journal of Cave and Karst Studies*, 73(1), 11–15.

Gnaneshwar, P., Shivaji, A., Srinivas, Y., Jettaiah, P., & Sundararajan, N. (2011). Very-low-frequency electromagnetic (VLF-EM) measurements in the Schirmacheroasen area, East Antarctica. *Polar Science*, 5(1), 11–19.

Herman, R. (2001). An introduction to electrical resistivity in geophysics. *American Journal of Physics*, 69(9), 943–952.

Jeng, Y., Lin, M. J., & Chen, C. S. (2004). A very low frequency-electromagnetic study of the geo-environmental hazardous areas in Taiwan. *Environmental Geology*, 46(6–7), 784–795.

McNeill, J. D., & Labson, V. F. (1991). Geological mapping using VLF radio fields. In *Electromagnetic Methods in Applied Geophysics: Volume 2, Application, Parts A and B* (pp. 521–640). Society of Exploration Geophysicists.

Papadopoulos, N. G., Yi, M. J., Kim, J. H., Tsourlos, P., & Tsokas, G. N. (2010). Geophysical investigation of tumuli by means of surface 3D electrical resistivity tomography. *Journal of Applied Geophysics*, 70(3), 192–205.

Paterson, N. R., & Ronka, V. (1971). Five years of surveying with the Very Low Frequency—Electro magnetics method. *Geoexploration*, 9(1), 7–26.

Ramesh Babu, V., Ram, S., & Sundararajan, N. (2007). Modeling and inversion of magnetic and VLF-EM data with an application to basement fractures: A case study from Raigarh, India. *Geophysics*, 72(5), B133–B140.

Santos, F. M., Mateus, A., Figueiras, J., & Gonçalves, M. A. (2006). Mapping groundwater contamination around a landfill facility using the VLF-EM method—a case study. *Journal of Applied Geophysics*, 60(2), 115–125.

Saydam, A. S. (1981). Very low-frequency electromagnetic interpretation using tilt angle and ellipticity measurements. *Geophysics*, 46(11), 1594–1605.

Selvadurai, A. P. S. (1996). The settlement of a rigid circular foundation resting on a half-space exhibiting a near surface elastic non-homogeneity. *International Journal for Numerical and Analytical Methods in Geomechanics*, 20(5), 351–364.

Sharma, S. P., & Baranwal, V. C. (2005). Delineation of groundwater-bearing fracture zones in a hard rock area integrating very low frequency electromagnetic and resistivity data. *Journal of Applied Geophysics*, 57(2), 155–166.

Singh, A., & Sharma, S. P. (2016). Interpretation of very low frequency electromagnetic measurements in terms of normalized current density over variable topography. *Journal of Applied Geophysics*, 133, 82–91.

Stummer, P., Maurer, H., & Green, A. G. (2004). Experimental design: Electrical resistivity data sets that provide optimum subsurface information. *Geophysics*, 69(1), 120–139.

Taiwo, S. M., Awoyemi, M. O., & Onyedim, G. C. (2016). Combined use of very low frequency electromagnetic (VLF-EM) and electrical resistivity survey for evaluation of groundwater potential of Modomo/Eleweran area, south western Nigeria. *African Journal of Environmental Science and Technology*, 10(7), 192–206.

Van Schoor, M. (2002). Detection of sinkholes using 2D electrical resistivity imaging. *Journal of Applied Geophysics*, 50(4), 393–399.

Index